Methods in Enzymology

Volume 122
VITAMINS AND COENZYMES
Part G

METHODS IN ENZYMOLOGY

EDITORS-IN-CHIEF

Sidney P. Colowick   Nathan O. Kaplan

Methods in Enzymology

Volume 122

Vitamins and Coenzymes

Part G

EDITED BY

Frank Chytil

DEPARTMENT OF BIOCHEMISTRY
VANDERBILT UNIVERSITY SCHOOL OF MEDICINE
NASHVILLE, TENNESSEE

Donald B. McCormick

DEPARTMENT OF BIOCHEMISTRY
EMORY UNIVERSITY SCHOOL OF MEDICINE
ATLANTA, GEORGIA

1986

ACADEMIC PRESS, INC.

Harcourt Brace Jovanovich, Publishers

Orlando   San Diego   New York   Austin
London   Montreal   Sydney   Tokyo   Toronto

COPYRIGHT © 1986 BY ACADEMIC PRESS, INC.
ALL RIGHTS RESERVED.
NO PART OF THIS PUBLICATION MAY BE REPRODUCED OR
TRANSMITTED IN ANY FORM OR BY ANY MEANS, ELECTRONIC
OR MECHANICAL, INCLUDING PHOTOCOPY, RECORDING, OR
ANY INFORMATION STORAGE AND RETRIEVAL SYSTEM, WITHOUT
PERMISSION IN WRITING FROM THE PUBLISHER.

ACADEMIC PRESS, INC.
Orlando, Florida 32887

United Kingdom Edition published by
ACADEMIC PRESS INC. (LONDON) LTD.
24–28 Oval Road, London NW1 7DX

LIBRARY OF CONGRESS CATALOG CARD NUMBER: 54-9110

ISBN 0–12–182022–X

PRINTED IN THE UNITED STATES OF AMERICA

86 87 88 89   9 8 7 6 5 4 3 2 1

Table of Contents

Section I. Ascorbic Acid

Section II. Thiamin: Phosphates and Analogs

Section III. Pantothenic Acid, Coenzyme A, and Derivatives

Section IV. Biotin and Derivatives

Section V. Pyridoxine, Pyridoxamine, Pyridoxal: Analogs and Derivatives

Section VI. Nicotinic Acid: Analogs and Coenzymes

Section VII. Flavins and Derivatives

Section VIII. Pteridines, Analogs, and Pterin Coenzymes (Folate)

Contributors to Volume 122

Article numbers are in parentheses following the names of contributors.
Affiliations listed are current.

R. KALERVO AIRAS (6), *Department of Biochemistry, University of Turku, SF-20500 Turku 50, Finland*

BRUCE M. ANDERSON (27, 29, 30), *Department of Biochemistry and Nutrition, Virginia Polytechnic Institute and State University, Blacksburg, Virginia 24061*

CONSTANCE D. ANDERSON (27, 30), *Department of Biochemistry and Nutrition, Virginia Polytechnic Institute and State University, Blacksburg, Virginia 24061*

BERTA ANDONDONSKAJA-RENZ (44), *Abteilung für Klinische Hämatologie der Gesellschaft für Strahlen- und Umweltforschung, D-8000 München 2, Federal Republic of Germany*

ADELBERT BACHER (32, 34), *Lehrstuhl für Organische Chemie und Biochemie, Technische Universität München, D-8046 Garching, Federal Republic of Germany*

MARK M. BASHOR (37), *Clinical Chemistry Division, Center for Environmental Health, Centers for Disease Control, Atlanta, Georgia 30333*

JAMES E. BISHOP (9), *University of Maryland Medical School, Department of Biochemistry, Baltimore, Maryland 21201*

ANDREW L. BOGNAR (55), *Department of Microbiology, University of Toronto, Toronto, Ontario, Canada M5S 1A8*

DELORES M. BOWERS-KOMRO (11, 19), *Department of Biochemistry, Emory University School of Medicine, Atlanta, Georgia 30322*

TOM BRODY (57), *Department of Biochemistry, University of Wisconsin—Madison, Madison, Wisconsin 53076*

GERDA C. CAERTELING (63), *Department of Microbiology, Faculty of Science, University of Nijmegen, Toernooiveld, 6525 ED Nijmegen, The Netherlands*

JANE L. CHASTAIN (11), *Department of Biochemistry, Emory University School of Medicine, Atlanta, Georgia 30322*

JOHN A. CIDLOWSKI (20), *Departments of Physiology and Biochemistry, University of North Carolina, Chapel Hill, North Carolina 27514*

JOAN E. CLARK (60, 61), *Department of Biochemistry, School of Medicine, Case Western Reserve University, Cleveland, Ohio 44106*

STEPHEN P. COBURN (17), *Department of Biochemistry, Fort Wayne State Developmental Center, Fort Wayne, Indiana 46815*

G. COHEN (13), *Department of Microbiology, George S. Wise School of Life Sciences, Tel-Aviv University, Ramat-Aviv, 69978 Tel-Aviv, Israel*

PAUL D. COLMAN (47), *Division of Hematology/Oncology, Children's Hospital of Los Angeles, Los Angeles, California 90054*

ROBERT J. COOK (40, 41), *Department of Biochemistry, Vanderbilt University School of Medicine, Nashville, Tennessee 37232*

JACK R. COOPER (4, 5), *Department of Pharmacology, Yale University School of Medicine, New Haven, Connecticut 06510*

MANASE M. DÂNŞOREANU (28), *Department of Biophysics, Institute of Medicine and Pharmacy, R-3400 Cluj-Napoca, Romania*

KARL DECKER (31), *Biochemisches Institut, Universität Freiburg, D-7800 Freiburg i. Br., Federal Republic of Germany*

M. T. DOIG (48), *Department of Chemistry, College of Charleston, Charleston, South Carolina 29424*

LANDIS W. DONER (1), *United States Department of Agriculture, Agriculture Research Service, Eastern Regional Research Center, Philadelphia, Pennsylvania 19118*

DALE E. EDMONDSON (39), *Department of Biochemistry, Emory University School of Medicine, Atlanta, Georgia 30322*

STEVEN M. FAYNOR (25), *International Clinical Laboratories, 5 Park Plaza, Nashville, Tennessee 37203*

JAMES H. FREISHEIM (45), *Department of Biochemistry, Medical College of Ohio, Toledo, Ohio 43699*

L. GOLDSTEIN (13), *Department of Biochemistry, George S. Wise School of Life Sciences, Tel-Aviv University, Ramat-Aviv, 69978 Tel-Aviv, Israel*

BEVERLY M. GUIRARD (23), *Department of Microbiology, The University of Texas, Austin, Texas 78712*

ROBERT W. GUYNN (59), *Psychiatry and Behavioral Sciences, The University of Texas Medical School, Houston, Texas 77225*

R. V. HAGEMAN (62), *Synergen, Inc., 1885 33rd Street, Boulder, Colorado 80301*

TORY M. HAGEN (19), *Department of Biochemistry, Emory University School of Medicine, Atlanta, Georgia 30322*

AMIYA K. HAJRA (9), *University of Michigan, Neuroscience Laboratory, Ann Arbor, Michigan 48109*

R. GAURTH HANSEN (7), *Department of Chemistry and Biochemistry, Utah State University, Logan, Utah 84322*

ROBERT P. HAUSINGER (33), *Department of Microbiology and Public Health, Michigan State University, East Lansing, Michigan 48824*

G. B. HENDERSON (42), *Division of Biochemistry, Scripps Clinic and Research Foundation, La Jolla, California 92037*

GAYLE C. HENEY (14), *Department of Molecular Pharmacology, Albert Einstein College of Medicine, Bronx, New York 10461*

KEVIN B. HICKS (1), *United States Department of Agriculture, Agriculture Research Service, Eastern Regional Research Center, Philadelphia, Pennsylvania 19118*

ARI HINKKANEN (31), *Biochemisches Institut, Universität Freiburg, D-7800 Freiburg i. Br., Federal Republic of Germany*

JOHN F. HONEK (33), *Department of Chemistry, Massachusetts Institute of Technology, Cambridge, Massachusetts 02139*

DONALD W. HORNE (43), *Department of Biochemistry, Vanderbilt University, Nashville, Tennessee 37232, and Veterans Administration Medical Center, Nashville, Tennessee 37203*

F. M. HUENNEKENS (42), *Division of Biochemistry, Scripps Clinic and Research Foundation, La Jolla, California 92037*

ARTHUR G. HUNT (8), *Department of Agronomy, College of Agriculture, University of Kentucky, Lexington, Kentucky 40546*

QUANG KHAI HUYNH (22), *Life Science Center, Biological Research Corporation, The Monsanto Company, St. Louis, Missouri 63198*

M. PERWAIZ IQBAL (54), *Department of Biochemistry, The Aga Khan University Medical School, Karachi 5, Pakistan*

B. A. KAMEN (53), *Department of Pediatrics, The University of Texas Health Science Center at Dallas, Southwestern Medical School, Dallas, Texas 75235*

MATTI T. KARP (24), *Department of Biochemistry, University of Turku, SF-20500 Turku 50, Finland*

PHILIP A. KATTCHEE (59), *Dental Branch, The University of Texas Dental School, Houston, Texas 77030*

TAKASHI KAWASAKI (3), *Department of Biochemistry, Hiroshima University School of Medicine, Hiroshima 734, Japan*

JAN T. KELTJENS (63), *Department of Microbiology, Faculty of Science, University of Nijmegen, Toernooiveld, 6525 ED Nijmegen, The Netherlands*

KHANDAN KEYOMARSI (47), *Department of Biochemistry, School of Medicine, University of Southern California, Los Angeles, California 90054*

PETER T. KISSINGER (46), *Department of Chemistry, Purdue University, West Lafayette, Indiana 47907*

JASON M. KITTLER (20), *Department of Molecular Biophysics and Biochemistry, Yale University, New Haven, Connecticut 06510*

A. ASHOK KUMAR (45), *Zymogenetics, Inc., 2121 N. 35th Street, Seattle, Washington 98103*

SANG-SUN LEE (38), *Department of Biochemistry, Emory University School of Medicine, Atlanta, Georgia 30322*

GARRY P. LEWIS (49), *Children's Medical Research Foundation, The Royal Alexandra Hospital for Children, University of Sydney, Sydney, Australia 2050*

TING-KAI LI (16), *Departments of Medicine and Biochemistry, Indiana University School of Medicine, and the Richard L. Roudebush Veterans Administration Medical Center, Indianapolis, Indiana 46223*

LEONARD F. LIEBES (2), *Department of Medicine, New York University School of Medicine, New York, New York 10016*

LARS G. LJUNGDAHL (60, 61), *Department of Biochemistry, University of Georgia, Athens, Georgia 30602*

PETER C. LOEWEN (51), *Department of Microbiology, University of Manitoba, Winnipeg, Manitoba, Canada R3T 2N2*

ALEC LUI (16), *Department of Medicine, Indiana University School of Medicine, and the Richard L. Roudebush Veterans Administration Medical Center, Indianapolis, Indiana 46223*

LAWRENCE LUMENG (16), *Departments of Medicine and Biochemistry, Indiana University School of Medicine, and the Richard L. Roudebush Veterans Administration Medical Center, Indianapolis, Indiana 46223*

CRAIG E. LUNTE (46), *Miami Valley Laboratory, Proctor and Gamble, Cincinnati, Ohio 45247*

J. DENNIS MAHUREN (17), *Department of Biochemistry, Fort Wayne State Developmental Center, Fort Wayne, Indiana 46815*

TOSHIO MATSUDA (4), *Department of Pharmacology, Faculty of Pharmaceutical Sciences, Osaka University, Osaka, Japan*

ROWENA G. MATTHEWS (52, 58), *Biophysics Research Division and Department of Biological Chemistry, The University of Michigan, Ann Arbor, Michigan 48109*

DONALD B. MCCORMICK (11, 19, 38), *Department of Biochemistry, Emory University School of Medicine, Atlanta, Georgia 30322*

NATALIE T. MEISLER (20), *Department of Biochemistry, University of Vermont, Burlington, Vermont 05405*

ALFRED H. MERRILL, JR. (18), *Department of Biochemistry, Emory University School of Medicine, Atlanta, Georgia 30322*

MARK S. MILLER (36), *Kraft, Inc., Research and Development, 801 Waukegan Road, Glenview, Illinois 60025*

RICHARD G. MORAN (47), *Division of Hematology/Oncology, Children's Hospital of Los Angeles; the Departments of Pediatrics and Biochemistry, School of Medicine and the School of Pharmacy, University of Southern California, Los Angeles, California 90054*

R. D. NARGESSI (12), *Triton Biosciences Inc., 1501 Harbor Bay Parkway, Alameda, California 94501*

COLIN J. NEWTON (25, 26), *Department of Clinical Pathology, Royal Liverpool Hospital, Liverpool L7 8XP, United Kingdom*

PETER NIELSEN (34), *Abteilung Medizinische Biochemie, Physiologisch-chemisches Institut, Universitätskrankenhaus Eppendorf, D-2000 Hamburg 20, Federal Republic of Germany*

KOHSUKE NISHINO (5), *Department of Hygiene, Faculty of Medicine, Kyoto University, Kyoto 606, Japan*

DEXTER B. NORTHROP (25, 26), *Division of Pharmaceutical Biochemistry, University of Wisconsin, Madison, Wisconsin 53706*

GEORGE A. ORR (14, 15), *Department of Molecular Pharmacology, Albert Einstein College of Medicine, Bronx, New York 10461*

D. G. PRIEST (48), *Department of Biochemistry, Medical University of South Carolina, Charleston, South Carolina 29425*

K. V. RAJAGOPALAN (62), *Department of Biochemistry, Duke University Medical Center, Durham, North Carolina 27710*

PRADIPSINH K. RATHOD (56), *Graduate Department of Biochemistry, Brandeis University, Waltham, Massachusetts 02154*

PETER RAUSCHENBACH (34), *Lehrstuhl für Organische Chemie und Biochemie, Technische Universität München, D-8046 Garching, Federal Republic of Germany*

PHILIP REYES (56), *Department of Biochemistry, The University of New Mexico School of Medicine, Albuquerque, New Mexico 87131*

SHELDON P. ROTHENBERG (54), *Division of Hematology/Oncology, Brooklyn Veterans Administration and Downstate Medical Centers, Brooklyn, New York 11209*

PETER B. ROWE (49), *Children's Medical Research Foundation, The Royal Alexandra Hospital for Children, University of Sydney, Sydney, Australia 2050*

BARRY SHANE (50, 55), *Department of Nutritional Sciences, University of California, Berkeley, California 94720*

TAKASHI SHIMAKATA (10), *Kawasaki Medical School, Kurashiki, Okayama 701-01, Japan*

ROBERT SILBER (2), *Department of Medicine, New York University School of Medicine, New York, New York 10016*

D. S. SMITH (12), *Department of Chemical Pathology, St. Bartholomew's Hospital, London EC1A 7HL, United Kingdom*

ESMOND E. SNELL (21, 22, 23), *Departments of Microbiology and Chemistry, The University of Texas, Austin, Texas 78712*

ROBERT L. STAHL (2), *The Emory Clinic, Section of Internal Medicine, 1365 Clifton Road N.E., Atlanta, Georgia 30322*

E. L. R. STOKSTAD (57), *Department of Nutritional Sciences, University of California at Berkeley, Berkeley, California 94720*

PAUL K. STUMPF (10), *Department of Biochemistry and Biophysics, University of California, Davis, California 95616*

MARIUS I. TELIA (28), *Department of Biophysics, Institute of Medicine and Pharmacy, R-3400 Cluj-Napoca, Romania*

JOHN W. THANASSI (20), *Department of Biochemistry, University of Vermont, Burlington, Vermont 05405*

JERRY ANN TILLOTSON (37), *Division of Comparative Medicine and Toxicology, Chemistry Branch, Letterman Army Institute of Research, Presidio of San Francisco, California 94129*

DACE VICEPS-MADORE (20), *Praxis Biologics, Inc., Rochester, New York 14623*

GODFRIED D. VOGELS (63), *Department of Microbiology, Faculty of Science, University of Nijmegen, Toernooiveld, 6525 ED Nijmegen, The Netherlands*

PAULI I. VUORINEN (24), *Department of Biomedical Sciences, University of Tampere, SF-33100 Tampere 10, Finland*

CONRAD WAGNER (40, 41), *Research Service, Veterans Administration Medical Center, Nashville, Tennessee 37203*

CHRISTOPHER WALSH (33), *Department of Chemistry, Massachusetts Institute of Technology, Cambridge, Massachusetts 02139*

ELAINE WANG (18), *Department of Biochemistry, Emory University School of Medicine, Atlanta, Georgia 30322*

HAROLD B. WHITE, III (35, 36), *Department of Chemistry, University of Delaware, Newark, Delaware 19716*

GARY WILLIAMSON (39), *Department of Biochemistry, Emory University School of Medicine, Atlanta, Georgia 30322*

SUSAN D. WILSON (43), *Research Service, Veterans Administration Medical Center, Nashville, Tennessee 37203*

N. WINICK (53), *Department of Pediatrics, The University of Texas Health Science Center at Dallas, Southwestern Medical School, Dallas, Texas 75235*

CARL T. WITTWER (7), *Department of Pathology, University of Utah Medical Center, Salt Lake City, Utah 84132*

BONITA W. WYSE (7), *Department of Nutrition and Food Science, Utah State University, Logan, Utah 84322*

S. A. YANKOFSKY (13), *Department of Microbiology, George S. Wise School of Life Sciences, Tel-Aviv University, Ramat-Aviv, 69978 Tel-Aviv, Israel*

DAVID A. YOST (29, 30), *Abbott Diagnostic Division, Abbott Laboratories, Abbott Park, North Chicago, Illinois 60064*

RON ZEHEB (14, 15), *Department of Internal Medicine and Human Genetics, University of Michigan Medical School, Ann Arbor, Michigan 48109*

HANS-JÖRG ZEITLER (44), *Abteilung für Klinische Hämatologie der Gesellschaft für Strahlen- und Umweltforschung, D-8000 München 2, Federal Republic of Germany*

Preface

Since Volumes 62, Part D, 66, Part E, and 67, Part F of "Vitamins and Coenzymes" were published as part of the *Methods in Enzymology* series considerable new information has become available. Advances continue to be made, rapid in some cases, on techniques and methodology attendant to assays, isolations, and characterization of vitamins and those systems responsible for their biosynthesis, transport, and metabolism. In some instances, additional coenzymic forms of principal groups, e.g., pterins, are now known to participate in roles not heretofore recognized, e.g., methane formation. In others, e.g., porphyrins and hemes, the coenzyme-like function of many warrants inclusion in current volumes.

Because of the new information on vitamins and coenzymes and the certainty that methods attendant to this area continue to provide impetus for further research, we have sought to provide investigators with more current modifications of earlier procedures as well as with those that have newly evolved. The collated information is divided into two new volumes that are the result of our efforts in soliciting contributions from numerous, active experimentalists who have published most of their findings in the usual, refereed research journals. Volume 122, Part G, covers most of the classically considered "water-soluble" forms including ascorbate, thiamin, pantothenate, and biotin. Volume 123, Part H, covers the "fat-soluble" groups of A, D, E, and K, and also covers B_{12}, carnitine, heme, and porphyrins.

We should like to express our gratitude to the contributors for their willingness to supply the information requested and for their tolerance of our editorial revisions. There has been an attempt to allow such overlap as would offer flexibility in the choice of method rather than presume any one is best for all laboratories. Omissions may be deliberate where modification of a "tried and true" technique is quite minor, but in some cases may be attributed to inadvertent oversight of the editors. We believe these volumes will provide useful and fairly replete addendums to the six earlier ones on this subject: 18 A, 18 B, 18 C, 62 D, 66 E, and 67 F.

We wish to acknowledge the encouragement of the founding editors of the *Methods in Enzymology* series, Dr. Nathan O. Kaplan and the late Dr. Sidney P. Colowick. It is with sadness that we join our colleagues in no longer having the kind and sapient advice of Dr. Colowick. Continued thanks are due to the staff of Academic Press for their help.

FRANK CHYTIL

DONALD B. MCCORMICK

METHODS IN ENZYMOLOGY

EDITED BY

Sidney P. Colowick and Nathan O. Kaplan

VANDERBILT UNIVERSITY
SCHOOL OF MEDICINE
NASHVILLE, TENNESSEE

DEPARTMENT OF CHEMISTRY
UNIVERSITY OF CALIFORNIA
AT SAN DIEGO
LA JOLLA, CALIFORNIA

I. Preparation and Assay of Enzymes
II. Preparation and Assay of Enzymes
III. Preparation and Assay of Substrates
IV. Special Techniques for the Enzymologist
V. Preparation and Assay of Enzymes
VI. Preparation and Assay of Enzymes (*Continued*)
Preparation and Assay of Substrates
Special Techniques
VII. Cumulative Subject Index

METHODS IN ENZYMOLOGY

EDITORS-IN-CHIEF

Sidney P. Colowick and Nathan O. Kaplan

VOLUME 86. Prostaglandins and Arachidonate Metabolites
Edited by WILLIAM E. M. LANDS AND WILLIAM L. SMITH

VOLUME 87. Enzyme Kinetics and Mechanism (Part C: Intermediates, Stereochemistry, and Rate Studies)
Edited by DANIEL L. PURICH

VOLUME 88. Biomembranes (Part I: Visual Pigments and Purple Membranes, II)
Edited by LESTER PACKER

VOLUME 89. Carbohydrate Metabolism (Part D)
Edited by WILLIS A. WOOD

VOLUME 90. Carbohydrate Metabolism (Part E)
Edited by Willis A. Wood

VOLUME 91. Enzyme Structure (Part I)
Edited by C. H. W. HIRS AND SERGE N. TIMASHEFF

VOLUME 92. Immunochemical Techniques (Part E: Monoclonal Antibodies and General Immunoassay Methods)
Edited by JOHN J. LANGONE AND HELEN VAN VUNAKIS

VOLUME 93. Immunochemical Techniques (Part F: Conventional Antibodies, Fc Receptors, and Cytotoxicity)
Edited by JOHN J. LANGONE AND HELEN VAN VUNAKIS

VOLUME 94. Polyamines
Edited by HERBERT TABOR AND CELIA WHITE TABOR

VOLUME 95. Cumulative Subject Index Volumes 61–74 and 76–80
Edited by EDWARD A. DENNIS AND MARTHA G. DENNIS

VOLUME 96. Biomembranes [Part J: Membrane Biogenesis: Assembly and Targeting (General Methods; Eukaryotes)]
Edited by SIDNEY FLEISCHER AND BECCA FLEISCHER

VOLUME 97. Biomembranes [Part K: Membrane Biogenesis: Assembly and Targeting (Prokaryotes, Mitochondria, and Chloroplasts)]
Edited by SIDNEY FLEISCHER AND BECCA FLEISCHER

Section I
Ascorbic Acid

[1] High-Performance Liquid Chromatographic Separation of Ascorbic Acid, Erythorbic Acid, Dehydroascorbic Acid, Dehydroerythorbic Acid, Diketogulonic Acid, and Diketogluconic Acid

By LANDIS W. DONER and KEVIN B. HICKS

L-Ascorbic acid (AA) and L-dehydroascorbic acid (DHAA) are the principal natural compounds with vitamin C activity, and chemical methods for their determination remain widely used. Interfering compounds, however, are often present in complex matrices such as biological tissues and foods. As a result, traditional approaches are being replaced by high-performance liquid chromatographic (HPLC) methods, which are more selective and sensitive. Recent comprehensive reviews[1-5] refer to various chemical and chromatographic methods available for determining AA, DHAA, and closely related compounds. In these reviews, their determination by HPLC on reversed-phase and ion-exchange stationary phases are discussed, as well as their detection by combinations of ultraviolet, electrochemical, and refractive index techniques. The importance of selecting appropriate extracting media and stabilizing solutions for these relatively unstable compounds is emphasized.

The primary basis for the physiological roles of vitamin C is the oxidation–reduction system between AA and DHAA. Hydrolysis of DHAA occurs outside a narrow acidic pH range and yields L-diketogulonic acid (DKGulA), with irreversible loss of vitamin C activity. D-Erythorbic acid (EA, i.e., isoascorbic acid) is epimeric to AA at C-5 and possesses little or no vitamin C activity; it may, in fact, be an antagonist of vitamin C.[6] EA forms an oxidation–reduction couple with D-dehydroerythorbic acid (DHEA), which readily and irreversibly hydrolyzes to D-diketogluconic

[1] J. R. Cooke and R. E. D. Moxon, in "Vitamin C (Ascorbic Acid)" (J. N. Counsell and D. H. Hornig, eds.), p. 167. Applied Science Publishers, London, 1981.

[2] H. E. Sauberlich, M. D. Green, and S. T. Omaye, in "Ascorbic Acid: Chemistry, Metabolism, and Uses" (P. A. Seib and B. M. Tolbert, eds.), Adv. Chem. Ser., No. 200, p. 199. Am. Chem. Soc., Washington, D.C., 1982.

[3] L. W. Doner, in "Trace Analysis" (J. F. Lawrence, ed.), Vol. 3, p. 113. Academic Press, New York, 1984.

[4] L. A. Pachla, D. L. Reynolds, and P. T. Kissinger, J. Assoc. Off. Anal. Chem. 68, 1 (1985).

[5] G. M. Jaffe, in "Handbook of Vitamins" (L. J. Machlin, ed.), p. 199. Dekker, New York, 1984.

[6] D. Hornig, Acta Vitaminol. Enzymol. 31, 9 (1977).

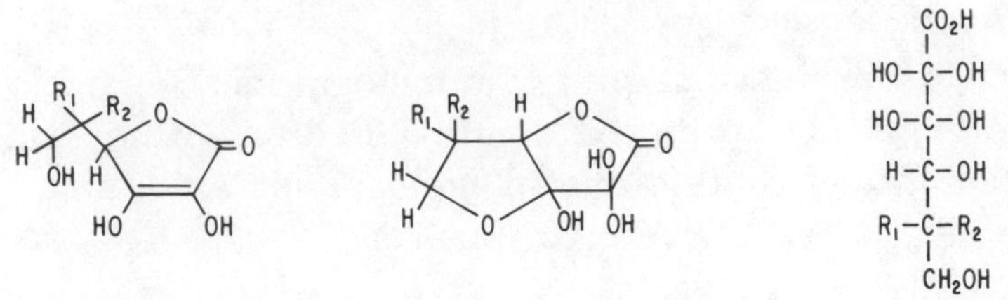

ASCORBIC ACID (AA) $R_1 = OH$, $R_2 = H$ | DEHYDROASCORBIC ACID (DHAA) $R_1 = OH$, $R_2 = H$ | DIKETOGULONIC ACID (DKGulA) $R_1 = OH$, $R_2 = H$

ERYTHORBIC ACID (EA) $R_1 = H$, $R_2 = OH$ | DEHYDROERYTHORBIC ACID (DHEA) $R_1 = H$, $R_2 = OH$ | DIKETOGLUCONIC ACID (DKGluA) $R_1 = H$, $R_2 = OH$

FIG. 1. Predominant structures in aqueous solution of separated compounds; DHAA and DHEA as hydrated bicyclic hemiketals, DKGulA and DKGluA as dihydrates.

acid (DKGluA). EA is widely used as an antioxidant by the food industry, since it is less expensive than AA.

In this report we emphasize the separation by weak anion-exchange HPLC of AA and EA, and these from their respective oxidation and hydrolysis products, extending an earlier study.[7]

Principle

The HPLC separation of AA, EA, DKGulA, and DKGluA on aminopropyl bonded-phase silica was accomplished by weak anion-exchange chromatography with acetonitrile–0.05 *M* KH_2PO_4 eluant. DHAA and DHEA are not truly organic acids, so their separation presumably results from partitioning between the mobile phase and the relatively water-enriched hydrated stationary phase. The conjugated enediol chromophore in AA and EA allows their detection at 270 nm; reduction of DHAA and DHEA with dithiothreitol (DTT) allows their indirect determination from the increase in AA and EA absorption. In aqueous solution, DHAA exists as a hydrated bicyclic hemiketal, and in DKGulA both ketone functions are hydrated. Analogous structures (Fig. 1) probably exist for DHEA and DKGluA. Although some carbonyl character is masked in these structures, they do exhibit some ultraviolet absorption. For many purposes, refractive index detection is the method of choice for the six compounds

[7] L. W. Doner and K. B. Hicks, *Anal. Biochem.* **115**, 225 (1981).

of interest in this study, especially with the recent development of highly sensitive detectors.

Preparation of Standard Compounds

DHAA and DHEA were readily prepared from AA and EA (both commercially available) as described earlier,[7,8] by air oxidation of ethanolic solutions containing activated charcoal. After filtration and removal of ethanol, pure syrups of DHAA and DHEA were obtained, as revealed by HPLC. These compounds are known only as syrups, but are conveniently stored as crystalline dimers, prepared[7,9] after dissolution of the monomers in nitromethane. The dimeric forms readily convert to DHAA and DHEA upon dissolving in acetonitrile–water (50 : 50, v/v).

Dimeric DHAA and DHEA were converted to DKGulA and DKGluA, respectively, by gradually titrating (in an ice bath over a period of 1 hr) aqueous solutions of 10 mg/ml with 0.5 N NaOH until the pH remained constant at 7.0. Solid samples of the sodium salts are obtained by lyophilization of these saponified solutions.

Preparation of Samples for HPLC

Standard mixtures of the title compounds were prepared just prior to analysis by HPLC (refractive index detection) by dissolving 32 mg each of AA, EA, and DHAA dimer, 64 mg of DHEA dimer, and 128 mg each of DKGulA and DKGluA (sodium salts) in 2 ml water. This solution was diluted with 2 ml acetonitrile and 20 μl injected.

For analysis of AA and EA by ultraviolet detection, 10 μg/ml solutions in acetonitrile–water (50 : 50, v/v) give high absorbances. The addition (1.0 mg/ml) of dithiothreitol (DTT) to one of duplicate samples at a final sample concentration of 1 mg/ml allows the indirect determination of DHAA and DHEA by increase in AA and EA absorption, since DTT quantitatively reduces these to AA and EA, respectively, within 15 min. To prevent reoxidation, DTT was also added to the mobile phase (1.0 mg/ml). Tyrosine serves well as internal standard for quantitation,[7] being well separated from the title compounds and having adequate ultraviolet absorption at 270 nm; it can be added to the sample diluent at a level of 30 mg/100 ml.

Orange juice samples were analyzed for AA by injecting after filtration (0.45-μm filter). Aqueous diluent containing tyrosine can be used for AA

[8] M. Ohmori and M. Takage, *Agric. Biol. Chem.* **42,** 173 (1978).
[9] H. Dietz, *Justus Liebigs Ann. Chem.* **738,** 208 (1970).

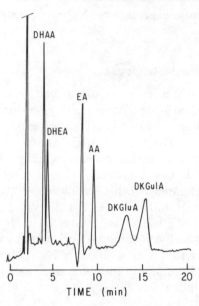

FIG. 2. Chromatography of standard mixture (20 μl injected) of dehydroascorbic acid (DHAA), dehydroerythorbic acid (DHEA), ascorbic acid (AA), erythorbic acid (EA), diketogluconic acid (DKGluA), and diketogulonic acid (DKGulA). Operating conditions: column, Zorbax NH$_2$; mobile phase, 3:1 acetonitrile–0.05 M KH$_2$PO$_4$; flow rate, 1.5 ml/min; refractive index detection, 16× attenuation; recorder chart speed, 0.25 cm/min.

quantitation, and dilution of a duplicate sample with both tyrosine and DTT (50 and 1.0 mg/ml, respectively) allows DHAA determination from the increase in AA absorbance. Likewise, urine samples (after adding an equal volume of acetonitrile and filtering) can be analyzed for AA and DHAA.

Instrument and Separation Procedure

The HPLC system consisted of a Hewlett-Packard HP 190 instrument equipped with a diode-array ultraviolet detector, 5- or 20-μl loop injectors, and an HP 85B computer. A Waters Associates[10] Model R401 refractive index detector also was used, sometimes in tandem with the ultraviolet detector. Separations were achieved by isocratic elution of a Zorbax NH$_2$ column (4.6 × 250 mm) with acetonitrile–0.05 M KH$_2$PO$_4$ (75:25,

---

[10] Reference to brand or firm name does not constitute endorsement by the U.S. Department of Agriculture over others of a similar nature not mentioned.

CHROMATOGRAPHIC EVALUATION OF SEPARATED
COMPOUNDS ON ZORBAX NH$_2$[a]

Compound	Retention time (min)	Capacity factor (k')	α[b]	R_s[c]
DHAA	3.6	0.91	0.22	
				1.04
DHEA	4.1	1.19	0.29	
				5.00
EA	8.1	3.32	0.82	
				1.80
AA	9.5	4.05	1.00	
				3.50
DKGluA	13.4	6.12	1.51	
				1.23
DKGulA	15.5	7.24	1.79	

[a] Chromatographic conditions as in Fig. 2.
[b] $\alpha = k'/k'_{AA}$.
[c] Resolution between adjacent peaks.

v/v) at a flow rate of 1.5 ml/min. Acetonitrile was added gradually with stirring to warmed 0.05 M KH$_2$PO$_4$, in order to avoid precipitate formation.

Discussion

The HPLC separation of the six compounds with a Zorbax NH$_2$ column is shown in Fig. 2, and the chromatographic characteristics are listed in the table. The resolution (R_s) between all adjacent pairs is >1.0, so the six compounds can be accurately quantified. If AA and EA are the analytes of interest, the flow rate (or polarity of eluant) can be increased; they remain efficiently separated at retention times of less than 5 min. Quantities injected (Fig. 2) were 160 μg AA, EA, and DHAA; 320 μg DHEA; and 640 μg DKGluA and DKGulA. The signals were quite attenuated (16×), and with highly sensitive refractive index detectors that are now available, detection of these compounds at 10–20% of these levels should be practical.

Figure 3 demonstrates the ability of tandem ultraviolet–refractive index detection to determine simultaneously AA and the various sugars in orange juice. The sugars are transparent to detection at 268 nm, while the level of AA is too low to be revealed by refractive index detection. Using tandem detection, all can be efficiently quantified, even though AA is not well separated from glucose, which is present in 50-fold excess of AA.

Measurements of DHAA have been made by others using ultraviolet detection; we find that the hydrated forms of DHAA and DHEA, as well as DKGulA and DKGluA, possess low extinction coefficients. Also, the

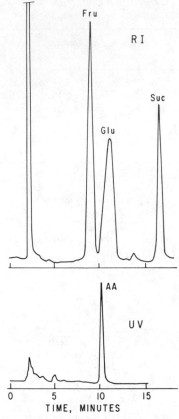

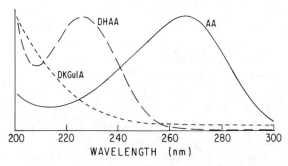

FIG. 3. Chromatogram of orange juice monitored by (below) tandem ultraviolet (UV, 268 nm) and (above) refractive index (RI, 8× attenuation) detection. Conditions as in Fig. 2. Fru, fructose; Glu, glucose; Suc, sucrose. (Reprinted with permission from Academic Press and Doner and Hicks.[7])

FIG. 4. Ultraviolet absorption spectra of AA, DHAA, and DKGulA.

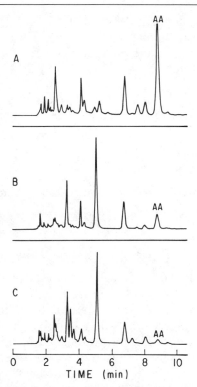

FIG. 5. Chromatogram of urine sample, monitored at (A) 270 nm, (B) 230 nm, and (C) 200 nm. Other conditions as in Fig. 2. Relative attenuations for (A), (B), and (C) are 4.7, 2.0, and 1.0, respectively.

maximum absorbance for DHAA and DHEA occurs at a wavelength of 227 nm, and for DKGulA and DKGluA <200 nm; other compounds which frequently occur in biological samples, however, also absorb at these low wavelengths. We, therefore, determined the dehydro forms indirectly, after reduction to their reduced forms, which possess high molar extinction coefficients (16,500) at 270 nm. Sensitive refractive index detection is suitable, in many cases, for the oxidized and hydrolyzed derivatives of AA and EA. Ultraviolet absorption spectra of AA, DHAA, and DKGulA are given in Fig. 4. Levels of AA and EA less than 5 ng can be detected, with good linearity for quantitation over a wide range.

Figure 5 shows that the chromatographic conditions described here can resolve many components of human urine and that monitoring the chromatogram at various wavelengths reveals significant differences. The AA peak corresponds to 26 mg/100 ml urine, and an ultraviolet scan of this peak verified its identity. DHAA can be indirectly quantified after reduction of a duplicate sample with DTT, which is added to the urine diluent along with tyrosine, the internal standard.[7]

Acknowledgment

We acknowledge David J. Hilber for excellent technical assistance.

[2] Glutathione Dehydrogenase (Ascorbate)

By ROBERT L. STAHL, LEONARD F. LIEBES, and ROBERT SILBER

Dehydroascorbic acid + 2GSH → ascorbic acid + GSSG

Glutathione dehydrogenase (ascorbate), i.e., dehydroascorbate reductase (glutathione : dehydroascorbate oxidoreductase, EC 1.8.5.1), catalyzes the glutathione (GSH)-dependent reduction of dehydroascorbic acid to ascorbic acid. The enzyme is widely distributed among plant tissues, where its presence was originally established using a dichlorophenol–indophenol titration technique.[1,2] Dehydroascorbate reductase has been detected in various animal cells using a coupled assay which monitors the oxidation of NADPH by glutathione reductase.[3] Some limitations of this method have been indicated.[4] Recently, the direct spectrophotometric assay for dehydroascorbate reductase described below was developed.[5]

Assay Method

Principle. The spectrum of ascorbic acid has a peak at 265 nm, while dehydroascorbic acid has no absorbance at that wavelength. The activity of dehydroascorbate reductase is assayed spectrophotometrically by measuring the change in absorbance at 265 nm associated with the generation of ascorbic acid.[5]

Reagents. Dehydroascorbic acid is made as follows by the oxidation of ascorbic acid according to Staudinger and Weis.[6] From a solution of

[1] B. M. Crook, *Biochem. J.* **35,** 226 (1941).
[2] M. Yamaguchi and M. A. Joslyn, *Arch. Biochem. Biophys.* **38,** 451 (1952).
[3] E. I. Anderson and A. Spector, *Invest. Ophthalmol.* **10,** 41 (1971).
[4] R. L. Stahl, L. F. Liebes, and R. Silber, *Biochim. Biophys. Acta* **839,** 119 (1985).
[5] R. L. Stahl, L. F. Liebes, C. M. Farber, and R. Silber, *Anal. Biochem.* **131,** 341 (1983).
[6] H. Staudinger and W. Weis, *Hoppe-Seyler's Z. Physiol. Chem.* **337,** 284 (1964).

20% Norit in 95% glacial acetic acid and 5% 0.2 N HCl, a 2-ml aliquot is taken and centrifuged for 1 min at 8000 g. The supernatant is removed and 0.2 ml glacial acetic acid is added to the charcoal. A solution of 25 mM ascorbic acid in H_2O is prepared and 1.8 ml is added to the charcoal. After vortexing for 3 min, the solution is separated from the charcoal by centrifugation, dried under reduced pressure, and stored in aliquots at $-20°$. Quantitative conversion to dehydroascorbic acid is documented by high-performance liquid chromatography according to the method of Farber *et al.*[7] in which dehydroascorbic acid is separated from ascorbic acid, reduced with dithiothreitol, and then measured as ascorbic acid following rechromatography by its absorbance at 254 nm.

Procedure. The standard reaction mixture contains 50 mM phosphate buffer (pH 6.8), 0.27 mM EDTA, 2.0 mM GSH, 1.0 mM dehydroascorbic acid, and 0.1 ml spinach extract in a total volume of 1 ml. The reaction is initiated by the simultaneous addition of dehydroascorbic acid and the enzyme preparation. Phosphate buffer is used instead of the cell extract in the blank. The absorbance at 265 nm is recorded at $37°$ for a 90-sec period and the increment over the nonenzymatic blank is taken as the measure of activity. The extinction coefficient of ascorbic acid ($\varepsilon_{mm} = 14.7$)[8] has been confirmed under assay conditions.

Units. A unit of enzyme activity is defined as the amount of enzyme which reduces 1 μmol of L-dehydroascorbic acid in 1 min at $37°$. Specific activity is expressed as units per milligram of protein. Protein concentration is measured by the method of Lowry *et al.*[9]

Purification Procedure

An extract from spinach leaves, used as a source of dehydroascorbate reductase, is prepared according to the method of Foyer and Halliwell.[10] Briefly, fresh mature deribbed spinach leaves are homogenized in 0.1 M KH_2PO_4 buffer (pH 6.3), for 1 min in a Waring blender. The homogenate is centrifuged at 20,000 g for 15 min. Solid $(NH_4)_2SO_4$ is slowly added to the supernatant to a saturation of 50% and the solution centrifuged at 20,000 g for 15 min. The $(NH_4)_2SO_4$ concentration of the supernatant is adjusted to 80%. Following another centrifugation at 20,000 g for 15 min, the supernatant is discarded and the pellet is redissolved in 10 mM

[7] C. M. Farber, S. Kanengiser, R. Stahl, L. Liebes, and R. Silber, *Anal. Biochem.* **134,** 355 (1983).
[8] C. Daglish, *Biochem. J.* **49,** 635 (1951).
[9] O. H. Lowry, N. J. Rosebrough, A. L. Farr, and R. J. Randall, *J. Biol. Chem.* **193,** 265 (1951).
[10] C. H. Foyer and B. Halliwell, *Phytochemistry* **16,** 1347 (1977).

KH$_2$PO$_4$ buffer (pH 6.3). The extract is dialyzed at 4° for 24 hr against 100 volumes 10 mM KH$_2$PO$_4$ buffer to remove GSH. This preparation serves as a source of dehydroascorbate reductase activity and is stable for at least 1 month after storage at $-20°$.

Properties. The enzyme obtained from the above purification procedure elutes from a Sephadex G-200 column as a symmetrical peak and has a MW of ~25,000.[10] The nonenzymatic reduction of dehydroascorbic acid by GSH occurs rapidly at pH 7.5–8.0; therefore despite the pH optimum of the enzymatic reduction of 7.5, the assays should be performed between pH 6.3 and 6.8. The K_m for GSH has been reported as 4.32 mM at pH 6.3 in the presence of 0.3 mM dehydroascorbate,[10] and as 3.7 mM at pH 6.8 with 0.5 mM dehydroascorbate.[4] The K_m for dehydroascorbic acid is much lower, reported as 0.34 mM at pH 6.3[10] and 0.47 mM at pH 6.8.[4]

Section II

Thiamin: Phosphates and Analogs

[3] Determination of Thiamin and Its Phosphate Esters by High-Performance Liquid Chromatography

By TAKASHI KAWASAKI

High-performance liquid chromatography (HPLC) of thiamin and its phosphate esters is the most sensitive method to determine these compounds at the subpicomole level within 10–20 min. This method is applicable to the analysis of thiamin and its phosphates not only in animal tissues but also in body fluids, foods, and pharmaceutical preparations. In addition, this method is useful in the study of intermediary metabolism of thiamin.[1]

Principle

In the chromatographic mode, two systems, straight-phase HPLC and reversed-phase HPLC, are used. In the former system, thiamin, ThMP,[2] ThDP, and ThTP are eluted from a column in this order. In the latter system, the elution order of thiamin phosphates is the reverse. When detection of thiamin phosphates is carried out fluorometrically, two different procedures are used: pre- and postcolumn derivatization methods. In the former method, thiamin and its phosphates in samples are first quantitatively converted to thiochrome and its phosphates, and then subjected to HPLC. In the latter method, thiamin and its phosphates are first chromatographed by HPLC, and then these compounds in the effluent are derivatized to thiochrome compounds followed by fluorometric detection. Although in the precolumn derivatization method, it is actually thiochrome and its phosphates that are chromatographed through a HPLC column, it is described in the text that thiamin and its phosphates are chromatographed by this method. The determination of thiamin and its phosphates by the precolumn derivatization method of the straight-phase and reversed-phase HPLC systems,[3,4] which have been developed in our laboratory, is described in this text.

[1] T. Kawasaki and H. Sanemori, in "Modern Chromatographic Analysis of the Vitamins" (A. P. DeLeenheer, E. E. S. Lambert, and G. M. De Ruyter, eds.), p. 385. Dekker, New York, 1985.
[2] Abbreviations used. HPLC, high-performance liquid chromatography; ThMP, thiamin monophosphate; ThDP, thiamin diphosphate; ThTP, thiamin triphosphate; Thc, thiochrome; TCA, trichloroacetic acid; DMF, dimethylformamide.
[3] K. Ishii, K. Sarai, H. Sanemori, and T. Kawasaki, Anal. Biochem. 97, 191 (1979).
[4] H. Sanemori, H. Ueki, and T. Kawasaki, Anal. Biochem. 107, 451 (1980).

METHODS IN ENZYMOLOGY, VOL. 122

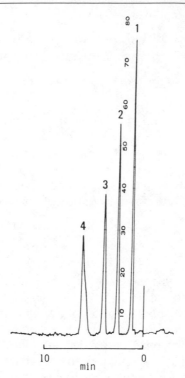

FIG. 1. Chromatogram of thiamin and its phosphates analyzed by straight-phase HPLC after precolumn derivatization of authentic compounds: 1, thiamin; 2, ThMP; 3, ThDP; 4, ThTP.

Reagents

Reagents used for conversion to thiochrome:
   BrCN, 0.3 M, prepared just before use
   NaOH, 1 M
Columns and solvents:
   LiChrosorb–NH$_2$ column for straight-phase HPLC:
      150 × 4.6 mm (i.d.), particle size 5 μm. The solvent is acetonitrile–90 mM potassium phosphate buffer (pH 8.4), 60 : 40 (v/v).
   TSK LS 410 (ODS) column for reversed-phase HPLC:
      300 × 4.6 mm (i.d.), particle size 10 μm. The solvent is 2.5% DMF–25 mM potassium phosphate buffer (pH 8.4) for elution of ThTP, ThDP, and ThMP, and 25% DMF–25 mM potassium phosphate buffer (pH 8.4) for subsequent elution of thiamin.
Instrumentation. The instrument used in our laboratory is a Jasco (Tokyo) component system composed of a pump (PG 350D) and a six-way

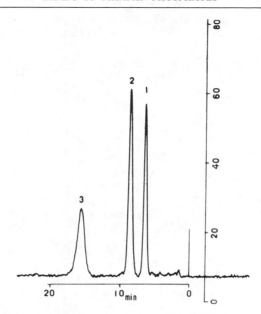

FIG. 2. Chromatogram of thiamin phosphates analyzed by reversed-phase HPLC after precolumn derivatization of authentic compounds: 1, ThTP; 2, ThDP; 3, ThMP.

injector (VL-611), which is connected to a spectrofluorometer (Jasco FP 110, flow cell volume 10 μl) and a LC data processor (Jasco DP-L220). Fluorometric detection is carried out at excitation wavelength of 365 nm and emission wavelength of 430 nm. Any commercially available instrumentation of HPLC equipped with a fluorometer may be used for the assay.

Procedure

To extract thiamin and its phosphates from animal tissues, tissue weighing 0.2–1 g is homogenized in 4 volumes of 10% TCA for 1 min by either a Teflon homogenizer or a Waring blender. The homogenate is centrifuged for 20 min at 15,000 g, and the supernatant is vigorously shaken with the same volume of water-saturated ethyl ether to remove TCA, followed by repeated treatment until pH of the water layer rises to 4.

Thiamin and its phosphates in the water layer of the extract or each authentic compound in standard solutions are derivatized to Thc and its phosphates. To 0.8 ml of sample solution, 0.1 ml of 0.3 M BrCN freshly prepared and then 0.1 ml of 0.1 M NaOH are added and mixed briefly. In

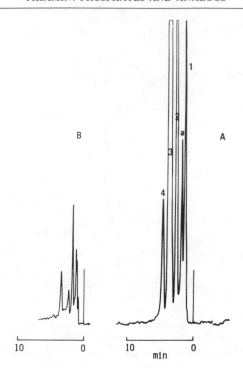

FIG. 3. Chromatogram of thiamin and its phosphates from rat liver by straight-phase HPLC after precolumn derivatization of the liver extract: (A) sample; (B) blank; 1, thiamin; 2, ThMP; 3, ThDP; 4, ThTP; and a, unidentified peak.

blank experiments, NaOH is first added to the water layer and mixed, followed by the addition of BrCN.

Ten microliters of the derivatized sample, standard solution, and the blank are subjected to HPLC. The flow rate of the effluent is 1.4 ml/min and the recording chart speed is 0.5 cm/min in both the straight-phase and reversed-phase HPLC methods. To elute thiamin by reversed-phase HPLC, the solvent containing 2.5% DMF and buffer (pH 8.4) is first used to elute ThTP, ThDP, and ThMP, and then is changed to that containing 25% DMF and buffer (pH 8.4). The effluent is then determined fluorometrically. The detection limit for thiamin and its phosphates is 0.05 pmol in straight- and reversed-phase HPLC.[1]

A chromatogram of authentic thiamin, ThMP, ThDP, and ThTP at each 1 pmol obtained by straight-phase HPLC is shown in Fig. 1, and one obtained by reversed-phase HPLC is shown in Fig. 2. Chromatograms of thiamin and its derivatives from rat liver, analyzed by straight-phase and the reversed-phase HPLC, are shown in Figs. 3 and 4, respectively. Blank

CONTENTS OF THIAMIN AND ITS PHOSPHATE ESTERS IN
RAT TISSUES[a]

Compound	Liver (nmol/g wet weight)	Brain (nmol/g wet weight)
Thiamin	1.45 ± 0.72 (4.8)	0.34 ± 0.13 (5.3)
ThMP	3.98 ± 1.87 (13.2)	1.27 ± 0.60 (19.6)
ThDP	24.02 ± 4.49 (79.7)	4.83 ± 0.49 (74.5)
ThTP	0.68 ± 0.1± (2.3)	0.04 ± 0.02 (0.6)
Total thiamin	30.13	6.48

[a] Values represent the means ± SD for five rats. The numbers
in parentheses refer to percentages of total thiamin.

sample peaks were subtracted from test sample peaks. Cellular concentrations of thiamin and its phosphates determined by the straight-phase HPLC were then calculated and are shown in the table.

Comments

1. All phosphate bonds remain intact when thiamin phosphates are converted to thiochrome phosphates by alkaline oxidation.[3]

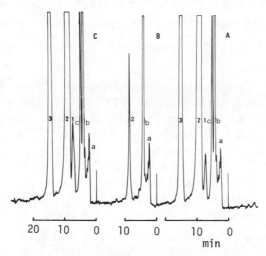

FIG. 4. Chromatogram of thiamin phosphates from rat liver by reversed-phase HPLC after precolumn derivatization of the liver extract: (A) sample; (B) blank; (C) sample plus 0.2 pmol of ThTP; 1, ThTP; 2, ThDP; 3, ThMP; a and b, peaks in the blank; and c, unidentified peak.

2. Standard curves should be prepared with each authentic thiamin, ThMP, ThDP, and ThTP, since the corresponding thiochromes show considerably different fluorescent intensity, as described previously.[3] Authentic thiamin phosphates should be checked by HPLC to determine the extent of contamination by the other phosphate forms of thiamin. For example, a commercially available ThDP preparation was found to contain 13.0% of ThMP and 1.25% of ThTP compared with 100% of ThDP.

3. The mobile phase used in the precolumn derivatization method of HPLC for thiamin and its phosphates should contain a buffer solution of pH above 8, since thiochrome fluoresces at pH values above 8.[3]

4. The postcolumn derivatization method of HPLC of thiamin and its phosphates has been reported both for straight-phase[5] and reversed-phase[6] systems. In this method, the effluent from the column was mixed with 0.01% potassium ferricyanide in 15% NaOH, which was sent to a mixing coil by a proportioning pump at the rate of 0.3 ml/min. Thiochrome fluorescence in the mixture was then measured fluorometrically. The detection limit of thiamin and its phosphates was 0.05 pmol.

[5] M. Kimura, T. Fujita, and Y. Itokawa, *J. Chromatogr.* **188**, 419 (1980).
[6] M. Kimura, B. Panijpan, and Y. Itokawa, *J. Chromatogr.* **245**, 141 (1982).

[4] Separation and Determination of Thiamin and Its Phosphate Esters by SP-Sephadex Chromatography

By JACK R. COOPER and TOSHIO MATSUDA

In recent years, the separation of thiamin, thiamin monophosphate (ThMP), thiamin pyrophosphate (ThPP), and thiamin triphosphate (ThTP) has been described using ion-exchange chromatography,[1] paper electrophoresis,[2] and high-performance liquid chromatography.[3–5] Because of interference from contaminants in tissue extracts, lack of separation of thiamin from ThMP requiring a separate assay after acid phosphatase treatment, and the lack of HPLC instrumentation, our laboratory developed a simple chromatographic procedure for brain extracts, which is reliable and capable of separating all forms of the vitamin.[6]

[1] J. M. Parkhomenko, A. A. Rybina, and A. G. Khalmuradoz, this series, Vol. 62, p. 59.
[2] H. K. Penttinen, *Acta Chem. Scand., Ser. B* **B32**, 609 (1978).
[3] K. Ishii, K. Sarai, H. Sanemore, and T. Kawasaki, *Anal. Biochem.* **97**, 191 (1979).
[4] B. C. Hemming and C. J. Gubler, *J. Liq. Chromatogr.* **3**, 1697 (1980).
[5] M. Kimura, T. Fujita, S. Nishida, and Y. Itokawa, *J. Chromatogr.* **188**, 417 (1980).
[6] T. Matsuda and J. R. Cooper, *Anal. Biochem.* **117**, 203 (1981).

Reagents

Citrate buffer, 0.1 M, pH 3.5
NaCl, 0.5 M in 0.1 M citrate buffer, pH 3.5
SP-Sephadex
Cyanogen bromide (bromine water, decolorized by the dropwise addition of 30% NaOH)

Procedure. Approximately 3 ml of a deproteinized brain extract is applied to a column (1.5 × 23 cm) of SP-Sephadex equilibrated with 0.01 M citrate buffer (pH 3.5). Stepwise elution at 0–4° is then carried out with 45 ml of 0.01 M citrate buffer to elute ThTP, 60 ml of 0.1 M citrate buffer for ThPP, and 90 ml of 0.5 M NaCl–0.1 M citrate buffer to elute ThMP and thiamin. The flow rate of the column is about 11 ml/cm²; 3-ml fractions are collected.

The thiamin compounds are assayed fluorometrically using the alkaline cyanogen bromide reaction of Fujiwara and Matsui.[7] To one aliquot of the fraction, 0.9 ml of cyanogen bromide is added, followed by 0.6 ml 30% NaOH. To the second aliquot, the order of addition is reversed in order to destroy thiamin and yield a blank value. The difference between the two readings is used to calculate the thiamin content. When assaying for ThPP and ThTP, readings are multiplied by 0.87 and 0.80, respectively, to correct for the quantal yield of the thiochrome fluorophore as previously reported.[8] In our laboratory, in order to ensure specificity in some extracts, thiaminase I was utilized.[9]

Comments

With this procedure, 10 pmol of the thiamin compounds could be reproducibly determined: linearity was obtained to 200 pmol. When 200 pmol of authentic radioactive thiamin and its phosphate esters were added to a brain extract, the recovery of the compounds was as follows: ThTP, 100.5%; ThPP, 98.3%; ThMP, 99.6%; thiamin 98.4%. However, as shown in Fig. 1 (lower panel) the fluorescence using the brain extract revealed six peaks (a–f) whereas only peaks a, c, e and f corresponded to authentic thiamin compounds. Although the excitation maximum (365 nm) was the same for all peaks and peaks c, e, and f exhibited the correct emission maximum for the vitamin (435–440 max), peak a showed not only an emission maximum at 435–440 nm but also a second one at 450 nm. Peaks b and d had emission maxima only at 450 nm. When the extracts were

[7] M. Fujiwara and K. Matsui, *Anal. Chem.* **25,** 810 (1953).
[8] H. K. Penttinen, this series, Vol. 62, p. 58.
[9] T. Matsuda and J. R. Cooper, *Proc. Natl. Acad. Sci. U.S.A.* **78,** 5886 (1981).

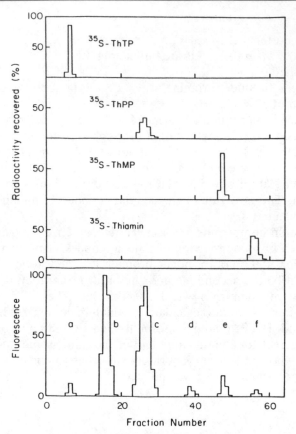

FIG. 1. Quantitative separation of thiamin and its phosphate esters from rat brain by SP-Sephadex column chromatography. The TCA extract from ~0.12 g wet weight of rat brain with (upper) or without (lower) each radioactive thiamin compound (0.2 nmol) was applied on a column of SP-Sephadex. The elution was carried out stepwise with 0.01 M citrate buffer (fraction 1–15), 0.1 M citrate buffer (fraction 16–35), and 0.1 M citrate buffer–0.5 M NaCl (fraction 36–65), and then radioactivity (upper) and fluorescence by alkaline cyanogen bromide (lower) of the eluates were measured as described under Procedure.

treated with thiaminase, peaks c, e, and f disappeared completely as would be expected, but peak a (ThTP) had a residual 13% fluorescence, indicating a contaminant. Accordingly, the value for ThTP was multiplied by 0.87. Peak b, which occurs between ThTP and ThPP, was not altered by thiaminase treatment. It should be noted that if acetate buffer is used instead of citrate, peaks a and b are not completely separated. We did not test d with thiaminase since it was of minor significance.

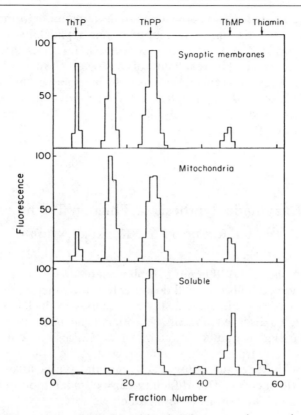

FIG. 2. Elution patterns of thiamin and its phosphate esters in synaptic plasma membrane, mitochondrial, and soluble fractions.

DISTRIBUTION OF THIAMIN AND ITS PHOSPHATE ESTERS IN RAT
BRAIN SUBFRACTIONS[a]

Fraction	Thiamin	ThMP	ThPP	ThTP
Membranes (3)	0	11.5 ± 1.0	71.9 ± 0.4	16.6 ± 1.3
Mitochondria (3)	0	10.9 ± 0.8	81.6 ± 2.1	7.5 ± 2.0
Soluble (2)	8.9	31.4	59.1	0.7

[a] The results are the means $\pm$ SEM where applicable. The number of separate experiments are shown in parentheses. The contents of total thiamin in membrane, mitochondrial, and soluble fractions were 15.1 ± 1.7, 149 ± 15, and 162 pmol/mg protein, respectively. Results are expressed as percentages in each fraction.

Since the procedure was developed to determine the forms of thiamin that occur in synaptic plasma membranes,[9] the applicability of the procedure was appraised after preparing subcellular fractions of rat brain as previously described. The results of this assay are shown in Fig. 2. After making the fluorescence corrections as described above, the final concentrations of thiamin and its phosphate esters are shown in the table. Using the procedure as described, only 40 mg of rat brain is required for a complete assay of all forms of the vitamin.

[5] Enzymatic Synthesis of Thiamin Triphosphate

By JACK R. COOPER and KOHSUKE NISHINO

Although the role of thiamin pyrophosphate (ThPP) as a coenzyme in intermediary metabolism is well documented, the function of thiamin triphosphate (ThTP) is still unclear. In nervous tissue, ThTP has been implicated in Leigh's disease (subacute necrotizing encephalomyelopathy) and considerable evidence, albeit circumstantial, suggests a role of this ester in conduction and transmission.[1] Until recently, the isolation and characterization of the enzyme system catalyzing the synthesis of ThTP have met with little success. This difficulty can be understood in light of this complex enzyme system as described below.

Assay Method

Principle. The substrate for this phosphoryltransferase (EC 2.7.4.15, thiamin-diphosphate kinase) is not free ThPP but ThPP that is bound to a protein. This substrate is prepared by injecting rats with [^{35}S]thiamin and subsequently isolating the [^{35}S]ThPP-protein from liver by ion-exchange and gel chromatography. After incubation of the labeled substrate with the enzyme and cofactor, [^{35}S]ThTP is isolated by column chromatography and the radioactivity determined.[2]

Reagents

Tris–HCl (pH 7.4), 0.5 *M*
ATP, 0.1 *M*

[1] J. R. Cooper and J. H. Pincus, *Neurochem. Res.* **4**, 223 (1979).
[2] K. Nishino, Y. Itokawa, N. Nishino, K. Piros, and J. R. Cooper, *J. Biol. Chem.* **258**, 11871 (1983).

$MgCl_2$, 0.1 M

Glucose, 0.1 M

[^{25}S]ThPP-protein substrate (equivalent to 4.0–6.0 μM [^{35}S]thiamin pyrophosphate)

Enzyme preparation

Trichloroacetic acid (TCA), 20%

Procedure. In a final volume of 100 μl are added, 10 μl of Tris–HCl (pH 7.4), 5 μl ATP, 5 μl $MgCl_2$, 10 μl glucose, 50 μl radioactive substrate, and 20 μl enzyme preparation. The reaction is initiated by the addition of the enzyme after the mixture was preincubated for 2 min at 37°. Following a 10 min incubation, 10 μl of TCA is added to terminate the reaction, followed by centrifugation. The supernatant is removed, extracted three times with 1.5 ml water–saturated ether to remove the TCA, and the residual ether in the supernatant removed by evaporation at 37° for 5 min. Approximately 70–80 μl of the supernatant is then applied to an AG-50-X8 (H$^+$) column (0.7 × 2.0 cm) and the ThTP eluted with 1 ml of water. The effluent, containing 90% of the radioactive ThTP that is synthesized, is collected in a scintillation vial and the radioactivity determined after the addition of scintillant. The amount of ThTP formed is calculated from the radioactivity and the previously calculated specific activity of the labeled substrate. Controls are carried out through the procedure both using zero time and with the omission of ATP from the incubation medium.

Preparation of [^{35}S]ThPP Substrate

After 6 weeks on a thiamin-deficient diet, Sprague–Dawley male rats are injected ip with 100 μCi of [^{35}S]thiamin–HCl (specific activity 40–200 mCi/mmol) dissolved in 0.3 ml of physiological saline twice at a 24-hr interval. Twenty-four hours after the second injection, rats are sacrificed, and the livers removed and homogenized with 4 volumes of 0.25 M sucrose containing 30 mM Tris–HCl (pH 8.6) and 0.1 mM EDTA. After centrifugation at 40,000 g for 20 min at 0–4°, the supernatant is removed and centrifuged at 105,000 g for 90 min. The resultant supernatant is dialyzed overnight against 30 mM Tris–HCl buffer (pH 8.6) at 0–4° and then applied to a DEAE-Sephadex A-25 column (6 × 40 cm) equilibrated with 30 mM Tris–HCl (pH 8.6). Elution with the same buffer yields the unabsorbed labeled substrate. These fractions are pooled, the pH adjusted to 7.4, and the pool is concentrated under a nitrogen atmosphere using an Amicon PM10 apparatus. The concentrate is then applied to a Sephadex G-100 column (4.2 × 150 cm) equilibrated with the same buffer. Radioactive fractions (Fig. 1) are collected and concentrated similarly to that described above. This fraction can be used as substrate for most

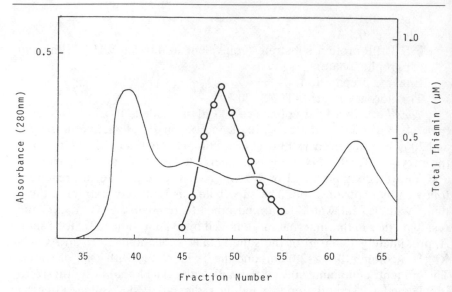

FIG. 1. Elution pattern of substrate protein on Sephadex G-100 column chromatography. The concentrated effluent containing substrate activity from DEAE-Sephadex A-25 chromatography was applied to a Sephadex G-100 column preequilibrated with 50 mM potassium phosphate buffer (pH 7.4), and eluted with the same buffer. Solid line, protein concentration; (○) total thiamin content of substrate protein.

experiments although it does contain a small amount of phosphoryltransferase. To block this activity, 1.0 M potassium phosphate buffer (pH 8.0), and 0.1 M iodoacetamide are added to the Sephadex G-100 eluant in a final concentration of 0.1 and 20 mM, respectively. The eluant is incubated at 37° for 30 min, centrifuged to remove the resultant precipitate, and then dialyzed against 20 mM potassium phosphate buffer (pH 7.4) overnight at 0–4°. The dialyzed sample is then concentrated with the Amicon PM10 membrane.

Purification of ThPP-ATP Phosphoryl Transferase

Step 1. Acetone Powder Preparation. Fresh bovine brain cortex, after the removal of meninges, is homogenized with 9 volumes of ice-cold 0.32 M sucrose containing 20 mM potassium phosphate buffer (pH 7.4), and 40 μM disodium EDTA in a Waring blender. After centrifugation at 1000 g for 10 min, the supernatant is removed and centrifuged at 10,800 g for 20 min. The supernatant is removed and the pellet is resuspended in 50 mM potassium phosphate buffer (pH 7.4), and recentrifuged for 20 min at 10,000 g. The resultant pellet (P_2) is resuspended in a small volume of the phosphate buffer and mixed with 9 volumes of acetone (−20°), and the

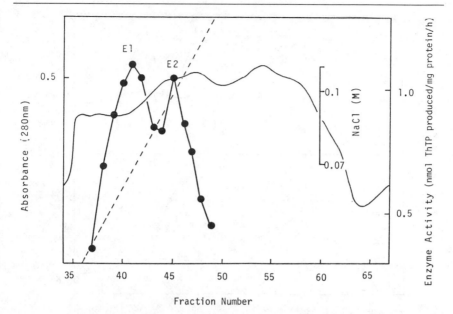

FIG. 2. Elution pattern of cellulose DE-52 column chromatography. The enzyme prepa-ration was applied to a DE-52 column and was eluted with a liner gradient of equal volumes of 20 mM potassium phosphate buffer (pH 7.6), containing 20% glycerol and 1.0 mM DTT, and 20 mM potassium phosphate buffer (pH 7.6), containing 20% glycerol, 1.0 mM DTT, and 1.0 M NaCl. Solid line, protein concentration; dotted line, NaCl concentration; (●) enzyme activity.

suspension stirred for 10 min at 0–4°. After filtration on a Büchner funnel through Whatman 3 MM filter paper, the acetone powder is washed with cold ether and air dried. The acetone powder is stored at −20° in a vac-uum desiccator.

Step 2. Extraction. The acetone powder is extracted with 30 volumes of 20 mM potassium phosphate buffer (pH 7.6), containing 1.0 mM dithiothreitol (DTT) via homogenization. After centrifugation at 105,000 g for 90 min, the supernatant is removed, the pellet reextracted and centri-fuged as above, and the two supernatants combined. Glycerol to a final concentration of 20% is then added to stabilize the enzyme.

Step 3. DE-52 Column Chromatography. The enzyme solution is ap-plied to a DE-52 column (6.5 × 30 cm) that has been preequilibrated with 20 mM potassium phosphate buffer (pH 7.6), containing 20% glycerol and 1.0 mM DTT. The column is washed with the same buffer and the enzyme eluted with a linear gradient of equal volumes (1500 ml) of the equilibrat-ing buffer and the buffer containing 1.0 M NaCl. From this procedure, as shown in Fig. 2, two peaks of enzyme activity were collected. Each

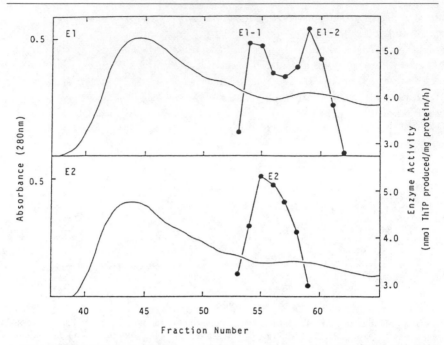

FIG. 3. Elution pattern of Sephacryl S-200 column chromatography. Sephacryl S-200 column (4.4 × 90 cm) was employed. Ten to 15 ml of the enzyme preparation of Step 3 (fractions E1 and E2) was applied separately and eluted with 50 mM potassium phosphate buffer (pH 7.2), containing 20% glycerol and 1.0 mM DTT. The elution speed was 4.0 cm/hr and about 12 1-ml fractions were collected. Solid line, protein; (●) enzyme activity.

enzyme fraction was individually concentrated under nitrogen by ultra-filtration using an Amicon PM10 filter membrane.

Step 4. Sephacryl S-200 Column Chromatography. The two enzyme fractions (E1 and E2) from Step 3 are separately applied to a column of Sephacryl S-200 (4.4 × 90 cm) equilibrated with 50 mM potassium phosphate buffer (pH 7.2), containing 20% glycerol and 1 mM DTT and eluted with the same buffer. As shown in Fig. 3, the E1 fraction gives two active fractions, referred to as E1-1 and E1-2, and the E2 fraction yields one peak of activity, referred to as E2.

Step 5. Chromatofocusing Chromatography. The active fractions from Step 4 were separately applied to a chromatofocusing column (1 × 37 cm) that was preequilibrated with 0.025 M histidine-HCl buffer (pH 6.2), containing 20% glycerol and 1.0 mM DTT. The eluant was polybuffer 74 HCl, pH 4.0 (0.0075 mmol/pH unit), containing 20% glycerol and 1.0 mM DTT. Active fractions were eluted at pH 5.1 ± 0.02. In order to eliminate polybuffer, which inactivates enzyme activity, the eluted fractions were immediately applied to a Sephadex G-75 column (3.2 × 10 cm) preequilibrated with 50 mM potassium phosphate buffer (pH 7.4) containing 20% glycerol, and elution carried out with the same buffer. To concentrate and retain enzyme activity, the eluants from the Sephadex chromatography were applied to an organomercurial agarose column (Affi-Gel 501, 0.7 × 4.0 cm) which was preequilibrated with 50 mM potassium phosphate buffer (pH 7.4), containing 20% glycerol and eluted with 3 ml of 50 mM potassium phosphate buffer (pH 7.4), containing 20% glycerol and 20 mM DTT.

Properties

Purity and Physicochemical Properties. The three active fractions (E1-1, E1-2, and E2) on disc gel electrophoresis showed a single protein band with the same R_f: the molecular weight of the enzyme was calculated to be 103,000. The pH optimum was 7.5, with Tris–HCl exhibiting slightly higher activity than phosphate buffer.

Dependencies of the Reaction. The reaction has an absolute dependence on ATP, Mg^{2+}, the cofactor (presumably glucose), and ThPP that is protein bound. When the bound substrate is replaced by free ThPP, no synthesis of ThTP is observed.

Inhibitions. p-Chloromercuribenzoate at $5 × 10^{-4}$ M inhibits activity by 79.5% and N-ethylmaleimide ($1 × 10^{-2}$ M) inhibits by 64%.

Miscellaneous. In addition to the substrate for the reaction being protein bound, ThTP is also synthesized bound to a protein from which it is released for assay on deproteinization. Both the protein-bound substrate and the enzyme are found in both brain and liver. We chose to prepare the substrate protein from liver because of the yield; we prepared the enzyme from brain since earlier work[3] had indicated that patients with Leigh's disease (subacute necrotizing encephalomyelopathy) excrete material which inhibits the synthesis of ThTP using a crude brain preparation but not the liver preparation.

[3] J. R. Cooper and J. H. Pincus, *J. Agric. Food Chem.* **20**, 490 (1972).

Section III

Pantothenic Acid, Coenzyme A, and Derivatives

[6] Pantothenase-Based Assay of Pantothenic Acid

By R. KALERVO AIRAS

The assay of pantothenic acid presented in this chapter is based on the exchange of radioactive β-alanine into preexisting pantothenic acid by pantothenase (EC 3.5.1.22).[1,2] The basic reaction of this enzyme is the hydrolysis of pantothenic acid to form β-alanine and D-pantoyl lactone. Catalysis occurs through an acyl enzyme intermediate. The first acylation step is fully reversible and fast enough to enable the exchange of the β-[14C]alanine into pantothenate.

$$\text{D-Pantothenate} + \text{E} \; \overset{\beta\text{-alanine}}{\underset{\beta\text{-[}^{14}\text{C]alanine}}{\rightleftharpoons}} \; \text{pantoyl-E} \rightleftharpoons \text{D-pantoyl lactone}$$

The rate of exchange of β-[14C]alanine depends on the amount of pantothenate at nonsaturating substrate concentrations, thereby making the assay possible.

Assay Procedure

The assay consists of extraction of pantothenate, the enzymatic reaction, separation of β-alanine and pantothenate by paper chromatography, and counting of the radioactivity of the pantothenate spot.

Extraction of pantothenic acid from various materials is somewhat limited by the pantothenase-based assay. The pantothenic acid should exist in concentrations of 10 μM or more. The extract should not contain organic solvents or high concentrations of chlorides or carboxylic acids, and the pH should be adjusted to 9.4. A simple acetone extraction of the pantothenic acid from biological materials was used in the original pantothenase-based method.[1]

The pantothenase is purified from *Pseudomonas fluorescens* NCIB 12017 (earlier *P. fluorescens* UK-1) as described previously.[2] A solution of pantothenase containing about 30 nkat/ml is prepared in 5 mM K$_2$SO$_4$ and stored frozen at $-20°$ in small portions. Repeated cycles of freezing and thawing inactivate the enzyme.

The reaction mixture is prepared by pipetting equal (e.g., 20 μl) volumes of the following five solutions: (1) the sample containing 10 μM–20

---

[1] R. K. Airas, *Anal. Biochem.* **134**, 122 (1983).
[2] R. K. Airas, this series, Vol. 62, p. 267.

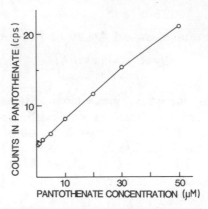

FIG. 1. Standard curve of the pantothenate assay. The reaction mixture (100 μl) contained 1 mM β-alanine with 3750 counts/sec of β-[^{14}C]alanine. Samples of 60 μl from the reaction mixture were drawn off at 10-min reaction times. No corrections for the background have been made.

mM pantothenate, adjusted to pH 9.4, (2) water or the pantothenate standard (e.g., 500 μM), (3) 500 mM diethanolamine–H$_2$SO$_4$ buffer (pH 9.4), (4) 5 mM β-alanine containing 500–5000 counts/sec per 20 μl of β-[1-^{14}C]alanine, and (5) the enzyme solution containing 0.5–1 nkat/20 μl. The concentrations in the reaction mixture are 100 mM diethanolamine–H$_2$SO$_4$ (pH 9.3), and 1 mM β-alanine. The reaction temperature is 25°.

Samples of 20–60 μl from the reaction mixture are pipetted onto Whatman No. 3 MM chromatography paper, and the chromatograms are developed for about 1.5 hr in 1-butanol–acetic acid–water (25 : 4 : 10 by volume). The paper strips containing the pantothenic acid spots, between R_f 0.6 and 0.9, are cut off and counted for radioactivity in a liquid scintillation counter. (For checking the R_f values, spots containing over 10 μg of β-alanine and pantothenate can be made visible by heating the chromatograms at 200° for about 5 min.)

The radioactivity in pantothenate is first increased during equilibration between β-alanine and pantothenate, and is then slowly decreased when the pantoyl lactone is released.[1] The samples for the assay should be taken before the maximum level of radioactivity in pantothenate has been attained, because corrections for inhibitions are then easier to make. Figure 1 shows the standard curve of the pantothenate assay. Lower β-alanine concentrations can also be used, but the assay becomes more sensitive to inhibiting compounds. Higher β-alanine concentrations should be used for a linear standard curve at high pantothenate concentrations. The background apparently varies depending on the isotope prepa-

ASSAY OF PANTOTHENATE FROM FOODS[a]

Food	Counts/sec	Pantothenate (μmol/kg)	Inhibition (%) caused by biological material
Potatoes	1.64	12	12
Beef	2.61	49	31
Mushrooms			
(*Lactarius rufus*)	2.51	36	15
Hen eggs	5.99	96	18
Wheat flour	1.57	19	26

[a] Pantothenate was extracted as described in Airas,[1] except the pH was adjusted with ammonia. The assay conditions were as in Fig. 1. The samples were assayed (a) without added pantothenate, (b) with 100 μM pantothenate added to the reaction mixture (to detect the inhibition), and (c) with 50 nmol of added pantothenate in the water phase of the extract. The counts from samples (a) and (c) were used to calculate the amounts of pantothenate.

ration used. In this case it was 0.18% of the total counts. As shown in the table, the assay has been applied to various food materials.

Comments

The transfer of β-alanine into pantothenate is inhibited by certain compounds present in biological samples. The chloride ion in concentrations as low as 50 mM causes 50% inhibition. Hydrochloric acid cannot, therefore, be used in preparing buffers. Other inhibiting compounds include glycine, carbonate, some carboxylic acids, and organic solvents. The presence of β-alanine in the samples also disturbs the assay. The assay can be done in spite of the inhibiting compounds if the samples are assayed both with and without a known amount of pantothenate.

The microbiological methods[3] for pantothenate assays are essentially vitamin assays and also measure other metabolites in the pathway from pantothenic acid to CoA. The pantothenase-based method measures only free pantothenic acid. This specificity can be either a limitation or an advantage depending on what is to be measured. This is also the case for the radioimmunoassay of pantothenate.[4]

[3] O. Solberg and I. K. Hegna, this series, Vol. 62, p. 201.
[4] C. Wittwer, B. Wyse, and R. G. Hansen, *Anal. Biochem.* **122,** 213 (1982).

[7] Enzymatic Hydrolysis of Pantetheine

By CARL T. WITTWER, BONITA W. WYSE, and R. GAURTH HANSEN

The enzymatic hydrolysis of pantetheine to pantothenate and cysteamine is the final common reaction in the degradation of the coenzymes phosphopantetheine and coenzyme A to the vitamin pantothenic acid. The enzymatic hydrolysis of pantetheine has been monitored by separation of radiolabeled substrate and product by electrophoresis,[1] pH-stat titration of reaction products,[1] and amino acid analysis of the N-ethylmaleimide adduct of cysteamine.[2] We had difficulty repeating the reported purification of this enzyme from horse kidney,[1] and subsequently developed four independent assays for the enzymatic hydrolysis of pantetheine,[3] and purified the enzyme from pig kidney.[4] An additional very useful assay has recently been reported.[5] As a continuous extension of the cysteamine assay,[4] it is based on the hydrolysis of the substrate, S-pantetheine-3-pyruvate. Presented below are our two most useful quantitative enzyme assays, a qualitative assay for visualization of enzyme activity on polyacrylamide gels, and our purification procedure of the enzyme from pig kidney. The radiolabeled assay is best used on tissue homogenates, whereas the continuous mercaptide assay is more convenient for partially purified preparations.

Assay Methods

Radiolabeled Assay[3]

Principle. Radiolabeled substrate and product are separated by a fast paper chromatography system and quantified by liquid scintillation counting.

Reagents

Tris–HCl buffer, 0.5 M, pH 8.1
Dithioerythritol, 0.1 M
Acetone

[1] S. Dupre and D. Cavallini, this series, Vol. 62, p. 262.
[2] S. Orloff, J. DeB. Butler, D. Towne, A. B. Mukherjee, and J. D. Schulman, *Pediatr. Res.* **15**, 1063 (1981).
[3] C. Wittwer, B. Wyse, and R. G. Hansen, *Anal. Biochem.* **122**, 213 (1982).
[4] C. T. Wittwer, D. Burkhard, K. Ririe, R. Rasmussen, J. Brown, B. W. Wyse, and R. G. Hansen, *J. Biol. Chem.* **258**, 9733 (1983).
[5] S. Dupre, R. Chiaraluce, M. Nardini, C. Cannella, G. Ricci, and D. Cavallini, *Anal. Biochem.* **142**, 175 (1984).

D-[^{14}C]Pantethine, 40 mM, 70 μCi/mmol. The following procedure is adapted from the 1-hydroxybenzotriazole-activated carbodiimide method of König and Gieger[6]: Dissolve 206 mg of dicyclohexylcarbodiimide and 135 mg of 1-hydroxybenzotriazole in 5.8 ml of dimethylformamide. Suspend 113 mg of cystamine–2HCl in the solution with magnetic stirring and cool the mixture on ice. Add 350 μl of aqueous sodium D-[1-^{14}C]pantothenate, 100 μCi/ml (specific radioactivity ~50 mCi/mmol), followed by 238 mg of calcium pantothenate (added slowly with stirring). All reactants dissolve within 30 min and dicyclohexylurea starts to precipitate within an hour. Keep on ice for a total of 2 hr and then overnight at room temperature. Rotary evaporate the reaction mixture below 30° and suspend in ~8 ml of H$_2$O. Filter off the precipitate in the cold and extract twice with an equal volume of ether. Rotary evaporate the aqueous phase below 30°, redissolve in H$_2$O, and pass through a 1.0 × 30-cm column of mixed-bed ion-exchange resin (e.g., AG-501-X8D), followed by ~250 ml of H$_2$O. Concentrate to ~10 ml and standardize the concentration by disulfide analysis.[7]

Procedure. Both substrate and enzyme are reduced before assay. Reduce the substrate by incubating [^{14}C]pantethine, dithioerythritol, Tris buffer, and H$_2$O (25:20:8:27) at 37° for at least 20 min. Reduce the enzyme at 37° for 20 min with 8 μl dithioerythritol and 8 μl Tris buffer in a total volume of 80 μl. The reaction is initiated by adding 20 μl of reduced substrate solution to 80 μl of the reduced enzyme.

The reaction is stopped after 15–30 min at 37° by the addition of 200 μl of acetone. After centrifugation, the supernatant is entirely applied near one end of a 2.5 × 22-cm strip of Whatman No. 4 filter paper with a hot-air dryer. With acetone–H$_2$O (90:10) as solvent, ascending chromatography in a rectangular chromatography jar (6 × 10 × 15 cm) *open to the atmosphere* separates labeled substrate and product. The solvent quickly travels up the paper, then slows and stops from evaporation. Pantetheine moves with the solvent front and pantothenate remains where spotted. After 15 min, the filter paper is removed, dried, and cut halfway between the origin and the solvent front. Each half is placed in a separate scintillation vial and quantified by liquid scintillation counting after the addition of a toluene-based cocktail.

Mercaptide Assay[3]

Principle. At a pH buffered around the pK value of the thiol groups of pantetheine (pK 9.9) and cysteamine (pK 8.1), the concentration of mer-

[6] W. König and R. Gieger, *Chem. Ber.* **103,** 788 (1970).
[7] W. L. Zahler and W. W. Cleland, *J. Biol. Chem.* **243,** 716 (1968).

captide ion increases as enzymatic hydrolysis proceeds. The mercaptide ion has a peak absorption around 235 nm of about 5000 M^{-1} cm^{-1}.

Reagents

Tris–HCl buffer, 0.5 M, pH 8.1

Tris–HCl buffer, 0.05 M, pH 8.1, deoxygenated by bubbling N_2 through the solution, and warmed to 37°

Dithioerythritol, 0.1 M

D-Pantetheine, 0.1 M. The following procedure is adapted from the method of Butler *et al.* for the reduction of disulfide-containing amines.[8] Dissolve 440 mg of D-pantethine and 370 mg of dithiothreitol in ~3.8 ml H_2O. Add 60 μl of 1 N NaOH and incubate for 30 min at 37°. Neutralize the solution with 60 μl of 1 N HCl and extract twice with 10 volumes of ethyl acetate saturated with H_2O, separating the phases by brief centrifugation. Rotary evaporate off excess ethyl acetate and standardize the concentration by sulfhydral analysis.[9] Alternatively, pantethine can be reduced to pantetheine with an excess of cysteine, which is later removed by ion exchange.[5]

Procedure. Reduce the enzyme at 37° for 20 min by adding 30 μl of 0.5 M Tris buffer and 30 μl of dithioerythritol and diluting the solution to a total volume of 300 μl. After reduction, add 2.685 ml of 0.05 M Tris buffer and follow the absorbance at 240 nm until a constant decreasing slope can be determined (~10 min). Add 15 μl of pantetheine and again follow the absorbance at 240 nm. The change in slope after substrate addition is proportional to the enzyme activity. The assay must be calibrated for each new batch of Tris buffer by running reactions to substrate exhaustion.

Visualization of Enzyme Activity on Polyacrylamide Gels[4]

Principle. Activity on nondenaturing polyacrylamide gels is visualized by local titration of cresol red. The enzymatic reaction produces acid at pH values intermediate to the pK values of pantetheine and cysteamine.

Reagents

D-Pantetheine, 100 mM

Dithiothreitol, 100 mM

Cresol red, 1 mM

Sodium hydroxide, 50 mM

[8] J. Butler, S. P. Spielberg, and J. D. Schulman, *Anal. Biochem.* **75,** 674 (1976).
[9] G. L. Ellman, *Arch. Biochem. Biophys.* **82,** 70 (1959).

Procedure. Apply protein with an enzyme activity of 10–100 nmol of pantetheine hydrolyzed per minute to 3-mm tube gels 10% in acrylamide and perform electrophoresis after Brewer and Ashworth.[10] Immediately after electrophoresis is completed, stir the gels in 500 volumes of deionized H_2O for 15 min. Repeat the washing for two additional 15-min periods. Bathe the gels in at least 2 volumes of pantetheine, dithiothreitol, cresol red, and H_2O (1 : 1 : 5 : 13) and add a minimum amount of sodium hydroxide to keep the solution red in color. After 10–60 min, bands of activity in the gels will turn yellow and can be monitored at 435 nm. This procedure has also been used with 1- or 3-mm slab gels. The activity stain is transient and slowly broadens secondary to diffusion.

Definition of Unit and Specific Activity. One unit hydrolyzes 1 μmol of pantetheine per minute when measured by the mercaptide assay. The relative activity of several commonly used assays has been compared.[3,5] Protein is determined by the method of Lowry *et al.*[11] Specific activity is given in units per milligram of protein.

Enzyme Purification

Fresh pig kidneys from a local slaughterhouse are cooled on ice and slices of cortex frozen until use. A total amount of 550 g of semisolid, thawed (4°, 8 hr) cortex is carried through the purification. Our equipment limited the scale of Steps 1 through 3 to ~275 g of cortex, so these steps were performed twice per purification. All steps are performed at 0–4° unless otherwise indicated.

Step 1. Homogenate. Kidney cortex (275 g) is homogenized with 4 volumes of 0.02 M Tris–HCl (pH 8.2), in a commercial Waring blender for 4 min at high speed. The suspension is centrifuged for 10 min at 10,000 g and the supernatant recovered by aspiration.

Step 2. Microsomal Agglutination. Microsomes in the supernatant are agglutinated at pH 4.2 by the addition of 1 M formic acid and the suspension centrifuged for 10 min at 20,000 g. The precipitate is resuspended in ~400 ml of deionized water, first by shaking in the original bottles, followed by homogenization in a Waring blender at high speed for 20 sec. The thick suspension is brought to pH 5.75 with 0.5 M Tris–HCl (pH 9.0).

Step 3. Solubilization by Treatment with 1-Butanol. 1-Butanol (1.5 volumes) is added to the microsomal suspension and the mixture homogenized in a Waring blender at low speed for 2 sec. After centrifugation for 10 min at 10,000 g, four layers become apparent from top to bottom: a

[10] J. M. Brewer and R. B. Ashworth, *J. Chem. Educ.* **46,** 41 (1969).

[11] O. H. Lowry, N. J. Rosebrough, A. L. Farr, and R. J. Randall, *J. Biol. Chem.* **193,** 265 (1951).

TABLE I
PURIFICATION OF A PANTETHEINE-HYDROLYZING ENZYME FROM
PIG KIDNEY

Step and fraction	Volume (ml)	Activity[a] (units)	Protein[b] (mg)	Specific activity (units/mg)
1. Homogenate	2700	250	99,000	0.0025
2. Microsomal agglutination	1100	170	43,000	0.0040
3. Butanol solubilization	990	150	3,900	0.038
4. Heat treatment	990	130	1,400	0.093
5. $(NH_4)_2SO_4$ fractionation	20	92	260	0.35
6. Hydrophobic chromatography	160	68	15	4.5
7. Hydroxyapatite chromatography	23	56	4.0	14

[a] As measured by the mercaptide assay. Samples from Steps 1 and 2 were measured by the radiolabeled assay and corrected for the difference between the radiolabeled and mercaptide assays. This difference was obtained by measuring later steps with both assays.

[b] Protein in Steps 1 and 2 was measured by the method of Lowry *et al.*[11] Later steps used an $E_{1 cm}^{1\%}$ (280) = 11.3.

yellow butanol layer, an intermediate zone of semisolid precipitate, a brown aqueous layer, and a small solid pellet. The aqueous phase is aspirated and immediately passed through a column (5.5 × 65 cm) of Sephadex G-25 (Pharmacia) previously equilibrated with 0.02 M phosphate (pH 7.0). The separation of protein (brown) from low molecular weight (yellow) compounds can be followed visually.

Step 4. Heat Treatment. Pool the void volume eluates of two runs through Step 3 and bring to pH 5.0 with 1 M formic acid. Heat in a 75° water bath up to 70° and hold at 70° for 2 min. Cool the suspension rapidly on ice and centrifuge for 10 min at 20,000 g, discarding coagulated protein.

Step 5. $(NH_4)_2SO_4$ Fractionation. The supernatant is saturated to 55% by adding 32.6 g of solid $(NH_4)_2SO_4$ per 100 ml and centrifuged for 15 min at 20,000 g. This supernatant in turn is saturated to 70% by adding 9.3 g $(NH_4)_2SO_4$ per 100 ml and centrifuged for 30 min at 20,000 g. The precipitate is dissolved in ~20 ml of buffer A [114 g of $(NH_4)_2SO_4$ added to 1 liter of 0.02 M phosphate and adjusted to pH 6.5 at 4°].

Step 6. Hydrophobic Chromatography. The protein from Step 5 is applied to a 2.5 × 23-cm column of octyl-Sepharose CL-4B (Pharmacia), previously equilibrated with buffer A. Wash the column with buffer A at a flow rate of ~150 ml/hr until the A_{280} of the eluate is less than 0.05. Elute the enzyme with a 20% ethylene glycol/80% buffer A (v/v) solution at the same flow rate until the A_{280} is less than 0.03. The entire protein peak is combined. The column can be regenerated by sequentially passing 1 column volume of 50 mM NaOH, 95% ethanol, deionized H_2O, and buffer A through the column.

Step 7. Hydroxyapatite Chromatography. The enzyme eluate from Step 6 is concentrated to ~20 ml by ultrafiltration and diafiltered into 0.001 M phosphate (pH 7.0) with an Amicon PM10 ultrafiltration membrane according to the directions from the manufacturer. The solution is applied at ~25 ml/hr to a 1.5 × 5-cm BioGel HT (Bio-Rad) hydroxyapatite column previously equilibrated with 0.001 M phosphate (pH 7.0). Enzyme is eluted at the same rate with 0.02 M phosphate (pH 7.0). Contaminating protein remains bound to the column.

The purification is summarized in Table I. The final preparation shows constant specific activity on elution from the hydroxyapatite column and only one band on nondenaturing and SDS electrophoresis with comigration of carbohydrate and enzyme activity. Microheterogeneity is apparent on isoelectrofocusing.

Properties

Stability and pH Optimum. At any step after enzyme solubilization (Step 3), the enzyme can be stored indefinitely in the frozen state without significant loss of activity. The pH profile of the pig kidney enzyme shows a broad plateau between pH 3 and 7 with an optimum pH of 9.0–9.5.[3] The horse kidney enzyme was initially reported to have a pH optimum between 4.0 and 5.5.[1] However, more recently, a similar basic pH optimum has been reported.[5]

Inhibitors and Activators. Similar to previous reports,[1] we have found the enzyme activated by thiols (e.g., 1–13 mM dithioerythritol),[3] and inhibited by thiol inhibitors (e.g., Hg^{2+}). Some inhibition by products is present,[1] but we have not observed any substrate inhibition if care is taken to reduce pantethine to pantetheine before initiation of the assay.

Substrate Specificity. The enzyme is very specific for the D-pantothenate portion of the molecule. Coenzyme A and β-alanylcysteamine are not hydrolyzed.[1,4] Analogs with only slight alterations in the pantothenate moiety, including L-pantetheine,[1] and analogs with a one-carbon increase

TABLE II
PANTETHEINE-HYDROLYZING ACTIVITY IN
VARIOUS RAT TISSUES[a]

Tissue	Specific activity (μunits/mg)
Kidney	470
Heart	420
Small intestine	390
Lung	150
Spleen	130
Large intestine	90
Testes	80
Liver	80
Salivary gland	70
Muscle	<50
Brain	<50

[a] We thank Mr. Sullivan Beck for running the radiolabled assay on these tissues.

or decrease in chain length[4] show little, if any, hydrolysis with the enzyme. The fully purified enzyme from pig kidney has insignificant activity against D-4'-phosphopantetheine (<0.5%) and can be used to assay specifically for D-pantetheine in the presence of phosphopantetheine.[12]

The enzyme is not specific for the cysteamine moiety.[4] S-Ethylpantetheine and pantothenyl-β-aminoethanol are readily hydrolyzed. The hydrolysis of N,N'-bis(pantothenyl)-1,6-diaminohexane suggests that pantethine is also a substrate. S-Pantetheine 3-pyruvate is hydrolyzed, and one of the products, S-cysteamine 3-pyruvate, spontaneously forms a cyclic compound with $\varepsilon = 6200$ at 296 nm.[5]

Kinetic Data. The apparent K_m of the pig kidney enzyme for D-pantetheine is 20 μM.[3] A value of 5 mM was originally reported for the horse kidney enzyme,[1] but a more recent estimate is 28 μM.[5] The maximum specific activity of the purified pig kidney enzyme is 14 μmol of pantetheine hydrolyzed/min · mg.[4] This compares to a specific activity of 0.4 μmol/min · mg for the reportedly homogeneous preparation from horse kidney.[1]

Physical Properties.[4] Estimated molecular weights from gel filtration and SDS electrophoresis are 54,000 and 60,000, respectively. The enzyme is a glycoprotein containing 11.8% carbohydrate by weight. Its extinction coefficient at 280 nm is $E_{1\,cm}^{1\%} = 11.3$, compared to $E_{1\,cm}^{1\%} = 15.2$ for the horse kidney preparation.[1]

[12] C. T. Wittwer, B. W. Wyse, and R. G. Hansen, *in* "Methoden der enzymatischen Analyse" (H.-U. Bergmeyer, ed.), 3rd ed., Vol. VII, p. 244. Verlag Chemie, Weinheim, 1985.

Distribution. When compared against subcellular markers, the enzyme is mainly, if not exclusively, microsomal.[4] Pantetheine hydrolase activity has been identified in human fibroblasts and white blood cells,[2] and in several organs of mammalian species.[1] Its distribution among rat tissues is given in Table II.

[8] Micromethod for the Measurement of Acetyl Phosphate and Acetyl Coenzyme A

By ARTHUR G. HUNT

Acetyl phosphate and acetyl coenzyme A (CoA) are key intermediates of fermentative and respiratory metabolism in bacteria. Moreover, they are obligative intermediates in the utilization of acetate by facultative aerobes such as *Escherichia coli.* I have developed a method to determine the concentrations of these metabolites in bacterial cultures.[1] This method, suitable for detecting levels as low as 0.02 nmol/mg protein in a growing bacterial culture, is described in detail here.

Reagents and Enzymes

[2,8-³H]Adenosine 5'-diphosphate, 25–40 Ci/mmol, from New England Nuclear

Acetate kinase (ATP : acetate phosphotransferase, EC 2.7.2.1), 200 units/mg, supplied as a suspension of 5 mg/ml in 3.2 M ammonium sulfate, from Boehringer Mannheim Biochemicals

Phosphotransacetylase (acetyl-CoA : orthophosphate acetyltransferase, EC 2.3.1.8), 100 units/mg, lyophilized, from Boehringer Mannheim Biochemicals

Acetyl phosphate, potassium–lithium salt, from Boehringer Mannheim Biochemicals

Acetyl coenzyme A, trilithium salt, from Boehringer Mannheim Biochemicals

Polyethyleneimine (PEI)-cellulose thin-layer chromatography plates, 0.1-mm layers, on plastic support, from Brinkmann

Solutions of acetyl phosphate and acetyl-CoA are spectrophotometrically calibrated as described by Hunt and Hong.[1]

[1] A. G. Hunt and J-S. Hong, *Anal. Biochem.* **108**, 290 (1980).

METHODS IN ENZYMOLOGY, VOL. 122

Methods

Summary

The method described here utilizes two enzymes involved with acetyl-CoA and acetyl phosphate metabolism in bacteria to measure these compounds. Basically, the acetyl-CoA and acetyl phosphate in perchloric acid extracts are converted to ATP, in the presence of [³H]ADP of high specific activity, using the enzymes phosphotransacetylase (PTA) and acetate kinase (AK):

$$\text{Acetyl-CoA} + P_i \xrightarrow{\text{PTA}} \text{acetyl phosphate} + \text{CoA} \qquad (1)$$

$$\text{Acetyl phosphate} + [^3\text{H}]\text{ADP} \xrightarrow{\text{AK}} \text{acetate} + [^3\text{H}]\text{ATP} \qquad (2)$$

Preparation of Cell-Free Extracts

Perchloric Acid Extraction. Cell-free extracts are routinely prepared by perchloric acid extraction, as described previously.[2,3] Care must be taken when neutralizing the extracts to avoid locally excessive concentrations of base, since acetyl phosphate is readily hydrolyzed in alkaline solutions.[4] This problem can be circumvented by neutralization with concentrated solutions of weak bases. In all instances, potassium salts are used so that perchlorate ions can be quantitatively removed as $KClO_4$. After neutralization, $KClO_4$ is removed by centrifugation and the extracts stored on ice or at $-20°$.

Treatment with Activated Charcoal. The sensitivity of the procedure described here is determined by the specific activity of the [³H]ADP in the cell-free extract. To obtain the greatest possible specific activity, endogenous ADP should be removed prior to the addition of the [³H]ADP. Also, since acetate kinase, the enzyme used here to convert acetyl phosphate to ATP (see below), catalyzes an ATP $\rightleftharpoons$ ADP exchange reaction, endogenous ATP should also be removed from the perchloric acid extracts before proceeding with this procedure. Removal of ATP and ADP is accomplished by treating neutralized extracts with activated charcoal. Forty to fifty milligrams of activated charcoal is added to 1 ml of extract, the mixture mixed briefly, and incubated on ice for 15 min. The bulk of the charcoal is then removed by centrifugation, and any remaining charcoal is eliminated by passing the extract through a glass wool plug in a Pasteur pipet.

² J-S. Hong, A. G. Hunt, P. S. Masters, and M. A. Lieberman, *Proc. Natl. Acad. Sci. U.S.A.* **76,** 1213 (1979).
³ B. R. Bochner and B. N. Ames, *J. Biol. Chem.* **257,** 9759 (1982).
⁴ E. R. Stadtman, this series, Vol. 3, p. 228.

This treatment quantitatively removes ATP from cell-free extracts without affecting the acetyl phosphate concentration of such extracts.[1] However, as much as 5% of the ADP in a sample remains after this treatment. For most purposes, this amount of ADP does not adversely affect the measurements described below. However, to measure very small quantities of acetyl phosphate, this residual ADP should be removed by a second charcoal treatment.

Determination of Acetyl Phosphate

Reaction of Acetyl Phosphate with [^{3}H]ADP. Acetyl phosphate is enzymatically converted to ATP in the presence of [^{3}H]ADP with acetate kinase. MgCl$_2$ and [^{3}H]ADP are added to perchloric acid extracts to final concentrations of 1 mM and 100 μCi/ml, respectively. The extract is then divided into 50-μl aliquots and known amounts of acetyl phosphate are added as internal standards. Five microliters of acetate kinase (50 μg/ml in 0.1 M potassium phosphate, pH 7.2, plus 1.0 mM MgCl$_2$; 200 units/mg) is added to each 50-μl aliquot and the samples incubated for 90 min at 30°. The samples are then placed on ice to stop the reaction.

Given the sensitivity of the method presented here, it is essential to make sure that no adenylate kinase activity is present either in the commercial acetate kinase preparation or in the perchloric acid extracts. This is best achieved by preparing two kinds of blank samples. First, extracts that have been heated to 100° for 5 min (to hydrolyze acetyl phosphate) are treated with acetate kinase and [^{3}H]ADP as described above, to determine the amount of contaminating adenylate kinase activity in the commercial acetate kinase preparation. Second, extracts containing [^{3}H]ADP are incubated without any enzyme. The amount of [^{3}H]ATP found in these samples is a measure of the amount of adenylate kinase present in the perchloric acid extract.

Acetate kinase has a rather broad pH optimum centered at 6.3.[5] Therefore, it is necessary to maintain the pH of the cell-free extract between 6.0 and 8.0. This is easily accomplished if the extract is buffered with a suitable buffer. Since bacterial minimal media are usually buffered with phosphate, and since these are most commonly used in metabolic studies, an additional buffer need not be added. However, rich media and tissue culture media may not be so buffered, and the buffer present after extraction may be an important consideration. If an additional buffer must be added, the activity of acetate kinase in this buffer should be checked independently.

[5] R. S. Anthony and L. B. Spector, *J. Biol. Chem.* **246**, 6129 (1971).

Large excesses of acetate kinase severely decrease the efficiency of the conversion of small amounts of acetyl phosphate to ATP.[1] This is presumably due to the ability of acetate kinase to phosphorylate itself, using ATP or acetyl phosphate.[6] Apparently, in the presence of large amounts of enzyme the phosphate moieties derived from acetyl phosphate remain bound to the enzyme and are not transferred to the [³H]ADP. To overcome this problem, very small quantities (250 ng) of acetate kinase are used.

Measurement of [³H]ATP. [³H]ATP produced by the reaction of [³H]ADP and acetyl phosphate is determined after separating [³H]ATP and [³H]ADP by thin-layer chromatography on PEI-cellulose layers.[7] Five microliters of a 50-μl sample is spotted onto a PEI-cellulose plate, 1 μl of 5 mM ATP spotted on top of each sample spot and the plates then developed in 4.0 M sodium formate (pH 3.4) at room temperature. Development is halted when the solvent has traveled roughly 15 cm (under these conditions, ATP has an R_f of 0.2, and ADP an R_f of 0.6). After drying, the ATP spots are located and marked under ultraviolet light. Each spot is scraped into a 5-ml liquid scintillation vial containing 0.5 ml of 4 M LiCl and the ATP eluted by vortexing briefly. Four milliliters of an appropriate aqueous-compatible scintillation fluid is added and the radioactive ATP determined by liquid scintillation counting. (The PEI-cellulose settles to the bottom of the vial and does not interfere with the determination of the radioactivity.)

Determination of Acetyl-CoA

Removal of Acetyl Phosphate and Reduced CoA. To facilitate the measurement of acetyl-CoA in cell-free extracts by this method, acetyl phosphate and reduced CoA should be completely removed from perchloric acid extracts. Removal of acetyl phosphate decreases the background seen in the measurements described below, and is essential in extracts in which acetyl phosphate concentrations are greater than acetyl-CoA concentrations. Acetyl phosphate is removed by heating extracts to 85° for 15 min. This destroys acetyl phosphate, but has no effect on acetyl-CoA.[1,8] Reduced CoA must be removed to ensure quantitative conversion of acetyl-CoA to acetyl phosphate by phosphotransacetylase. Reduced CoA is oxidized as described by McDougal and Dargar.[9] Fifty microliters of 20 mM N-ethylmaleimide is added to 1 ml of extract and the mixture incubated for 5 min at 37°. Excess N-ethylmaleimide is subsequently removed

[6] R. S. Anthony and L. B. Spector, *J. Biol. Chem.* **245,** 6739 (1970).
[7] E. Randerath and K. Randerath, *J. Chromatogr.* **16,** 126 (1964).
[8] E. R. Stadtman, *J. Biol. Chem.* **196,** 535 (1952).
[9] D. B. McDougal, Jr. and R. V. Dargar, *Anal. Biochem.* **97,** 103 (1979).

with 50 μl of 40 mM dithiothreitol and an additional 5-min incubation at 37°.

Conversion of Acetyl-CoA to Acetyl Phosphate. To measure acetyl-CoA, this compound must be converted to acetyl phosphate prior to treatment of perchloric acid extracts with activated charcoal. Acetyl-CoA is stoichiometrically converted to acetyl phosphate by treating extracts with phosphotransacetylase. Extracts that have been heated to 85° and treated with N-ethylmaleimide are brought to 0.1 M in potassium phosphate (pH 7.2) and 1 mM in MgCl$_2$. Phosphotransacetylase (2.5 mg/ml in 0.1 M potassium phosphate pH 7.2, plus 1.0 mM MgCl$_2$; 1000 units/mg) is then added to 2.5 units/ml and the mixtures incubated for 60 min at 37°. These extracts are then treated with activated charcoal and the acetyl phosphate formed in the phosphotransacetylase reaction determined as described above.

Calculation of Acetyl Phosphate and Acetyl-CoA Contents

In order to determine the concentration of acetyl phosphate and acetyl-CoA in a sample, the specific activity of the [³H]ADP and the counting efficiencies for each sample must be known. These parameters can be determined empirically for each sample by performing reactions containing varying amounts of added acetyl phosphate as internal standards. Also, the contribution from residual ATP must be known. This is estimated by preparing blank samples as described above. The radioactivity from these blank samples is regarded as background. Samples with relatively large background values should be discarded, since high backgrounds are indicative of substantial adenylate kinase activity and prohibit an accurate determination of acetyl phosphate and acetyl-CoA levels. The background value is subtracted from the values of the radioactivity of the sample reactions, with or without internal standards. By plotting the radioactivity versus picomoles of added acetyl phosphate, the specific activity of the [³H]ADP and the counting efficiency together are reflected in the slope of the resulting line. The acetyl phosphate or acetyl-CoA content is then simply the value of the intercept divided by the slope of the line.

It is important to take into account the various manipulations described above when calculating the acetyl phosphate and acetyl-CoA contents of the original extract. For example, a bacterial suspension is usually diluted by 20% in the course of perchloric acid extraction and neutralization, and another 10% in removing reduced CoA. These dilutions affect the final values of acetyl phosphate and acetyl-CoA concentrations, as well as the sensitivity of this method.

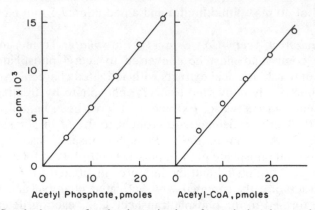

FIG. 1. Standard curves for the determination of acetyl phosphate and acetyl-CoA. Various amounts of acetyl phosphate or acetyl-CoA, in 50 μl of 0.1 M potassium phosphate (pH 7.2) plus 1 mM MgCl$_2$, were reacted with [^{3}H]ADP (100 μCi/ml, 25–40 Ci/mmol). The acetyl phosphate was reacted by adding 0.05 units of acetate kinase. Acetyl-CoA was reacted by adding 0.125 units of phosphotransacetylase and 0.05 units of acetate kinase. Reactions were incubated for 90 min at 37°. [^{3}H]ATP was separated from [^{3}H]ADP and quantitated as described under Methods. (Taken, with permission, from Hunt and Hong.[1])

Discussion

Standard curves for the determination of acetyl phosphate and acetyl-CoA by this procedure are given in Fig. 1. Given the slopes of these lines (620 and 550 cpm/pmol, respectively) and the observation that, using 0.5 μCi of radioactivity on PEI-cellulose plates obtained from Brinkmann, the background radioactivity is usually approximately 500 cpm, the limits of sensitivity of the methods described are roughly 1 pmol per 0.05 ml of appropriately treated cell-free extract.

Acetyl phosphate is a relatively labile compound.[4] However, perchloric acid extraction does not significantly affect acetyl phosphate concentrations.[1] Acetyl-CoA is likewise not affected by perchloric acid extraction.[10] The procedure described here is therefore well suited for measurements of these compounds in bacterial suspensions or other instances in which their concentrations in cell-free extracts are expected to be very low (between 0.02 and 1 μM).

Since this technique requires the use of acetate kinase for the conversion of acetyl phosphate and ADP to acetate and ATP, and since others have noted that this enzyme is capable of using propionyl-CoA as a substrate,[11] it is possible that this procedure is not specific for acetyl phos-

[10] W. Prinz, W. Schoner, U. Haag, and W. Seubert, *Biochem. Z.* **346**, 206 (1966).
[11] I. A. Rose, M. Grunberg-Manago, S. R. Korey, and S. Ochoa, *J. Biol. Chem.* **211**, 737 (1954).

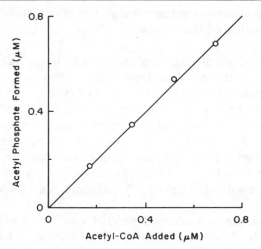

FIG. 2. Efficiency of conversion of acetyl-CoA to acetyl phosphate. Solutions of acetyl-CoA (0.2–0.8 μM) were heated and treated with N-ethylmaleimide, DTT, and phosphotransacetylase as described under Methods. The acetyl phosphate formed was then measured, also as described in Methods. (Taken, with permission, from Hunt and Hong.[1])

phate. However, under the conditions described for the acetate kinase reactions, less than 2% of added propionyl phosphate is converted to ATP.[1] Therefore, this method is specific for acetyl phosphate (and acetyl-CoA). (The failure to see any substantial reaction with propionyl phosphate here probably reflects vast differences in the rates with which propionyl phosphate and acetyl phosphate are converted to ATP by acetate kinase.[11])

Since the equilibrium of the phosphotransacetylase reaction favors the formation of acetyl-CoA (K_{eq} = 134 ± 20 at 27°, pH 7.6[12]), it is necessary to ascertain whether the measures taken to remove reduced CoA are adequate to drive the conversion of acetyl-CoA to acetyl phosphate, in the presence of 0.1 M phosphate, to completion. The data presented in Fig. 2 indicate that this conversion is indeed quantitative, at least over the range of acetyl-CoA concentrations examined.

As mentioned above, this procedure is able to detect levels of acetyl phosphate or acetyl-CoA as low as 20 nM in a cell-free extract. In principle, this sensitivity can be enhanced even further. [α-^{32}P]ADP can be used instead of [^{3}H]ADP. Large volumes can be applied to the thin-layer chromatography plates, and two-dimensional thin-layer chromatography can be used to facilitate the separations for such larger volumes. However, the somewhat larger background radioactivity that results from replacing

[12] T. Nojiri, F. Tanaka, and I. Nakayama, *J. Biochem. (Tokyo)* **69**, 789 (1971).

^{32}P with ^{3}H or from applying larger quantities of radioactivity to the thin-layer chromatography plates partially offsets the expected increases in sensitivity.

Another, more promising, approach would be to measure the ATP formed in the acetate kinase reactions with purified preparations of luciferin and luciferase. This alternative would obviate the need for [^{3}H]ADP of high specific activity, an essential feature of the method presented here. The inherent sensitivity of the luciferin–luciferase assay would extend the range of detection of acetyl phosphate and acetyl-CoA to the subnanomolar range. However, this modification can only be made with highly purified preparations of luciferase, since even small amounts of adenylate kinase will interfere with the measurement of acetyl phosphate and acetyl-CoA.

I have used the procedure described here to measure the acetyl phosphate levels of different strains of *E. coli* under a variety of conditions.[13] I have observed concentrations ranging from less than 0.2 nmol/mg cell protein (in arsenate-treated wild-type cells) to greater than 5 nmol/mg cell protein (in glucose-fed suspensions of an acetate kinase-deficient strain of *E. coli*). These values probably represent the limits of the acetyl phosphate contents of *E. coli*, and thus serve as a guide for other such studies dealing with bacteria that possess the acetate kinase–phosphotransacetylase pathway.

[13] A. G. Hunt, Ph.D. Thesis, Brandeis University, Waltham, Massachusetts (1982).

[9] Preparation of Radioactive Acyl Coenzyme A

By AMIYA K. HAJRA and JAMES E. BISHOP

Principle

$$R—^{14}COOH + (COCl)_2 \longrightarrow R—^{14}C(=O)—Cl + CO + CO_2 + HCl \qquad (1)$$

$$R—^{14}C(=O)—Cl + CoASH \xrightarrow[\text{tetrahydrofuran–water}]{\text{pH 8.8}} R—^{14}C(=O)\text{\textasciitilde}SCoA + HCl \qquad (2)$$

In this procedure, long-chain ^{14}C-labeled fatty acid is first converted to the acyl chloride by oxalyl chloride,[1] which is then condensed with

[1] R. Adams and L. H. Ulich, *J. Am. Chem. Soc.* **42**, 599 (1920).

CoASH to form the corresponding long-chain acyl-CoA. The condensation method is based on the procedure originally described by Seubert,[2] which was scaled down and modified to conserve the expensive radioactive fatty acid.[3] The modifications consist of (1) using excess CoASH instead of excess acyl chloride to ensure high conversion of the radioactive acyl chloride and (2) using a buffered reaction mixture rather than adjusting the pH continuously during the reaction. The acyl-CoAs are purified by precipitating with $HClO_4$, thus removing the excess CoASH, and then washing the precipitate with acetone and diethyl ether to remove the unconverted fatty acid.

Reagents

Coenzyme A Na or Li salt, 15 μmol (~14 mg of 90% purity)
[1-^{14}C]Palmitic acid (5 μmol, 250 μCi)
Oxalyl chloride
Dry benzene, prepared by distilling benzene from CaH_2. Store in a desiccator over anhydrous $CaSO_4$
Tetrahydrofuran, freshly distilled. Precaution should be taken not to distill it to dryness to avoid the explosion of peroxides, if present.
Acetone, dried over anhydrous Na_2SO_4
Anhydrous diethyl ether (commercial, freshly opened can)

Method

[1-^{14}C]Palmitic acid was dried three times in a dry 15-ml Corex tube (Corning Glass, No. 8441) with dry benzene under a gentle stream of extra dry N_2.[4] Dry benzene (0.8 ml) and oxalyl chloride (0.4 ml) were then added to the fatty acid residue, and the tube was flushed with N_2, stoppered, and incubated for 1 hr at 37°. Liquids were evaporated off by a stream of nitrogen and the oily residue was dried three times with N_2, with aliquots of dry benzene added each time to remove the last traces of oxalyl chloride. The acyl chloride was used immediately.

To the dry acyl chloride, 1.3 ml of a solution containing 15 μmol CoASH in tetrahydrofuran–aqueous 150 mM $NaHCO_3$ buffer (2.2:1.0, adjusted to pH 8.8 with NaOH) was added. The mixture was mixed quickly by vortexing, and the tube was flushed with N_2, stoppered, and incubated at 37° for 30 min. The reaction was stopped by adding 20 μl of

---

[2] W. Seubert, *Biochem. Prep.* **7**, 80 (1960).
[3] J. E. Bishop and A. K. Hajra, *Anal. Biochem.* **106**, 344 (1980).
[4] Extra dry nitrogen should be used to dry the fatty acid and acid chloride. Hydrolysis of the acid chloride occurs, thus lowering the yield, if N_2 in water-pumped cylinder is used.

10% $HClO_4$ and most of the tetrahydrofuran was evaporated off under a gentle stream of N_2.

To the semisolid residue, 2.5 ml of 1.3% $HClO_4$ was added, followed by mixing and then chilling on ice. The mixture was centrifuged at 18,000 g for 15 min at 4°, the supernatant was removed, and the precipitate was washed again with 1.3% $HClO_4$ by mixing and centrifuging as above. The residue was further washed successively in a similar manner once with 4 ml acetone[5] and twice with 4 ml anhydrous ether. The last traces of ether were removed from the fluffy residue by incubating the tube in a shaking water bath at 37°. The overall yield was 70–80% with a purity >90%. The final product was dissolved in 10 mM phosphate buffer (pH 6.0) and stored under N_2 at −20°. The product is found to be stable for more than a year when stored under such conditions (see below).

Comments

The above method describes the preparation of [1-^{14}C]palmitoyl-CoA. However, radioactive mono- and polyunsaturated long-chain acyl-CoAs can also be prepared in a similar manner. To protect these unsaturated fatty acids from oxidation, an equimolar amount of the antioxidant butylated hydroxytoluene (BHT) should be added to the fatty acids during the preparation of the acyl chlorides. For polyunsaturated acyl-CoAs, excess BHT (0.5%) should also be included with the solution in which the acyl-CoAs are dissolved, for prolonged storage. Under such conditions, the acyl-CoAs (such as linoleoyl-CoA) remain stable for at least 6 months. The original report[3] that the polyunsaturated acyl-CoAs were not stable to storage was due to the use of an insufficient amount of BHT during storage.

The concentration and purity of the acyl CoAs can be determined spectrophotometrically by measuring the absorbances at 260 nm (E_{260} = 1.54×10^4 M^{-1} cm^{-1}) and 232 nm (E_{232} = 8.7×10^3 M^{-1} cm^{-1}).[3,6] The purity can also be determined by thin-layer chromatography on Merck Silica Gel 60 plate using butanol–water–acetic acid (50:30:20) as the developing solvent.[3] The acyl-CoAs have an R_f of 0.46 in this system compared to 0.86 for fatty acids.

This method can also be adapted to prepare larger amounts of nonradioactive acyl-CoAs. For this purpose equimolar amounts of acyl chloride

---

[5] Reagent-grade acetone contains variable amounts of water which interferes with pelleting down the acyl CoA. Therefore, acetone should be dried, as indicated, before use.

[6] E. R. Stadtman, this series, Vol. 3, p. 931.

and CoASH are used. Up to 25 μmol of each component can be used in 1.5 ml of the above reaction mixtures and the product can be purified in the same way. For larger quantities it is necessary to increase the volume and also to maintain the pH at 8.8 by adding 0.5 M Na$_2$CO$_3$ during the reaction.

[10] Purification of Plant Acetyl-CoA:Acyl Carrier Protein Transacylase

By Takashi Shimakata and Paul K. Stumpf

Acetyl-CoA:acyl carrier protein transacylase (acetyl transacylase) catalyzes the transfer of the acetyl group from the SH group of CoA to that of acyl carrier protein (ACP):

$$CH_3\text{-}CO\text{-}S\text{-}CoA + HS\text{-}ACP \rightleftharpoons CH_3CO\text{-}S\text{-}ACP + CoA\text{-}SH$$

Unlike animal and yeast enzymes which are integrated in the multifunctional polypeptide chain of their fatty acid synthetases (FAS), the *Escherichia coli* and plant acetyl transacylases can be easily separated from the other component enzymes of FAS by conventional purification procedures.[1-6] The counterpart in *E. coli* has already been purified partially and characterized.[1,3] Since plant acetyl transacylase has as yet not been purified and characterized, its partial purification, characterization, and function are described in this chapter.

Assay Method

Principle. Acetyl transacylase is assayed by counting the radioactivity of [1-^{14}C]acetate transferred to ACP from CoA. The reaction is stopped by addition of cold perchloric acid and the mixture is poured on a Millipore filter. After unreacted [^{14}C]acetyl-CoA was washed out with cold perchloric acid, the filter containing acid-insoluble [^{14}C]acetyl-ACP is

[1] J. J. Volpe and P. R. Vagelos, *Annu. Rev. Biochem.* **42**, 21 (1973).
[2] S. J. Wakil, J. K. Stoops, and V. C. Joshi, *Annu. Rev. Biochem.* **52**, 537 (1983).
[3] I. P. Williamson and S. J. Wakil, *J. Biol. Chem.* **241**, 2326 (1966).
[4] T. Shimakata and P. K. Stumpf, *Arch. Biochem. Biophys.* **217**, 144 (1982).
[5] T. Shimakata and P. K. Stumpf, *Plant Physiol.* **69**, 1257 (1982).
[6] T. Shimakata and P. K. Stumpf, *J. Biol. Chem.* **258**, 3592 (1983).

transferred to a scintillation vial and counted to determine the amount of acetyl-ACP present.

Reagents

Tris–HCl, 1 M, pH 8.1

Dithiothreitol (DTT), 0.1 M, freshly prepared

Acyl carrier protein, 1 mM, purified from *E. coli* (Grain Processing Company, Iowa) by the method of Rock and Cronan,[7] and reduced fully with DTT

[1-^{14}C]Acetyl-CoA (26.1 mCi/mmol), 191.2 μM, purchased from New England Nuclear and diluted with nonlabeled acetyl-CoA

Ice-cold perchloric acid (PCA), 5% (v/v)

Millipore filter (25-mm diameter, pore size 0.45 μm)

Phase Combining System (PCS) (Amersham)

Xylene

Procedure. The following reagents are added for each assay (50 μl): Tris–HCl, 5 μl; DTT, 2.5 μl; ACP, 2.5 μl; [1-^{14}C]acetyl-CoA, 7.85 μl; and acetyl transacylase (with variable units of activity) plus water, 32.15 μl. Acetyl transacylase is preincubated with the above reagents without [^{14}C]acetyl-CoA for 15 min at 33° with vigorous shaking and then the reaction is started by addition of [^{14}C]acetyl-CoA. After incubation for 5 min at 33°, the reaction is stopped by addition of 0.25 ml of cold PCA (5%), and the tubes are placed in the ice bath for 15 min to ensure complete precipitation of [^{14}C]acetyl-ACP. The resulting precipitates are retained on Millipore filters and washed three times with ~25-ml aliquots of cold PCA. Each filter is transferred to a scintillation vial containing 10 ml of mixture of PCS–xylene (2 : 1, v/v) and counted in a Beckman LS-230 counter after standing for 10 min. The reaction rate is nearly linear, both with time for 5 min and with enzyme concentration to 10 μg protein.

Purification Procedure

Materials

Fresh spinach leaves (*Spinacia oleracea* L. var. Viroflay), obtained from local market

Potassium phosphate buffer, 1 M, pH 7.5

KDG buffer, containing indicated concentration of potassium phosphate buffer (pH 7.5), 0.5 mM DTT, and 20% glycerol

Dithiothreitol (DTT), freshly prepared

---

[7] C. O. Rock and J. E. Cronan, Jr., *Anal. Biochem.* **102,** 362 (1980).

Glycerol
$(NH_4)_2SO_4$, finely powdered
KOH, 2 M
KCl, 1 M
KSCN, 1 M
Cheesecloth
Amicon PM30 membrane filter
Sephacryl S-300 (Pharmacia), equilibrated with 0.05 M KDG buffer
Cibacron blue–agarose (Bio-Rad, 100–200 mesh), equilibrated with
    0.05 M KDG buffer
Sephadex G-25, equilibrated with 0.05 M KDG buffer
DEAE-cellulose, equilibrated with 0.05 M KDG buffer

Procedure

Crude Extract. Washed spinach leaves (500 g) with the stems removed
are homogenized for 5 min in a Waring blender with 250 ml of 0.05 M
potassium phosphate buffer (pH 7.5) containing 0.5 mM DTT. The homogenate is centrifuged at 10,000 g for 20 min. By carefully decanting the
contents of the centrifuge tube through several layers of cheesecloth, the
middle extract layer can be separated readily from the top lipid layer and
the pellet. The filtrate is again centrifuged at 78,000 g for 60 min. The
resulting supernatant fraction is identified as crude extract, and stored at
−20° until needed. These extracts can be stored for at least 3 months
without appreciable loss of activity.

Ammonium Sulfate Fractionation. All solutions are kept at 4°. Glycerol (50 ml) and 2 mM DTT are added to 200 ml of spinach crude extract
(1728 mg protein). After the mixture is stirred for 20 min, ammonium
sulfate (powder) is added gradually with stirring to 40% saturation, and
the pH of the solution is adjusted to 7.5 with 2 M KOH. The solution is
stirred for an additional 30 min before centrifugation at 10,000 g for 30
min. Additional ammonium sulfate is added to the resulting supernatant
until 80% saturation is obtained. The precipitate formed between 40 and
80% ammonium sulfate saturation is dissolved in 20 ml of 0.05 M KDG
buffer and dialyzed overnight against 2 liters of 0.05 M KDG buffer.

First Sephacryl S-300 Chromatography. The dialyzed solution is applied to a Sephacryl S-300 column (4 × 41 cm) equilibrated with 0.05 M
KDG buffer. The column is eluted with the same buffer and fractions of
4.85 ml are collected. Acetyl transacylase is eluted between fractions 64
and 83. These fractions were pooled.

Affi-Gel Blue Column Chromatography. The pooled solution is applied to an Affi-Gel Blue column (1.4 × 12 cm) equilibrated with 0.05 M

KDG buffer at a flow rate of 10 ml/hour. The column is washed with 50 ml of 0.2 M KDG buffer and then with 50 ml of 0.2 M KDG buffer containing 1 M KCl. After washing, the column is eluted with 0.2 M KDG buffer containing 1 M KSCN and fractions of 1.89 ml each are collected. Acetyl transacylase is eluted into this fraction together with β-hydroxyacyl-ACP dehydrase and malonyl transacylase. KSCN is removed by passing the pooled fractions through a Sephadex G-25 column (a suitable size) equilibrated with 0.05 M KDG buffer and acetyl transacylase is excluded.

Second Sephacryl S-300 Chromatography. Fractions containing acetyl transacylase are pooled, concentrated to 2 ml with a PM30 membrane filter, and applied to a second Sephacryl S-300 column (1.5 × 47.5 cm) equilibrated with 0.05 M KDG buffer. The column is eluted with the same buffer, and fractions of 1.05 ml each are collected. This procedure separates essentially all the β-hydroxyacyl-ACP dehydrase protein from the acetyl transacylase fractions.

First and Second DEAE-Cellulose Chromatographies. Fractions (46–51) containing acetyl transacylase activity are pooled and applied to a first DEAE-cellulose column (1.4 × 12 cm) equilibrated with 0.05 M KDG buffer. After washing the column with the same buffer (most of malonyl transacylase was passed through), a 100-ml linear gradient from 0.05 to 0.2 M KDG buffer is employed. Acetyl transacylase is eluted at about 0.07 M potassium phosphate. The acetyl transacylase fractions still contain low levels of malonyl transacylase activity. These fractions are therefore pooled, concentrated with a PM30 membrane filter, and applied to a Sephadex G-25 column equilibrated with 0.05 M KDG buffer. The excluded fractions are now applied to a second DEAE-cellulose column (1.4 × 12 cm) equilibrated with 0.05 M KDG buffer and fractionated again as described for the first DEAE-cellulose procedure. Fractions containing acetyl transacylase completely free of malonyl transacylase activity are collected, concentrated with a PM30 membrane filter, and stored at −70° without any loss of activity for several months. Table I summarizes the purification protocol of the transacylase.[8]

Properties

Miscellaneous Properties. Although the final preparation was not pure, it was completely free from the other spinach FAS enzymes. The molecular weight of the purified acetyl transacylase was estimated to be 48,000 by Sephacryl S-300 gel filtrations.[6]

[8] M. M. Bradford, *Anal. Biochem.* **72**, 248 (1976).

TABLE I
PURIFICATION OF SPINACH ACETYL-CoA : ACP TRANSACYLASE

Step	Protein[a] (mg)	Total activity[b] (nmol/min)	Specific activity (nmol/min·mg)	Purification (-fold)
Crude extract	1204	39.0	0.032	1
$(NH_4)_2SO_4$	797	25.8	0.032	1
Sephacryl S-300	191	21.4	0.112	3.5
Affi-Gel Blue	9.76	9.14	0.936	29
Sephacryl S-300	2.03	6.22	3.06	96
DEAE-cellulose I	0.71	4.36	6.14	192
DEAE-cellulose II	0.45	2.52	5.60	175

[a] Protein was determined by the method of Bradford,[8] using bovine serum albumin as a standard.

[b] The activity was measured with acetyl-CoA and ACP as described in the text.

pH Activity Profile. The acetyl transacylase was maximally active at pH 8.1, with the half-maximum activity at pH 7.5, and about 78% of the maximum activity was retained at pH 9.1.[6]

Specificity. The purified acetyl transacylase was relatively specific for acetyl-CoA like the *E. coli* acetyl transacylase,[3] and it could transacylate butyryl-CoA, hexanoyl-CoA, and octanoyl-CoA at a rate of 35, 18, and 7% of that of acetyl-CoA, respectively.[6] The K_m and V_{max} values for the transacylase are summarized in Table II.

TABLE II
K_m AND V_{max} FOR SUBSTRATES OF PURIFIED
SPINACH ACETYL TRANSACYLASE[a]

Substrate	K_m (μM)	V_{max} (nmol/min·mg protein)
Acetyl-CoA	8.00	6.18
Butyryl-CoA	8.31	2.17
Hexanoyl-CoA	8.62	1.18
Octanoyl-CoA	—	0.45
Malonyl-CoA	—	0

[a] The kinetic parameters of spinach acetyl transacylase for each substrate were obtained from double-reciprocal plots of substrate concentration data (not shown here).

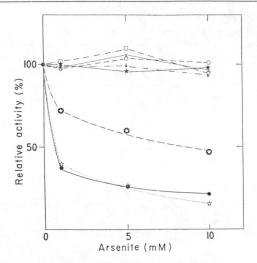

FIG. 1. Arsenite inhibition of fatty acid synthesis, and all the component enzymes of the FAS system in crude spinach extracts. An aliquot of arsenite solution (1.0 M) which was freshly dissolved and adjusted to pH 7.5 was added to each reaction mixture (without substrate) containing fresh crude spinach extract (0.34 mg protein) to give a final concentration as indicated and preincubated for 15 min at 25°. Then each reaction was initiated by addition of substrate. *De novo* fatty acid synthesis (●); acetyl transacylase (☆); malonyl transacylase (□); β-ketoacyl-ACP synthetase I (○) and II (◐); β-ketoacyl-ACP reductase (★); β-hydroxyacyl-ACP dehydrase (▲); enoyl-ACP reductase (×).

Inhibitors. The acetyl transacylase was inhibited almost completely by 5 mM arsenite or 5 mM p-chloromercuribenzoate and about 50% by 5 mM N-ethylmaleimide.[6]

Function of Acetyl Transacylase

Rate-Limiting Step in the Spinach FAS System

Because of its lowest specific activity among seven individual activities of FAS system in plant crude extracts, acetyl transacylase could be expected to be a candidate for the rate-limiting step for fatty acid synthesis from acetyl-CoA and malonyl-CoA.[4–6,9] It was found that when acetyl-ACP or hexanoyl-ACP was added as substrate, thereby bypassing the acetyl transacylase reaction, the level of malonyl-CoA incorporation was increased considerably.[6] In addition, inhibitory studies with arsenite (Fig. 1) supported the conclusion that the acetyl transacylase should be consid-

[9] P. K. Stumpf and T. Shimakata, *in* "Biosynthesis and Function of Plant Lipids" (W. W. Thomson, J. B. Mudd, and M. Gibbs, eds.), p. 1. Am. Soc. Plant Physiol., Rockville, MD.

ered as a rate-limiting enzyme.[6] The other individual enzymes were found not to be rate limiting, based on the inhibitory experiments with antisera and cerulenin.[6]

Addition of Purified Acetyl Transacylase to Spinach Crude Extract or Reconstituted FAS System

Evidence has already been presented that elevated levels of acetyl transacylase may shift the synthesis of fatty acid from the normal C_{16}–C_{18} pattern to shorter chain fatty acids.[4–6,9] Therefore, this enzyme, together with β-ketoacyl-ACP synthetase II which was required for the conversion of palmitic to stearic acid,[10] was added to crude spinach extract or reconstituted FAS system consisting of highly purified enzymes in increasing amounts. In both cases, medium-chain fatty acids increased with increasing amounts of added acetyl transacylase with a concomitant decrease in stearate, whereas the increase of stearate was related to increasing levels of synthetase II.[6]

Conclusion

The purification procedures and some characteristics of spinach acetyl-CoA : ACP transacylase were described. The role of the acetyl transacylase was explored by addition of varying concentration of the transacylase either to spinach leaf extract or to a completely reconstituted FAS system. The results suggested that (1) acetyl transacylase was rate-limiting step in plant FAS system; (2) increasing concentration of this enzyme markedly increased the levels of the medium-chain fatty acids. These results suggested that modulation of the acetyl transacylase activity may have important implications in the type of fatty acids synthesized, as well as the amount of fatty acids formed.

[10] T. Shimakata and P. K. Stumpf, *Proc. Natl. Acad. Sci. U.S.A.* **79**, 5808 (1982).

Section IV

Biotin and Derivatives

[11] Separation of Biotin and Analogs by High-Performance Liquid Chromatography

By DELORES M. BOWERS-KOMRO, JANE L. CHASTAIN, and
DONALD B. MCCORMICK

Biotin and its analogs including catabolites have been isolated and characterized by several investigators, as cited in the review by McCormick and Wright.[1] Typical isolation techniques involve gravity-flow, anion-exchange chromatography with gradient elution from water to dilute formic acid or 1 M ammonium formate. Once isolated, biotin and its analogs have been differentiated on paper or TLC sheets in various solvent systems[2,3] and quantitated at the microgram level using the colorimetric reaction with p-dimethylaminocinnamaldehyde.[4] Desbene *et al.*[5] have demonstrated that biotin, dethiobiotin, the *d*- and *l*-sulfoxides, and sulfone can be separated by C_{18} normal-phase chromatography once they are derivatized to esters using p-bromophenacyl bromide for UV detection (at 254 nm) and 4-bromomethylmethoxycoumarin for fluorimetric detection (excitation filter with maximum transmittance at 360 nm and emission filter with cutoff at 410 nm). Roders *et al.*[6] have reported the HPLC determination of biotin in pharmaceutics. To our knowledge HPLC separation of underivatized biotin, its analogs, and metabolites has not been reported, even though HPLC is a common method used for separation of other vitamins, coenzymes, and their analogs (see citations in this and earlier volumes of "Vitamins and Coenzymes" in this series). Using reverse-phase or anion-exchange HPLC we can differentiate biotin from those analogs that possess variations in the side chain, thiophane ring, and ureido ring.

Instrumentation

Any commercially available HPLC system can be used as long as it is capable of gradient elution and has a variable-wavelength detector with low-UV capabilities. The results presented here were obtained with a

[1] D. B. McCormick and L. D. Wright, *in* "Comprehensive Biochemistry" (M. Florkin and E. H. Stotz, eds.), Vol. 21, p. 81. Am. Elsevier, New York, 1971.
[2] H. M. Lee, L. D. Wright, and D. B. McCormick, *J. Nutr* **102**, 1453 (1972).
[3] H. Ruis, R. N. Brady, D. B. McCormick, and L. D. Wright, *J. Biol. Chem.* **243**, 547 (1968).
[4] D. B. McCormick and J. A. Roth, *Anal. Biochem.* **34**, 226 (1970).
[5] P. L. Desbene, S. Coustal, and F. Frappier, *Anal. Biochem.* **128**, 359 (1983).
[6] E. Roder, U. Engelbert, and J. Troschutz, *Fresenius Z. Anal. Chem.* **319**, 426 (1984).

METHODS IN ENZYMOLOGY, VOL. 122

Waters Associates HPLC system (Milford, MA) consisting of a 450 variable-wavelength detector, a U6K injector, a 6000A and M45 solvent delivery system with 720 controller, and a data module. The elution patterns were monitored at 220 nm with solvent selection to give minimum background drift.

Columns and Solvent System

Reverse-phase separations were achieved using a RP-C_{18} guard column (4.6 × 30 mm) (Brownlee Labs Inc., Santa Clara, CA) followed by a 10 μm μ-Bondapak C_{18} column (4.6 × 250 nm) (Waters Associates). The chromatographic conditions giving the best separation were provided by a 35-min gradient from aqueous trifluoroacetic acid (0.05%, adjusted to pH 2.5 with KOH) to aqueous trifluoroacetic acid (0.05%, adjusted to pH 2.5 with KOH)–acetonitrile (70 : 30, v/v), followed by holding at 70 : 30 ratio for 5 min at a flow rate of 1 ml/min. It was necessary to reequilibrate the column 5 min between each run to achieve maximum reproducibility.

Anion-exchange separations were achieved using an Aquapore AX-300 column (4.6 × 250 mm) preceded by its respective guard column (4.6 × 30 mm) (Brownlee Labs, Inc., Santa Clara, CA). Chromatographic conditions giving the best separation were provided by a 30-min linear gradient from 50 mM Tris chloride (pH 4.5) to 50 mM Tris chloride (pH 6.5) at a flow rate of 1 ml/min. It was necessary to reequilibrate the column at least 10 min between each run to achieve maximum reproducibility. Other solvent systems were tried, such as isoprotic aqueous buffers (pH 4.5, 5.5, 6.5) with gradients that increased ionic strength to 0.5 M using sodium chloride; however, these were less effective.

All solvents were prepared from glass-distilled deionized water, filtered through a 0.45-μm filter, and degassed before use. Acetonitrile (HPLC grade) was obtained from J. T. Baker Chemical Co. (Phillipsburg, NJ).

Compounds

Biotin and the analogs presented here were obtained from synthetic procedures[7] and/or isolated from cultures[1] and were in the purest state available. The polar compounds were dissolved in 1 mM NaOH, whereas the nonpolar compounds were dissolved in water–acetonitrile (1 : 1, v/v).

[7] L. H. Sternbach, in "Comprehensive Biochemistry" (M. Florkin and E. H. Stotz, eds.), Vol.11, p. 66. Am. Elsevier, New York, 1963.

HPLC RETENTION TIMES OF BIOTIN, ANALOGS, AND METABOLITES

Compound	$pK_a{}^a$	Retention time (min)	
		C_{18} μ-Bondapak[b]	Aquapore AX-300[c]
Side-chain variations			
Biotin	6.22	26.0	17.7
Homobiotin	6.29	32.2	18.4
Bisnorbiotin	5.86	15.5	16.7
Tetranorbiotin	4.85	6.1	15.3
α-Dehydrobiotin		25.2	22.3
Biotinol		27.2	3.9
Biocytin		19.0	3.7
Biotin methyl ester		37.0	3.8
Thiophane ring variations			
Biotin l-sulfoxide	6.05	13.3	13.5
Biotin d-sulfoxide	6.04	14.1	12.1
Biotin sulfone	6.10	4.8	14.8
Dethiobiotin	6.35	30.2	14.0
Ureido ring variations			
2'-Thiobiotin		25.8	23.7
2'-Iminobiotin		4.5–5.0	4.3

[a] pK_a values of side-chain carboxyls taken from Sigel et al.[8]

[b] Linear gradient elution (35 min) from aqueous trifluoroacetic acid (0.05%, neutralized to pH 2.5 with KOH) to aqueous trifluoroacetic acid (0.05%, neutralized to pH 2.5 with KOH)–acetonitrile (70 : 30, v/v) followed by holding at 70 : 30 ratio for 5 min at a flow rate of 1 ml/min.

[c] Linear gradient elution (30 min) from 50 mM Tris chloride (pH 4.5) to 50 mM Tris chloride (pH 6.5) at a flow rate of 1 ml/min.

Individual solutions as well as mixtures were injected to determine the resolution being achieved.

Applications

Side-Chain Variations

Using reverse-phase chromatography, biotin is easily separated from analogs with varying side-chain lengths (see table). There is a linear relationship (correlation coefficient 0.9997) between the number of methylene carbons in the chain and the retention time. Separation on the anion-exchange column shows a similar linearity for homobiotin, biotin, and bisnorbiotin with five, four, and two methylene groups within the alkanoic chains; however, tetranorbiotin, with a carboxyl group alpha to the thio function, falls below the line produced by the three homologs. This may reflect the interaction of the carboxylic acid proton with the thiophane ring sulfur as reflected by the relatively low pK_a.[8] Although these compounds were isolated using anion-exchange chromatography (gravity flow), the separation achieved between tetranorbiotin and homobiotin is not as pronounced on the anion-exchange column Aquapore AX-300 (difference is 3.1 min) as it is on the C_{18} μ-Bondapak (difference is 28.1 min). Thus reverse-phase chromatography seems to be the best method for separation.

The unsaturated α-dehydrobiotin is unresolved from biotin in a mixture containing the two compounds in the C_{18} column but is completely resolved from biotin on an anion-exchange column. Changing the terminal function of the side chain from acid to alcohol, namely biotin to biotinol, increases the retention slightly on the C_{18} column (26.0 to 27.2 min) and has a marked influence on the retention on the Aquapore AX-300 (17.73 to 3.92 min). The methyl ester of biotin is well retained, whereas biocytin is not retained well on a C_{18} column; neither can be adequately resolved on the anion-exchange column.

Thiophane Ring Variations

Oxidation of the sulfur in the thiophane ring of biotin to d- and l-sulfoxides and the sulfone increases the polarity of the molecule and shortens the retention time from 26.0 to 14.1, 13.3, and 4.8 min, respectively, on the reverse-phase C_{18} column. Removing the sulfur atom from the ring, i.e., in dethiobiotin, makes the compound less polar; thus, it is retained longer (30.2 min) than biotin. Using an anion-exchange column, the separation profile closely reflects the differences in the pK_a values (the higher the pK_a, the longer the retention time). The small pK_a difference in the sulfoxide isomers (0.01 units) would not appear to be sufficient to account for a partial resolution observed when a mixture is injected. Changing the flow rate and gradient profile might allow for quantitation of these two components if the mixture were not very complex.

Ureido Ring Variations

Removing the carbonyl function from the ring, as in the diaminocarboxylic acid of biotin, lowers the UV absorbancy so that its detection was not observed. Changing the 2'-position oxygen to sulfur, i.e., 2'-

[8] H. Sigel, D. B. McCormick, R. Griesser, B. Prijs, and L. D. Wright, *Biochemistry* **8**, 2687 (1969).

thiobiotin, does not significantly change the characteristics of the molecule with respect to reverse-phase chromatography; however, the guanidino-like 2'-iminobiotin is protonated ($pK_a \approx 12$) and this analog elutes fairly rapidly. Changes in the 2' position do affect the ionic nature of the molecule enough to provide quite different retention times on the anion-exchange column.

[12] Fluorometric Assays for Avidin and Biotin

By R. D. NARGESSI and D. S. SMITH

Several methods have been described for the determination of biotin, a growth factor. These include microbiological, colorimetric, and enzymatic methods.[1-4] Competitive protein binding assays which exploit the biotin-binding property of avidin and employ a labeled-biotin ligand as tracer have also been developed. Radioisotopes,[2-6] coenzymes,[7,8] enzymes,[9] and fluorogenic enzyme substrates[10] have been used for labeling of biotin. A spectrophotometric method[11] involves biotin displacement of a dye from the avidin binding site. Biotin-induced changes in the absorption[11] or fluorescence[12] properties of avidin form the basis of particularly simple biotin assays. Avidin, the egg-white protein, can easily be determined by a variety of binding assay techniques.[1,2,7,13]

There has been much interest in exploitation of the avidin–biotin interaction in immunochemical assay systems.[14,15] When working with fluores-

[1] D. B. McCormick and L. D. Wright, this series, Vol. 18, Part A, p. 379.
[2] D. B. McCormick and L. D. Wright, this series, Vol. 62, Part D, p. 279.
[3] K. Gröningsson and L. Jansson, *J. Pharm. Sci.* **68**, 364 (1979).
[4] S. Haarasilta, *Anal. Biochem.* **87**, 306 (1978).
[5] T. Horsburgh and D. Gompertz, *Clin. Chim. Acta* **82**, 215 (1978).
[6] R. Rettenmaier, *Anal. Chim. Acta* **113**, 107 (1980).
[7] R. J. Carrico, J. E. Christner, R. C. Boguslaski, and K. K. Yeung, *Anal. Biochem.* **72**, 271 (1976).
[8] H. R. Schroeder, R. J. Carrico, R. C. Boguslaski, and J. E. Christner, *Anal. Biochem.* **72**, 283 (1976).
[9] C. R. Gebauer and G. A. Rechnitz, *Anal. Biochem.* **103**, 280 (1980).
[10] J. F. Burd, R. J. Carrico, M. C. Fetter, R. T. Buckler, R. D. Johnson, R. C. Boguslaski, and J. E. Christner, *Anal. Biochem.* **77**, 56 (1977).
[11] N. M. Green, this series, Vol. 18, Part A, p. 418.
[12] H. J. Lin and J. F. Kirsch, *Anal. Biochem.* **81**, 442 (1977).
[13] N. M. Green, *Adv. Protein Chem.* **29**, 85 (1975).
[14] S. M. Costello, R. T. Felix, and R. W. Giese, *Clin. Chem. (Winston-Salem, N.C.)* **25**, 1572 (1979).
[15] J. L. Guesdon, T. Ternynck, and S. Avrameas, *J. Histochem. Cytochem.* **27**, 1131 (1979).

METHODS IN ENZYMOLOGY, VOL. 122

cein-labeled avidin, we noticed that the fluorescence of the labeled protein was enhanced upon binding of its specific ligand, biotin. This observation enabled the development of simple assays for avidin and biotin that are more sensitive and practical than the previously described fluorometric assays based on the intrinsic fluorescence of avidin.[12]

Materials and Methods

Fluorescein isothiocyanate isomer I (FITC), avidin, and d-biotin were obtained from Sigma (Poole, Dorset, United Kingdom). Triton X-100 was obtained from BDH (Poole, Dorset, United Kingdom).

Preparation of Fluorescein-Labeled Avidin

Avidin and FITC were reacted in sodium bicarbonate buffer (50 mM, pH 9.0) at molar ratios of $1:1$, $1:4$, $1:6$, and $1:10$ by adding 250-μl aliquots of avidin solution (10 g/liter) to 25, 100, 150, and 250 μl of FITC solution (580 mg/liter), respectively. Reaction mixtures were incubated at room temperature overnight, applied to columns of Sephadex G-25 fine grade (1.2 × 20 cm), and eluted with bicarbonate buffer. The entire labeled protein peak, identified by its color, was collected from each column, and was well separated from minimal amounts of unreacted FITC.

Labeled proteins were designated according to the molar ratio of FITC to protein in the reaction mixture, assuming a molecular weight of 66,000 for avidin.[13] Thus, $FITC_1$–avidin, $FITC_4$–avidin, $FITC_6$–avidin, and $FITC_{10}$–avidin were prepared and stored at $-18°$ until use. Concentrations of labeled proteins are given on the basis of their protein content, assuming total recovery from the original reaction mixtures.

Fluorometry

A Perkin-Elmer Model 1000 ratio-recording filter fluorometer was used, which was equipped with broad-band interference filters types FITC (440–490 nm bandpass) in the excitation path, and DB2 (510–600 nm bandpass) in the emission path. Both filters were from Barr and Stroud (Anniesland, Glasgow, United Kingdom). Because of the high light throughput of the FITC filter, it was necessary to limit the excitation beam to avoid overloading the reference detector of the fluorometer; this was achieved by placing a black plastic plate with a central circular aperture of 10-mm diameter in the excitation filter holder. The continuous wavelength filter of the Model 1000 was wound out of the emission beam and not used.

All fluorometric measurements in this fluorometer were made using disposable polystyrene test tubes (75 x 11 mm, No. 55.478 from Walter Sarstedt, Leicester, United Kingdom). To enable measurements on test tubes, an adapter unit with the same external dimensions as a conventional fluorometer cuvette was made. A 45-mm length of half-inch (12.7 mm) square aluminum rod was drilled out to a diameter of 11 mm down most of its long axis. Slits (30 x 5 mm) were cut in each face to allow entry and exit of light beams. The adapter was anodized black, and then permanently placed in the cuvette holder of the fluorometer. To accommodate the height of the test tubes, the normal sample compartment lid of the fluorometer was replaced with a standard Perkin-Elmer accessory unit (Ittrich sample cover).

Total fluorescence signals were recorded in arbitrary units and included a constant buffer background of 4.5 units. On the intensity scale chosen, a 10 nmol/liter solution of fluorescein gave a total signal of 125 units.

Relationship between Fluorescence Intensity of Avidin-Labeled Products and Molar Labeling Ratio

Fluorescein-labeled avidin products were diluted in phosphate–Triton buffer (100 mM sodium phosphate, pH 7.5, containing 0.1% v/v Triton X-100) to a protein concentration of 4 mg/liter. To 100 μl of each in assay tubes was added 1.1 ml of phosphate–Triton buffer, the tube contents mixed, and fluorescence intensities were then measured.

As demonstrated in Fig. 1, fluorescence of fluorescein-labeled avidin products did not show a linear relationship with the degree of labeling and was maximal at a molar labeling ratio, of FITC to avidin, of 6:1. The FITC$_4$–avidin-labeled product was chosen for further studies.

Assay of Biotin

To duplicate tubes containing biotin solutions (600 μl) of various concentrations in phosphate–Triton buffer was added 600 μl of fluorescein-labeled avidin solution (1 mg protein/liter) in the same buffer. After 15 min incubation at room temperature, fluorescence was measured by placing each tube into the fluorometer. Preliminary kinetic studies had demonstrated that binding reactions were complete within 5 min.

The fluorescence of fluorescein-labeled avidin was enhanced by biotin to a maximum extent of about 2-fold, which enabled construction of a standard dose–response curve for the assay of biotin (Fig. 2). The mini-

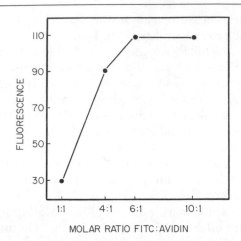

FIG. 1. Dependence of fluorescence of fluorescein-labeled avidin on molar labeling ratio.

mal detectable dose at the 95% confidence level[16] was 0.5 ng of biotin. Similar fluorescence enhancement was noted using labeled protein preparations obtained by reacting FITC and avidin at molar ratios of 1 : 1 or 6 : 1.

Assay of Avidin

To duplicate tubes containing avidin solutions (300 μl) of various concentrations in phosphate–Triton buffer was added 600 μl of fluorescein-labeled avidin solution (1 mg protein/liter), followed by 300 μl of biotin solution (16 μg/liter) in the same buffer. After 15 min incubation at room temperature, fluorescence was measured as above. The assay established for avidin was based on competition between the labeled and unlabeled protein for binding of a limited amount of biotin (Fig. 3). The minimal detectable dose was 0.04 μg of avidin.

Discussion

Since there is no gross change in the structure of avidin upon binding of biotin,[13] the relatively large increase in fluorescence of fluorescein-labeled avidin was unexpected. However, biotin binding to dansyl–avidin has been found[13] to lead to 40% reduction in emission intensity; this observation, together with fluorescence polarization results, was interpreted as suggesting local displacement of the dansyl groups into a more

[16] D. Rodbard, *Anal. Biochem.* **90,** 1 (1978).

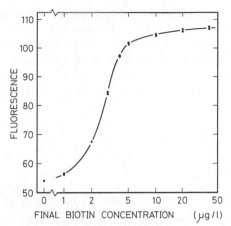

FIG. 2. Typical standard curve for fluorometric assay of biotin, utilizing biotin-induced enhancement of the fluorescence of fluorescein-labeled avidin.

aqueous environment where they had greater rotational freedom and less interaction with the protein structure. This finding would be consistent with a biotin-induced enhancement in the fluorescence of fluorescein groups attached to similar sites on the avidin molecule.

The quenching of the protein fluorescence of avidin upon binding of biotin has been used as the basis of existing fluorometric assays for avidin and biotin.[12] The biotin assay is comparatively insensitive, with a usefully applicable range between 20 and 120 ng of the analyte (a 6-fold span).

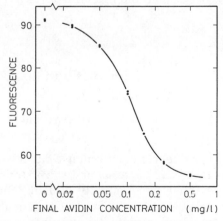

FIG. 3. Typical standard curve for fluorometric assay of avidin, utilizing competition between the labeled and unlabeled avidin for binding of biotin.

Proteins or other materials fluorescing in the ultraviolet may cause interference. The use of fluorescein-labeled avidin offers several advantages. These include increased sensitivity due to high quantum efficiency of fluorescein; absence of interference from proteins, since fluorometry is performed at visible wavelengths; and finally the fact that end point measurement can easily and directly be made in disposable plastic tubes used throughout the assay, again because of the visible-wavelength fluorometry.

The labeled avidin fluorometric assays offer a sensitivity about 20-fold better than the protein fluorescence method, and comparable with that of radio-assays using tracers of low specific activity,[1,2] and with that of the binding assays employing other types of nonisotopic labeled reactants.[7-10] Although microbiological assays,[1,12,13] enzymatic assays,[4] and radioassays employing tracers of high specific activity[2,5,6] can be between 10 and 100 times more sensitive, they are generally more complex than the described fluorometric methods.

Similar to many binding assays, the fluorometric assays provide a comparatively limited useful range. The biotin assay covers about a 4-fold useful range between 1.5 and 6 ng of analyte (Fig. 2) and the avidin assay about a 10-fold range between 0.06 and 0.6 μg (Fig. 3).

All radioassays[1,2,5,6] involve a separation step in which avidin-bound and free labeled biotin are separated by a variety of techniques. In the nonisotopic competitive-binding assays,[7-10] such a step is not required since binding to avidin changes the signal from the labeled biotin. The labeled avidin fluorometric assays described above provide an unusual instance of a nonisotopic nonseparation binding assay system in which the binding protein, rather than the ligand, is labeled.

[13] Solid-Phase Assay for *d*-Biotin and Avidin on Cellulose Disks

By L. GOLDSTEIN, S. A. YANKOFSKY, and G. COHEN

Most biotin assay procedures in current use either exploit the absolute growth requirement of certain bacterial and yeast strains for *d*-biotin, or else depend on the extremely high affinity of avidin for this vitamin.[1-15] Although the above two basic approaches to biotin quantitation are comparable in terms of specificity, sensitivity, and reproducibility, the binding assays, particularly those employing immobilized avidin, are faster and generally easier to perform.[1-15] Nevertheless, when avidin is covalently linked to particulate supports (e.g., Sepharose beads), it becomes difficult to distribute accurately constant amounts of the immobilized protein among different samples, and equally difficult to avoid losing some of the biotin–immobilized-avidin complex while centrifuging or filtering particles to remove unbound biotin. Both of these problems can be overcome by changing the geometry of the avidin-carrying support. Specifically, filter paper disks of uniform size, and loaded with essentially constant amounts of immobilized avidin, are easily transferable from one aqueous environment to another. Such avidinylated disks of known biotin-binding capacity form the basis of our sequential competition assay in which primary immersion of the disks in a sample of unknown biotin content is followed by exposure to an excess of radioactive biotin.[16] In the event that disks can bind more biotin than test samples contain, decreased disk capacity for the labeled ligand becomes a direct measure of sample biotin content.

Covalent linkage of avidin to filter paper disks is effected by first introducing isonitrile (isocyanide, $-C{\equiv}N$) functional groups onto the cellulosic matrix and then attaching the protein via its carboxyl groups in a four-component condensation (4CC) reaction.[17-21] This type of reaction generates a stable peptide bond between support and ligand. Analogously,

[1] E. A. Bayer and M. Wilchek, *Methods Biochem. Anal.* **26,** 1 (1980).
[2] K. Daksinamurti and S. P. Mistry, *J. Biol. Chem.* **238,** 294 (1963).
[3] L. D. Wright and H. R. Skeggs, *Proc. Soc. Exp. Biol. Med.* **117,** 95 (1944).
[4] N. M. Green, *Biochem. J.* **94,** 23c (1965).
[5] N. M. Green, *Biochem. J.* **89,** 585 (1963).
[6] R. D. Wei and L. D. Wright, *Proc. Soc. Exp. Biol. Med.* **117,** 17 (1964).
[7] D. Dakshinamurti, D. Landman, A. D. Ramamurti, and R. J. Constable, *Anal. Biochem.* **61,** 225 (1974).
[8] R. L. Hood, this series, Vol. 62, p. 279.
[9] K. Dakshinamurti and L. Allan, this series, Vol. 62, p. 284.
[10] H. L. Lin and J. F. Kirsch, this series, Vol. 62, p. 287.
[11] M. S. Kulomaa, H. A. Elo, and P. J. Tuohimaa, *Biochem. J.* **175,** 685 (1978).
[12] H. A. Elo, P. J. Tuohimaa, and O. Janne, *Mol. Cell. Endocrinol.* **2,** 203 (1975).
[13] H. A. Elo and P. J. Tuohimaa, *Biochem J.* **140,** 115 (1974).
[14] H. A. Elo and P. J. Tuohimaa, this series, Vol. 62, p. 290.
[15] J. A. Swack, G. L. Zander, and M. F. Utter, *Anal. Biochem.* **87,** 114 (1978).
[16] S. A. Yankofsky, R. Gurevitch, A. Niv, G. Cohen, and L. Goldstein, *Anal. Biochem.* **118,** 307 (1981).
[17] L. Goldstein, A. Freeman, and M. Sokolovsky, *Biochem. J.* **143,** 497, (1974).
[18] A. Freeman, M. Sokolovsky, and L. Goldstein, *J. Solid-Phase Biochem.* **1,** 261 (1976).
[19] A. Freeman, R. Granot, M. Sokolovsky, and L. Goldstein, *J. Solid-Phase Biochem.* **1,** 275 (1976).
[20] A. Freeman, M. Sokolovsky, and L. Goldstein, *Biochim. Biophys. Acta* **571,** 127 (1979).
[21] L. Goldstein, *J. Chromatogr.* **215,** 31 (1981).

it is possible to attach biotin covalently to the cellulosic matrix through its carboxyl group to give biotin–cellulose disks. This forms the basis of a simple and rapid avidin assay. Furthermore, since the avidin–biotin linkage is almost as strong as a covalent bond,[22-25] the product of interaction between biotin–cellulose disks and avidin, namely avidin–(biotin–cellulose) disks, is just as useful for biotin assay as the avidin–cellulose disks already mentioned.

Preparation of Avidin–Cellulose, Biotin–Cellulose, and Avidin–(Biotin–Cellulose) Disks

Isonitrile–Cellulose Disks

Filter-paper disks are chemically modified to contain isonitrile functional groups according to the procedure of Freeman *et al.*[20] The method involves partial ionization of polysaccharide hydroxyl groups, followed by nucleophilic attack of such polymeric alkoxide ions on an isonitrile containing a good leaving group in the ω position: 1-tosyloxy-3-isocyanopropane [p-CH$_3$-C$_6$H$_4$-SO$_2$-O-(CH$_2$)$_3$NC].

Preparation of Reagents. Dimethyl Sulfoxide. Dry dimethyl sulfoxide is used throughout. To this effect, analytical grade, anhydrous dimethyl sulfoxide (Merck) is stored over molecular sieve.

Sodium tert-Butoxide. Metallic sodium (1.15 g, 0.05 mol) and *tert*-butanol (100 ml) are warmed gently with magnetic stirring under reflux until complete dissolution and made up to 1 liter with dry dimethyl sulfoxide. Small volumes of the resulting 0.05 M solution are stored in tightly stoppered dark bottles at $-18°$.

1-Tosyloxy-3-isocyanopropane. The reagent is prepared from 3-aminopropanol via the N-formylaminopropanol[20] derivative as described below, or, alternatively, can be purchased from Fluka (Buchs, Switzerland).

N-Formylaminopropanol. An equivalent amount of ethyl formate (26.4 ml, 0.33 mol) is added dropwise to strongly stirred 1-amino-3-propanol (25 ml, 0.33 mol). Stirring is continued for 1 hr at room temperature. The ethanol formed in the reaction is removed in a rotatory evaporator and the residue distilled *in vacuo*. The N-formylaminopropanol yield is about 26 g (77%), b.p. 115–118° (10^{-2} mm Hg).

1-Tosyloxy-3-isocyanopropane. A pyridine solution (50 ml) of p-toluenesulfonyl chloride (38.2 g, 0.2 mol) is added dropwise over 30 min to a

[22] N. M. Green, *Adv. Protein Chem.* **29,** 85 (1975).
[23] N. M. Green, *Biochem. J.* **89,** 599 (1963).
[24] N. M. Green, *Biochem. J.* **89,** 609 (1963).
[25] N. M. Green and E. J. Toms, *Biochem. J.* **133,** 687 (1973).

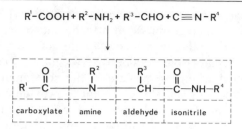

$$R^1\text{-COOH} + R^2\text{-NH}_2 + R^3\text{-CHO} + C \equiv N - R^4$$

Fig. 1. Four-component condensation between amine, carboxyl, aldehyde, and isonitrile.

vigorously stirred, ice-cooled solution of N-formylaminopropanol (10.3 g, 0.1 mol) in pyridine (50 ml). The reaction mixture is stirred over ice for 1 hr. Cold water (100 ml) is then added and the mixture extracted with three 50-ml portions of diethyl ether–hexane (5 : 1, v/v). The combined extract is washed with cold water and dried over sodium sulfate. The solvent is removed by evaporation and the residue dissolved in diethyl ether–hexane (3.5 : 1, v/v). The solution is left at $-18°$ and the white crystalline solid which forms is collected on a filter, washed with 10 ml of ice-cold hexane, and air dried. The yield is 7 g (25%), m.p. 37–38°.

Preparation of Isonitrile–Cellulose Disks.[16,20] Disks of 1-cm diameter are cut from sheets of Whatman filter paper (Grade 542, ashless, hardened) and then modified to contain isonitrile groups as follows. About 700 disks (5 g) are suspended in 100 ml of dry dimethyl sulfoxide and allowed to swell with gentle stirring at 40° for 30 min in a stoppered flask. Sodium *tert*-butoxide (0.05 M) in dimethyl sulfoxide (50 ml) is added dropwise with stirring, and ionization is allowed to proceed to equilibrium for 15 min. A dimethyl sulfoxide solution (50 ml) of 1-tosyloxy-3-isocyanopropane (2.4 g) is then added, and stirring at 40° is continued for 4 hr. The disks are removed, resuspended in methanol, exhaustively washed in a Büchner funnel under suction with methanol followed by diethyl ether, and air dried. Disks are stored in closed vials at 4° and remain fully active under these conditions for at least 1 year.

Avidin–Cellulose Disks

As illustrated in Fig. 1, 4CC reactions involve the simultaneous participation of amine, carboxyl, aldehyde, and isonitrile and lead to the formation of a stable, N-substituted peptide bond between the carboxyl (R^1-COOH) and amine (R^2-NH$_2$) moieties, in which the aldehyde and isonitrile components (R^3-CHO and R^4-NC) appear as the side chain attached to the amide nitrogen. Due to their polyampholyte nature, proteins such as avidin can serve as either amine or carboxyl donors in 4CC reactions, in which the actual mode of attachment to the support (R^4- in Fig. 1) is determined by the overall composition of the reaction mixture. In the presence of water-soluble aldehyde (acetaldehyde), amine-mode coupling is obtained when the protein is the sole amine donor in the reaction and an excess of acetate is present to provide carboxyl groups. Conversely, addition of acetaldehyde and excess amine (e.g., Tris, trishydroxymethylaminomethane) allows the protein to undergo covalent attachment through its carboxyl groups. Carboxyl-mode attachment gives rise to avidin–cellulose filters with greater biotin-binding capacity than in the case of amine-mode coupling. Procedural details for covalent attachment of commercial avidin (Sigma, chromatographically purified, 10–15 U/mg protein) to isonitrile–cellulose disks are given below.

Procedure. About 700 isonitrile–cellulose disks (5 g) are suspended in 200 ml of cold (4°) 0.1 *M* Tris (pH 7.4) buffer containing 40 μg/ml of avidin. Based on its known molecular weight, approximately 0.15 nmol of avidin tetramer is available per disk. The mixture, in a glass-stoppered flask, is magnetically stirred or gently shaken in a rotatory shaker. Redistilled, ice-cold acetaldehyde (1 ml) is then added to make the mixture 0.09 *M* in aldehyde. After stirring overnight at 4° with the stopper tightly in place, the reaction mixture is decanted off and the disks are washed several times with 5–10 volumes of water in a Büchner funnel under suction, before soaking in 0.1 *M* phosphate at pH 7 in the cold for several hours to remove absorbed protein (20–30% of total disk protein). Disks are then washed extensively with phosphate buffer and again with distilled water before storage at −18° in water suspension. Repeated cycles of freezing and thawing of avidin–cellulose disks have no deleterious effect on their biotin-binding ability, and they are also stable in water (with 10 m*M* sodium azide present to suppress the growth of microorganisms) at room temperature for up to 5 days. However, care must be taken to keep disks moist at all times, because drying invariably leads to significant inactivation of the immobilized avidin. Providing that it was stored at 4° in a tightly stoppered bottle, the avidin–Tris–acetaldehyde supernatant can be used twice more to avidinylate isonitrile–cellulose filters without significant dimunition of final biotin-binding capacity.

Biotin–Cellulose Disks

Procedure. Isonitrile–cellulose disks (7 g) are suspended in 80 ml of ice-cold 0.1 *M* Tris (pH 7.0) buffer containing 30 mg *d*-biotin. To the tightly stoppered vessel is next added 0.3–0.5 ml of cold acetaldehyde (redistilled), and the reaction is allowed to proceed in the cold for 18–24 hr with either gentle mechanical stirring or rotatory shaking. Care must be taken to keep the rate of agitation below the point at which attrition of the paper disks becomes significant. Removal of all traces of absorbed biotin from biotinylated disks is crucial and the terminal washing procedure must therefore be exhaustive. In our experience, continuous washing under a gentle stream of tap water for 24 hr achieves this latter end. Disks are now sequentially washed with methanol and diethyl ether in a suction funnel and air-dried. Dry disks are stable indefinitely at room temperature.

Avidin–(Biotin–Cellulose) Disks

The source of avidin for preparation of avidin–(biotin–cellulose) disks can be (1) commercial avidin or (2) egg white.

Procedure 1. From Commercial Avidin. Chromatographically purified avidin (tetramer) is dissolved to a final concentration of 100–200 μg/ml in 0.5 M phosphate (pH 7.0) buffer made 5 mM in cetyltrimethylammonium-bromide (CTAB) and 10 mM in sodium azide. Approximately 20 biotin–cellulose disks are added for every milliliter of this solution, and the mixture then allowed to incubate with occasional shaking for 3–4 hr at 37°. The mixture of polyvalent anion plus cationic detergent minimizes nonspecific adsorption of avidin (a glycoprotein) to the cellulosic matrix of the disks. Greater than 90% of the avidin-binding capacity of the disks is satisfied under the above conditions.

Procedure 2. From Egg White. Egg white from one fresh egg (about 30 ml) is slowly mixed with 1 volume of 2% Triton X-100 while magnetically stirring at room temperature. The resulting clear homogenate is then further diluted with 2 volumes of 0.5 M phosphate (pH 7.0) buffer containing 5 μmol/ml CTAB and centrifuged at 8000 g and room temperature for 20 min to clear. After discarding the gel-like, transparent precipitate, the clear supernatant (about 120 ml) can be used without further treatment as the source of avidin tetramer for preparation of avidin–(biotin–cellulose) disks. The combination of the high phosphate concentration and cationic detergent (CTAB) serves to minimize unspecific absorbtion of avidin and other macromolecules to the cellulosic matrix. Incubation of 100 disks per 100 ml of detergent-treated homogenate for 48 hr at room temperature with occasional stirring produces avidin–(biotin–cellulose) disks of sufficiently high biotin-binding capacity to be useful as biotin assay vehicles. Alternatively, 6–8 hr at 37° yields the same result. The egg white–detergent mixture can be reused at least twice more to avidinylate additional batches of biotin–cellulose disks. It is also crucial to employ eggs which have not been held in cold storage for too long. Prolonged storage appears to promote blocking of biotin-binding sites, presumably as a consequence of biotin leakage from yolk to white.

Avidinylated disks prepared by either method are washed free of excess reaction mixture prior to use. However, since ligation of avidin is carried out under conditions which minimize nonspecific adsorption of protein, there is no need for an extensive washing procedure at this stage.

Biotin Assay Procedure

This two-step procedure involves (1) exhaustive uptake of biotin from a sufficiently diluted aliquot of the unknown sample by exposure to a fixed number of avidin–cellulose or avidin–(biotin–cellulose) disks, and (2) immersion of the washed disks in a solution of [14C]biotin or [3H]biotin containing an excess of the labeled ligand.

Measurement of the Biotin-Binding Capacity of Avidinylated Disks by Saturative Exposure to Radioactive Biotin

Up to 100 avidin–cellulose or avidin–(biotin–cellulose) disks are immersed in 30 ml of radioactive biotin solution (150 ng/ml) made 10 mM in sodium azide and incubated with occasional shaking for 30 min (or longer) at any temperature between 4 and 37° The disks are then transferred with forceps into a beaker containing 200–300 ml of water, spread over the surface of a circular piece of filter paper in a large Büchner funnel, and successively washed six times apiece with water, 0.05 M NaHCO$_3$, and water under suction. After drying in air for about 20 min, the disks are individually immersed in Hydro Luma fluor mixture (Lumac Systems AG, Basel, Switzerland) and counted in a liquid scintillation spectrometer at appropriate settings.

Depending on the capacity of the first batch of avidinylated disks for biotin, it is generally possible to reuse the same radioactive biotin solution to saturate at least one more batch of disks. In any case, adding enough [14C]- or [3H]biotin to the standard solution to maintain a saturating level of the ligand allows efficient use of radioactive material.

Exhaustive Uptake of Biotin from Standard or Unknown Samples by Avidinylated Disks

The routine procedure employed is to immerse five avidin–cellulose disks, or the same number of avidin–(biotin–cellulose) disks, in 5 ml of biotin-containing sample and then incubate them at room temperature with gentle rotatory or reciprocal shaking for 2–3hr. It may be noted that each disk is marked with a soft pencil to facilitate its later identification. Providing that the amount of biotin in the sample is less than the overall

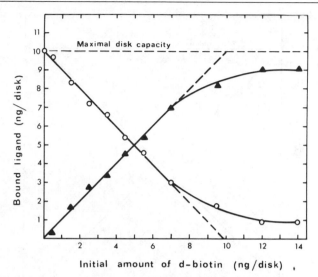

FIG. 2. Stoichiometry of uptake of *d*-biotin by avidin–cellulose disks. Initial binding reactions with the indicated amounts of *d*-biotin dissolved in the basal salts mixture of Davis and Mingioli[26] were carried out for 3 hr at 25° with six avidin–cellulose disks present per 6 ml of sample. Disks were then washed, saturated with [¹⁴C]biotin (280 cpm/ng), washed again, dried, and counted. (○) [¹⁴C]biotin bound per disk; (▲) calculated amount (ng) *d*-biotin bound per disk; values obtained by subtracting experimental values from the maximal disk capacity (horizontal dashed line).

capacity of the disks to accept biotin, the above procedure removes greater than 90% of all biotin present from solution. Individual disks are then extensively washed and saturated with radioactive biotin as already detailed.

The biotin-binding capacity of individual disks in a given batch may vary by as much as 20–30% from the batch average. Consequently, single-disk determinations are inherently imprecise and replicate assays per sample become necessary. Averaging the results of five single-disk determinations provides the requisite assay precision. Moreover, given the manipulative ease of the assay, such multiple-disk operations do not greatly add to the time and effort expended in determining the biotin content of even large numbers of samples. Indeed, the most important factor in determining how many samples can be assayed at once is the need to run simultaneously a series of dilutions from each unknown sample to ensure that the condition of disk subsaturation will be met.

Representative calibration curves for sequential competition assay of *d*-biotin with avidin–cellulose and avidin–(biotin–cellulose) disks are presented in Figs. 2 and 3, respectively. In the case in which avidin is cova-

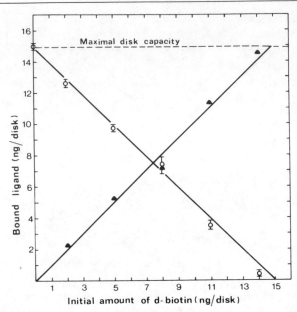

FIG. 3. Stoichiometry of uptake of d-biotin by avidin–(biotin–cellulose) disks. The biotin–cellulose disks were used in this case were avidinylated by immersion in detergent-treated egg-white solution as detailed in text. Each datum point represents the average of five replicate determinations. Assay details and plot symbols as in the legend to Fig. 2.

lently bound to the cellulosic matrix (Fig. 2),[26] the range of the linear response does not extend beyond 70% saturation of available biotin-binding sites. The observed departure from linearity, presumably due to the existence of avidin sites with reduced affinity for biotin, probably results from chemical damage to avidin molecules during the covalent coupling reaction. If so, noncovalent attachment of avidin to biotin–cellulose disks should avoid this complication. As can be seen in Fig. 3, the amount or radioactive biotin ultimately bound to avidin–(biotin–cellulose) disks is inversely proportional to d-biotin input throughout. It is also evident that the effective upper limit of avidin–cellulose disks for biotin is only about 8 ng per disk (Fig. 2), whereas that of avidin–(biotin–cellulose) disks is approximately 15 ng per disk (Fig. 3). Disks of the latter type with even higher capacities for biotin can be prepared by raising the level of d-biotin in the 4CC reaction mixture. Resulting biotin–cellulose disks have a greater surface density of covalently bound biotin and are consequently able to ligate more avidin than those prepared in a standard reaction mixture.

[26] B. D. Davis and E. S. Mingioli, *J. Bacteriol.* **60**, 17 (1950).

The smallest amount of biotin per disk which can be measured with reasonable accuracy is about 1 ng. Accordingly, the lowest concentration of biotin assayable by the standard procedure described above is about 1 ng/ml. The standard disk assay is therefore comparable in sensitivity to bioassay procedures. Nevertheless, by increasing the volume of the test sample and prolonging the time of incubation one may accurately measure the biotin content of solutions much more dilute than 1 ng/ml. For example, exposing each disk to 20 ml of sample for 30 hr will permit accurate assay of a 50 pg/ml solution of biotin. Conversely, sample volumes as small as 0.1 ml (per disk) and incubation times as short as 30 min will suffice to determine the biotin content of sufficiently concentrated solutions.

Assay of Avidin on Biotin–Cellulose Disks

The ability of biotin–cellulose disks to remove avidin from crude mixtures forms the basis for the simple assay procedure described below.

Pretreatment of Disks and Glassware

Well-washed biotin–cellulose disks are immersed in a solution containing 4 mg/ml each of lysozyme (Sigma, egg white) and bovine serum albumin (Sigma, Type V) dissolved in PTA buffer (0.5 M phosphate, 1% Triton X-100, 10 mM sodium azide, pH 7.0). Uptake of avidin from solution by disks treated in this way is linearly related to avidin concentration, presumably because nonspecific adsorption of the glycoprotein in question to the cellulose is minimalized. Similarly, presoaking of glassware and micropipet tips in a relatively dilute (5-fold) solution of detergent-treated egg white prevents loss of sample avidin to the walls of dilution tubes and pipets.

Exhaustive Uptake of Avidin from Solution by Biotin–Cellulose Disks

Biotin–cellulose disks are placed in the wells of a tissue culture microplate of appropriate size (one disk per well). Avidin-containing test solutions are diluted 1 : 1 with double-strength PTA buffer (pH 7.0) and 200 μl of diluate placed on each appropriately marked disk. The plates are covered, incubated at 37° for 3 hr, washed successively with PTA buffer and water, and then immersed in a standard solution of radioactive biotin to saturate the biotin-binding sites of avidin taken up from the test solution. Providing that uptake of avidin is exhaustive, and that filter-associated avidin molecules possess uniform biotin-binding capacity, the above procedure leads to direct proportionality between the initial avidin concen-

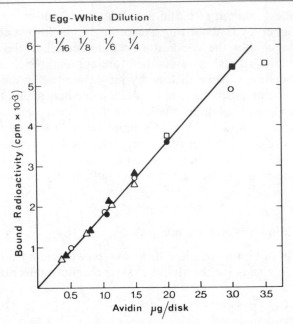

FIG. 4. Stoichiometry of uptake of avidin by biotin–cellulose disks. All data points represent the average of five single-disk determinations in 200 μl of test sample as detailed in text. Top abscissa denotes final dilution of egg white from one egg, and bottom abscissa pertains to chromatagraphically purified commercial avidin. Open and closed symbols signify different batches of biotin–cellulose disks. (○, ●) Commercial avidin; (△, ▲) diluted egg white; (□, ■) avidin-enriched egg white.

tration of the test sample and the final amount of radioactive biotin immobilized on the disk. This linear relationship is demonstrated in Fig. 4 for solutions of known commercial avidin content, for serially diluted solutions of egg white, and for solutions of egg white enriched with known amounts of commercial avidin. It should, however, be noted that about 3.5 μg of avidin per 200-μl aliquot is the highest concentration at which a linear response is still obtained. Consequently, test solutions falling outside the linear range of the assay must undergo retesting after dilution.

[14] Purification of Avidin and Its Derivatives on 2-Iminobiotin-6-aminohexyl-Sepharose 4B

By GEORGE A. ORR, GAYLE C. HENEY, and RON ZEHEB

Principle

The avidin–biotin complex possesses an extremely low dissociation constant (K_d) of approximately 10^{-15} M.[1] The tightness of this interaction has formed the basis for the use of the complex in several techniques in membrane and molecular biology.[2,3] In many of these studies, it would be desirable if an affinity isolation system were available for the isolation of biologically active avidin derivatives that were uncontaminated by damaged avidin or unconjugated reporter molecules. Cuatrecasas and Wilchek have reported the isolation of avidin on a biocytin (biotin-ε-N-lysyl)-Sepharose 4B column.[4] The conditions required for the elution of the specifically bound protein were a combination of low pH (1.5) and 6 M guanidine-HCl; either alone did not effect elution. Although avidin is not inactivated by these harsh elution conditions, it is uncertain whether certain avidin derivatives, e.g., horseradish peroxidase or alkaline phosphatase conjugated avidin, would retain their desired biological activity.

We have found,[5] in agreement with the observations of Green,[1] that the free base form of 2-iminobiotin forms a stable complex with avidin, but that the salt form interacts poorly with its binding protein (Fig. 1). Our studies indicate that the decrease in affinity observed at neutral and acidic pH values is due to the combined protonation of the cyclic guanidino group of 2-iminobiotin and the ionization of some residue on avidin.[5] We have used this pH-dependent alteration in binding to develop an efficient affinity isolation procedure for avidin and its derivatives.[6]

Synthesis of 2-Iminobiotin Hydrobromide

The two-step synthesis involves the alkaline hydrolysis of biotin with barium hydroxide at 140° yielding 5-(3,4-diaminothiophan-2-yl)pentanoic

[1] N. M. Green, Adv. Protein Chem. 29, 85 (1975).
[2] E. A. Bayer and M. Wilchek Trends Biochem. Sci. 3, N257 (1978).
[3] E. A. Bayer and M. Wilchek, this series, Vol. 62, p. 308.
[4] P. Cuatrecasas and M. Wilchek, Biochem. Biophys. Res. Commun. 33, 235 (1968).
[5] G. A. Orr, J. Biol. Chem. 256, 761 (1980).
[6] G. Heney and G. A. Orr, Anal. Biochem. 114, 92 (1981).

METHODS IN ENZYMOLOGY, VOL. 122

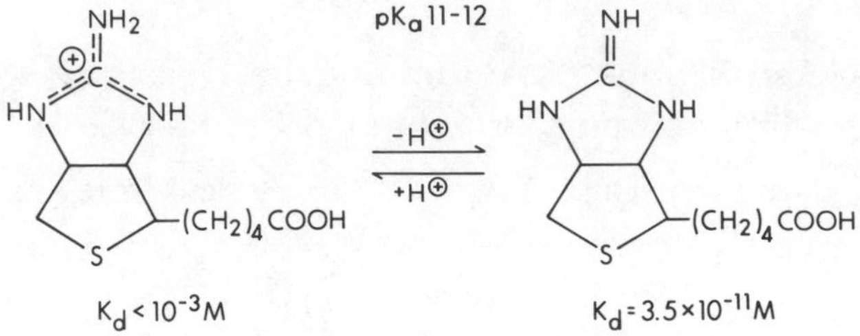

FIG. 1. Structure of 2-iminobiotin showing ionization of the cyclic guanidino group.

acid.[7] 2-Iminobiotin is subsequently prepared from this diaminocarboxylic acid derivative of biotin by reaction with cyanogen bromide.[8]

5-(3,4-Diaminothiophan-2-yl)pentanoic acid. Biotin (1 g) and barium hydroxide (3 g) are mixed together and placed into a Pyrex hydrolysis tube (15 × 250 mm). Water (7 ml) is added and the tube sealed under mild vacuum (house vacumm line, approx 30 mm Hg). After heating at 140° for 21 hr, the tube contents are removed, CO_2 is bubbled into the suspension, and the insoluble precipitate ($BaCO_3$) is removed by filtration. The filtrate is acidified to pH 4 using 2 N H_2SO_4, filtered, and concentrated *in vacuo*. Addition of methanol induces crystallization of the diaminocarboxylic acid sulfate and this process is allowed to continue overnight at 40° (56% yield). Thin-layer chromatography on silica gel ($CHCl_3$: H_2O : concentrated NH_4OH, 75 : 25 : 5) reveals the presence of a single I_2- and ninhydrin-positive spot of R_f 0.6.

2-Iminobiotin. The diaminocarboxylic acid sulfate (2.5 g) is treated with barium carbonate (5 g) and dissolved in 30 ml of hot H_2O, and the solution immediately filtered through a prewarmed fritted glass funnel. The precipitate is washed with an additional 10 ml of hot H_2O. Cyanogen bromide (1.9 g) is added to the filtrate, and crystals of 2-iminobiotin (free base form) start to form within 5 min. The reaction mixture is left at 4° for 12 hr. The yield of 2-iminobiotin is 1.8 g (m.p. > 260° decomp.). To convert the free base into the hydrobromide salt, 2-iminobiotin is suspended in H_2O (40 ml) and heated to approximately 45°, and 1% HBr is added dropwise until all of the 2-iminobiotin dissolves. The solution is concentrated *in vacuo* and the residue recrystallized from hot 2-propanol–methanol (6 : 4, v/v). 2-Iminobiotin hydrobromide has a melting point of 222–223°. Thin-layer chromatography on silica gel eluting with $CHCl_3$: H_2O : concentrated NH_4OH(70 : 25 : 5, v/v) reveals the presence of a single I_2-positive spot with an R_f of 0.38.

[7] K. Hofmann, D. B. Melville, and V. Du Vigneaud, *J. Biol. Chem.* **141,** 207 (1941).
[8] K. Hofmann and A. E. Axelrod, *J. Biol. Chem.* **187,** 29 (1950).

Purification of Avidin

2-Iminobiotin is coupled to 6-aminohexyl-Sepharose 4B using a water-soluble carbodiimide. Although 2-iminobiotin contains a potentially reactive guanidino group (pK_a 11–12), this does not complicate the coupling, since the reaction is carried out at pH 4.8, where the group is fully protonated. Purified avidin is bound by the affinity matrix at pH 11 and is eluted as a sharp peak when the pH is lowered to 4. Biotin-treated avidin is not retained by the affinity column, indicating that binding is specific to the 2-iminobiotin moiety. For efficient binding to immobilized 2-iminobiotin, the pH must be greater than 9.0. We have used this affinity matrix for the isolation of avidin from homogenized egg whites in a single step. However, for the purification of large amounts of avidin, we have found it more convenient to carry out a preliminary ammonium sulfate fractionation.

2-Iminobiotin-6-aminohexyl-Sepharose 4B. 2-Iminobiotin (400 mg, 1.23 mmol) is added to 100 ml of 6-aminohexyl-Sepharose 4B (40 ml of packed resin in H_2O), prepared by the method of Poráth,[9] and the pH adjusted to 4.8 with HBr (1%, v/v). 1-Cyclohexyl-3-(2-morpholinoethyl)-carbodiimidemetho-p-toluene sulfonate (4.24 g, 10 mmol) is added portionwise over a period of 10 min. The pH is kept at 4.8 throughout the reaction by the addition of 1% HBr and is constant after 3–5 hr. The resin is washed with 1 M NaCl (2 liters) and H_2O (2 liters) and packed into a column. The binding capacity of the affinity matrix, using these coupling conditions is 0.75 mg purified avidin/ml of swollen gel.

Affinity Isolation of Avidin. Homogenized egg whites from 24 fresh jumbo eggs are diluted with H_2O (2 : 1, v/v) and the solution is brought to 70% saturation with ammonium sulfate (enzyme grade, Schwartz/Mann) at 4°. After stirring for 2 hr, the mixture is centrifuged (8000 g, 20 min), the supernatant brought to 100% saturation, and the mixture left stirring at 4° overnight. After centrifugation (8000 g, 30 min), the pellet is dissolved in H_2O (40 ml) and dialyzed against H_2O (3 × 2 liters). The pH of the dialysate is adjusted to 11 with 1 N NaOH and NaCl (1 M) added. The crude avidin solution is applied to 2-iminobiotin-6-aminohexyl-Sepharose 4B (40 ml) which had previously been equilibrated with 50 mM sodium carbonate (pH 11) containing 1 M NaCl. The column is washed with equilibrating buffer (20 ml/hr) until the absorbance at 282 nm returns to baseline. Avidin is eluted from the column with 50 mM ammonium acetate (pH 4) containing 0.5 M NaCl. Protein content is measured by absorbance at 282 nm and avidin content by its ability to bind either 4-hy-

[9] J. Poráth, this series, Vol. 18 Part B, p. 13.

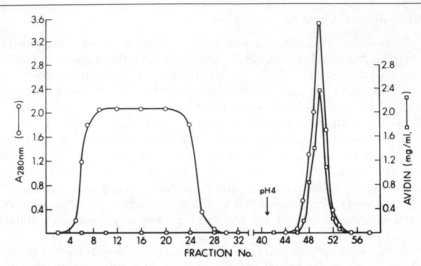

FIG. 2. Affinity purification of avidin on 2-iminobiotin-6-aminohexyl-Sepharose 4B. Crude avidin was loaded onto the column at pH 11 and specifically eluted by application of pH 4 buffer.

droxyazobenzene-2'-carboxylic acid or [^{14}C]biotin.[10] The appropriate fractions are pooled, dialyzed against H_2O, and lyophilized.

As can be seen from Fig. 2, avidin is eluted as a sharp peak after application of the low pH buffer. Greater than 90% of the crude avidin applied to the column is recovered in the specifically eluted fractions, and the yield of avidin from 24 eggs is in the range of 15–20 mg. We have used the same affinity column for several such avidin preparations with no apparent loss in activity. The avidin obtained by this procedure is pure as judged by its ability to bind 14.4 μg of [^{14}C]biotin/mg of protein. Literature values for pure avidin range from 13.8 to 15.1.[1,4] SDS–polyacrylamide gel electrophoresis of the specifically eluted fractions reveals a single polypeptide with an apparent molecular weight slightly larger than the hemoglobin monomer (16,000). Native avidin (68,000) is composed of four identical subunits.[1] As a further indication of purity, if the specifically eluted protein is treated with biotin and rechromatographed on the affinity column, no protein is retained and eluted at pH 4.

Conclusions

The pH-dependent interaction between 2-iminobiotin and avidin has enabled us to develop an affinity isolation procedure for avidin which

[10] N. M. Green, this series, Vol. 18 Part A, p. 418.

overcomes the harsh elution condition associated with immobilized biotin columns. We have also used this system for the purification of both [125]I- and rhodamine-labeled avidin.[6] This method should also prove suitable for the purification of enzyme-conjugated avidin derivatives, e.g., horseradish peroxidase and alkaline phosphates conjugates.

Acknowledgments

This research was supported in part by Grants GM 27851, GM 34029, and Cancer Core Grant P30-CA13330. G.A.O. is a recipient of a Research Career Development Award (HD 00577). R.Z. was supported in part by NIH Training Grant 5T32-GM 07260.

[15] Use of Avidin–Iminobiotin Complexes for Purifying Plasma Membrane Proteins

By RON ZEHEB and GEORGE A. ORR

Principle

An analytical and preparative approach which provides information concerning the organization and function of cell surface components without the prior isolation of plasma membranes has been developed.[1,2] The basis of the technique is the covalent attachment of compounds containing 2-iminobiotin, the cyclic guanidino analog of biotin, onto cell surface proteins. The "tagged" species are then isolated by virtue of the unique interaction between the covalently attached ligand and its binding protein, avidin. The pH-dependent interaction of 2-iminobiotin with avidin makes recovery possible. At high pH, the free base form of 2-iminobiotin retains the high-affinity specific binding to avidin characteristic of biotin ($K_d = 10^{-11}\ M$) whereas at acidic pH values, the salt form of the analog interacts poorly with avidin ($K_d > 10^{-3}\ M$). By chemically modifying the valeric acid side chain of 2-iminobiotin (Fig. 1), it is possible to create a series of molecules capable of covalently "tagging" membrane proteins and glycoproteins with 2-iminobiotin. The specific site of covalent attachment can be varied depending on the congener of 2-iminobiotin used and the choice of prior chemical or enzymatic treatment of the target membranes. Additionally, the method permits labeling under native conditions in which cell membrane integrity is maintained.

[1] G. A. Orr, *J. Biol. Chem.* **256,** 761 (1981).
[2] R. Zeheb, V. Chang, and G. A. Orr, *Anal. Biochem.* **129,** 156 (1983).

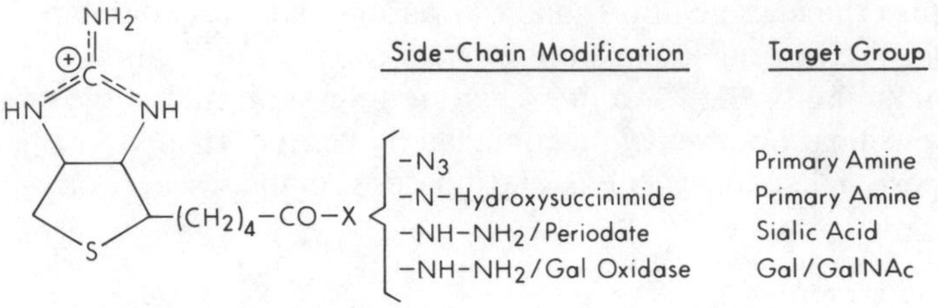

NH₂

$\overset{\oplus}{\underset{HN}{C}}\overset{\parallel}{\underset{NH}{}}$

$(CH_2)_4-CO-X$

	Side-Chain Modification	Target Group
	$-N_3$	Primary Amine
	$-N-$Hydroxysuccinimide	Primary Amine
	$-NH-NH_2$ / Periodate	Sialic Acid
	$-NH-NH_2$ / Gal Oxidase	Gal / GalNAc

FIG. 1. Some methods for selectively attaching 2-iminobiotin onto specific cell surface targets.

Synthesis of 2-Iminobiotin and Derivatives

2-Iminobiotin. The synthesis of 2-iminobiotin is described in the previous chapter of this volume.[3]

2-Iminobiotin N-Hydroxysuccinimide Ester. The N-hydroxysuccinimide ester of 2-iminobiotin is prepared by dissolving 2-iminobiotin hydrobromide (324 mg, 1 mmol) and N-hydroxysuccinimide (115 mg, 1 mmol) in dry N,N-dimethylformamide (3 ml) at 4°. Dicyclohexylcarbodiimide (206 mg, 1 mmol) is added and the reaction allowed to proceed for 1 hr at 4° and then overnight at room temperature. The dicyclohexylurea is removed by filtration and the organic solvent taken off *in vacuo*. The residue is recrystallized from hot 2-propanol, yielding the N-hydroxysuccinimide ester (58% yield, m.p. 160–161°). Analysis, calculated for $C_{14}H_{21}N_4O_4SBr$: C 39.91, H 5.02, N 13.30, S 7.61%; found, C 40.00, H 5.20, N 13.20, S 7.34%.

2-Iminobiotin Methyl Ester Hydrochloride. 2-Iminobiotin is converted to its methyl ester hydrochloride by treatment with methanolic HCl at 4° for 15 hr. Removal of the solvent gives the desired compound in quantitative yield (m.p. 156–157°). Analysis, calculated for $C_{11}H_{20}N_3O_2SCl$: C 44.97, H 6.86, N 14.30, S 10.91, Cl 12.07%; found, C 44.97, H 7.09, N 14.31, S 10.66, Cl 12.09%.

2-Iminobiotin Hydrazide Hydrochloride. 2-Iminobiotin methyl ester hydrochloride (0.5 g) is dissolved in methanol (15 ml), and hydrazine hydrate (1 ml) added. After 5 hr at room temperature, the reaction mixture is taken to dryness *in vacuo* and washed several times with diethyl ether, and the residue recrystallized from 2-propanol–diethyl ether (55% yield, m.p. 209.5–210.5°). Analysis, calculated for $C_{10}H_{20}N_5OSCl$: C 40.88, H 6.86, N 23.84, S 10.91, Cl 12.07%; found, C 40.81, H 6.73, N 23.62, S 11.07, Cl 12.23%.

2-Iminobiotin Acyl Azide Hydrochloride. The acyl azide of 2-imino-

[3] G. A. Orr, G. C. Heney, and R. Zeheb, this volume, [14].

biotin is formed by the reaction of 2-iminobiotin hydrazide (5 mM in 100 mM HCl) with an equal volume of sodium nitrite (200 mM in 100 mM HCl) for 30 min at 4°. Both solutions should be prechilled to 4° prior to mixing. The acyl azide is extracted into benzene by three successive washes of 200 μl each. The benzene washes are pooled and delivered to a glass test tube, and the benzene evaporated under a stream of nitrogen. The reactive azide should be used immediately (MW 304).

Preparation of Other Materials

Surface Labeling of Cells with 2-Iminobiotin

Reaction with Sialic Acid Residues. Cells are washed in PBS (10 mM sodium phosphate, 140 mM NaCl; pH 7.4, containing 10 mM benzamidine) and resuspended at a concentration of 1 × 10⁸ per ml. The cell suspension is reacted with an equal volume of 2 mM sodium metaperiodate (the appropriate concentration of sodium metaperiodate may vary for different cell types and should be determined experimentally) in PBS, and the reaction mixture left on ice for 15 min with occasional shaking. The reaction is stopped by the addition of 10 ml of ice-cold PBS, and the cells centrifuged and washed three times. The chemically oxidized cells are resuspended in PBS containing 5 mM 2-iminobiotin hydrazide. Incubation is for 1–3 hr at 37° with occasional shaking, after which the cells are centrifuged and washed three times.

Reaction with Galactose/N-Acetylgalactosamine Residues. Cells are washed in phosphate-buffered saline (pH 6.8, containing 10 mM benzamidine) and incubated with galactose oxidase (20 U per 1 × 10⁸ cells per ml) for 30 min at 37°. The cells are then centrifuged, washed three times with PBS, and incubated with 2-iminobiotin hydrazide as above.

Reaction with Primary Amino Groups. Cells suspended in 1 ml of PBS (pH 7.4, 1 × 10⁸ per ml) are labeled by the addition of 2-iminobiotin N-hydroxysuccinimide ester (25 μl of a 200 mM stock solution in dimethyl sulfoxide, 5 mM final concentration) at 4° for 30 min. After incubation, cells are centrifuged and washed as above.

As with the N-hydroxysuccinimide ester, the acyl azide derivative of 2-iminobiotin reacts with primary amino groups. Cells suspended in PBS (10⁸/ml, 100 μl) are added directly to tubes which have been plated with the acyl azide as described.

Solubilization of 2-Iminobiotin-Derivatized Cells

Prior to retrieval of 2-iminobiotin-derivatized proteins, cells are subject to detergent solubilization. The detergent type and concentration must be determined for the particular cell system under study. We have found the following protocol to be generally satisfactory.

2-Iminobiotin "tagged" cells are solubilized using 500 μl of solubilization buffer (50 mM borate, pH 8.0, 0.5 mM phenylmethylsulfonyl fluoride, 10 mM benzamidine, and 2%, (w/v), Triton X-100) for 100 μl packed washed cells. Solubilization is allowed to proceed for 30 min at 4° with occasional vortexing. The solution is centrifuged at 40,000 g for 30 min to remove undissolved material, and the clear supernatant retained. Detergent concentrations of up to 4% (w/v) of Triton X-100, Nonidet P-40, or Lubrol PX do not interfere with the subsequent recovery of 2-iminobiotinylated proteins by immunoprecipitation (Fig. 2C) or by affinity chromatography on immobilized avidin.[1,2]

Preparation of Anti-Avidin Serum

Anti-avidin serum is raised in rabbits (New Zealand White, female) by subcutaneous injecions of a mixture of avidin (1 mg/ml in PBS) and Freund's adjuvant (1 : 1, v/v) at 2-week intervals for 6–12 weeks. The first immunization is with complete adjuvant, while subsequent boosts are with incomplete Freund's adjuvant. When antiserum titers (measured by the ability to precipitate ^{125}I-labeled avidin)[4] have reached sufficiently high levels (>1 : 1,000), the animals are bled by cardiac puncture and serum isolated by centrifugation (30 min at 3000 g, 4°). Serum is stored in aliquots at -70° until used.

Preparation of Avidin-Sepharose 4B

Avidin-Sepharose 4B is prepared according to the method of Bodanszky and Bodanszky.[5]

Recovery of 2-Iminobiotinylated Proteins by Immunoprecipitation

Immunoprecipitation of detergent-solubilized 2-iminobiotin "tagged" proteins is carried out in buffers (50 mM sodium borate, pH 8.5–9.0 or 50 mM glycine, pH 9.5–10) containing 1 mg/ml BSA, 300 mM NaCl, and 0.5% (w/v) Triton X-100. Solubilized proteins, in the above buffer, are reacted with a slight molar excess of avidin. Anti-avidin antiserum is added (approximate final dilution of 1 : 200) and the mixture incubated at 37° for 90 min. Twenty-five microliters of a 10% suspension of insoluble protein A (IgGSorb, The Enzyme Center, Malden, MA) is used to bring

[4] G. Heney, and G. A. Orr, *Anal. Biochem.* **114,** 92 (1981).
[5] A. Bodanszky and G. Bodanszky, *Experientia* **26,** 327 (1970).

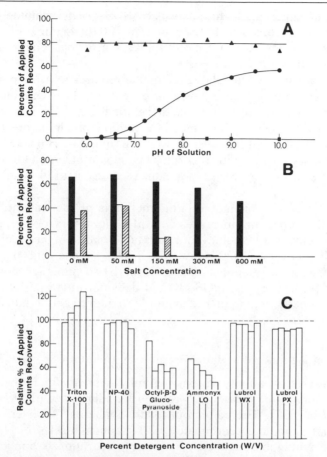

FIG. 2. Percentage of applied counts recovered by (A) immunoprecipitation of labeled BSA from solutions buffered at increasing pH. (▲) Biotinyl ^{125}I-labeled BSA; (●) 2-iminobiotinyl ^{125}I-labeled BSA; (■) ^{125}I-labeled BSA. Results are corrected for counts precipitated by preimmune serum which never exceed 4%. (B) Immunoprecipitation of 2-iminobiotinyl ^{125}I-labeled BSA at pH 9.0, as a function of increasing salt concentration. Immunoprecipitations were carried out in the presence of (from left to right) (i) avidin and immune serum, (ii) avidin and preimmune serum, (iii) avidin and no serum, (iv) no avidin and no serum. (C) Immunoprecipitation of 2-iminobiotinyl ^{125}I-labeled BSA at pH 9.0, as a function of increasing detergent concentration. From left to right, 0.1, 0.5, 1.0, 2.0, and 4.0% (w/v) of each detergent shown. Results are shown referenced to a no-detergent control, indicated by the dotted line.

down the antibody (30 min at 37°). The insoluble protein A is pelleted by centrifugation and washed twice with immunoprecipitation buffer, and twice with 10 mM Tris buffer (pH 8.0) to remove unbound material and excess salt, respectively. Immunoprecipitated proteins can be released back into the supernatant, for analysis by SDS–polyacrylamide gel electrophoresis, by incubation of the pellet at pH 4.0 for 15 min, or by incubation in the presence of 1 mM biotin for 15 min, or by boiling in sample application buffer for 2 min.

Each application of the above technique requires a titration of avidin to determine a molar excess. Once an excess is achieved, adding more avidin is not harmful so long as neither antibody nor IgGSorb becomes limiting. Avidin excess is determined by multiple precipitations (in triplicate), keeping all variables constant while varying the avidin content in 2-fold increments starting with 1 μg/ml. The amount of avidin above which further increases do not result in greater recovery is used in subsequent experiments.

Since the avidin–2-iminobiotin interaction is pH dependent, the efficiency of this procedure increases with pH (Fig. 2A). Good recovery of labeled proteins can be achieved at solution conditions between pH 7.5 and 9.5. No recovery occurs below pH 6.5. The presence of 300 mM NaCl in the immunoprecipitation buffer is essential to minimize nonspecific binding of proteins to the IgGSorb, and does not interfere with specific binding of either 2-iminobiotin to avidin, or avidin to anti-avidin antibody (Fig. 2B).

Recovery of 2-Iminobiotinylated Proteins by Column Chromatography

Recovery is by affinity chromatography over avidin-Sepharose 4B. Detergent-solubilized membrane proteins are adjusted to pH 9.5–11, and made 1 M in NaCl. The column of avidin-Sepharose 4B is equilibrated with 50 mM ammonium carbonate (pH 11) containing 1.0 M NaCl. The solubilized 2-iminobiotinylated membrane preparation is applied to the affinity column, and the column is washed with the high pH buffer (15 ml/hr) until protein and/or radioactivity (if using a radiolabeled preparation) returns to background levels. Specifically retained proteins are eluted with 50 mM ammonium acetate (pH 4.0) containing 0.1 M NaCl. Specifically bound proteins can also be eluted, at any pH, with buffers containing 1 mM biotin. However, this procedure can only be used once on any single column since the avidin–biotin complexes which are formed cannot be broken under nondenaturing conditions. Fractions are collected and assayed for protein content or radioactivity. All chromatographic buffers should contain 0.5% Triton X-100 or other suitable nonionic detergent. If protein is measured by absorbance at 280 nm, then the use of Lubrol PX in place of Triton X-100 is required since the latter absorbs strongly at this wavelength.

Detection of 2-Iminobiotinylated Proteins

Proteins are resolved by SDS–polyacrylamide gel electrophoresis[6] and transferred to nitrocellulose paper by the method of Burnette[7] (Western blot). Nonspecific binding sites are blocked by incubating the nitrocellulose in blocking buffer (10 mM Tris–HCl, 140 mM NaCl, pH 7.4, containing 5% IgG free bovine serum albumin) for 2 hr at 37°. After this step all subsequent incubations and washes are carried out with 50 mM sodium borate (pH 9.5) containing 1% IgG free bovine serum albumin (SBB). The blocked nitrocellulose replica is incubated with avidin (10 μg/ml in SBB) for 15 min followed by several washes with SBB to remove unbound avidin. Nitrocellulose sheets are then incubated with anti-avidin antibody (1 : 1000 dilution in SBB) for 2 hr at room temperature. Visualization of 2-iminobiotinylated proteins can be achieved in either of two ways. In the first method, the nitrocellulose is incubated with ^{125}I-labeled protein A (5 × 10^5 cpm/ml in SBB for 45 min), washed, and dried, and exposed to X-ray film in order to produce an autoradiographic image of the labeled bands. The second method involves the incubation of the nitrocellulose with peroxidase-conjugated goat anti-rabbit immunoglobulins (1 : 1000 dilution in SBB) for 1 hr, washing with SBB, and incubating with substrate. The substrate is prepared by adding 1 part of stock solution (3 mg 4-chloro-1-naphthol/ml methanol) to 4 parts buffer (50 mM sodium borate, pH 9.5) and adding 5 μl of 30% hydrogen peroxide to 10 ml of the above solution. Nitrocellulose sheets are incubated with substrate for 15–30 min or until bands become visible. The sheets are then washed with buffer and dried. The inclusion of 5 mM biotin in the initial avidin incubation serves as a control.

Conclusions

A major advantage of this technique is the versatility of the labeling reaction. By modification to the valeric acid side chain, 2-iminobiotin can be used in a variety of chemical and enzymatic labeling reactions. Some of the methods that we have developed include the attachment of 2-iminobiotin to (1) Gal/GalNAc by galactose oxidase and 2-iminobiotin hydrazide; (2) sialic acid by sodium metaperiodate and 2-iminobiotin hydrazide; and (3) primary amino groups by the N-hydroxysuccinimide ester, or the acyl azide of 2-iminobiotin. The procedure allows one to study the surface composition of intact cells without the prior isolation and

[6] U. K. Laemmli, Nature (London) 227, 680 (1970).
[7] W. N. Burnette, Anal. Biochem. 112, 195 (1981).

purification of the plasma membrane, and may serve as a powerful first step in the purification of surface proteins for structural or functional studies and for the preparation of antibody.

Acknowledgments

This research was supported in part by Grants GM 27851, GM 34029, and Cancer Core Grant P30-CA13330. G.A.O. is a recipient of a Research Career Development Award (HD 00577). R.Z. was supported in part by NIH Training Grant 5 T32 GM 07260.

Section V

Pyridoxine, Pyridoxamine, Pyridoxal: Analogs and Derivatives

[16] Preparation of [^{14}C]Pyridoxal 5'-Phosphate

By ALEC LUI, LAWRENCE LUMENG, and TING-KAI LI

Radiolabeled pyridoxal 5'-phosphate (PLP) is useful as a substrate for metabolic experiments[1] and a chemical probe for membrane studies.[2] Over the years, several approaches have been used to synthesize radiolabeled PLP. Stock *et al.*[3] reported the incorporation of tritium into PLP by reducing it with ^{3}H-labeled sodium borohydride to form [^{3}H]pyridoxine-P and then oxidizing the latter back to [^{3}H]PLP with manganese dioxide. This method can be used for the synthesis of tritium-labeled PLP only. Since the isotope is incorporated into the carbonyl group of PLP, a disadvantage of this method of labeling PLP is the susceptibility of the radioactive coenzyme to lose its tritium label when applied to metabolic studies. Cabantchik *et al.*[4] produced [^{32}P]PLP by first synthesizing [^{32}P]pyridoxamine-P from pyridoxamine plus anhydrous ortho[^{32}P]phosphoric acid. The [^{32}P]pyridoxamine-P was then oxidized to [^{32}P]PLP by manganese dioxide. Eger and Rifkin[2] synthesized [^{32}P]PLP enzymatically by reacting pyridoxal with [γ-^{32}P]ATP in the presence of *Escherichia coli* pyridoxal kinase. Because the radioactive label of [^{32}P]PLP can be rapidly removed by the high PLP-phosphohydrolase activity of many tissues, the [^{32}P]PLP synthesized by these methods is not suitable for many metabolic experiments. Additionally, [^{32}P]PLP has the disadvantage of short isotopic half-life. While the method described by Eger and Rifkin[2] can be modified to synthesize other kinds of radiolabeled PLP, e.g., [^{14}C]PLP, the low yield (less than 20%) of the method is its major weakness.

The disadvantages of [^{3}H]- and [^{32}P]PLP described above do not pertain to [^{14}C]PLP. In order to prepare [^{14}C]PLP, we have modified the procedure previously described by Eger and Rifkin.[2] [4,5-^{14}C]Pyridoxine is commercially available or can be prepared simply by condensation of [2,3-^{14}C]diethyl fumarate with 4-methyl-5-ethoxyoxazole followed by lithium aluminum hydride reduction of the adduct.[5] [^{14}C]Pyridoxine is oxidized to [^{14}C]pyridoxal with MnO$_2$. [^{14}C]Pyridoxal, is isolated by cation-

---

[1] A. Lui, L. Lumeng, and T.-K. Li, *J. Nutr.* **113**, 893 (1983).

[2] R. Eger and D. B. Rifkin, *Biochim. Biophys. Acta* **470**, 70 (1977).

[3] A. Stock, F. Ortanderl, and G. Pfleidener, *Biochem. Z.* **344**, 353 (1966).

[4] I. Z. Cabantchik, M. Balshin, W. Breuer, and A. Rothstein, *J. Biol. Chem.* **250**, 5130 (1975).

[5] C. E. Colombini and E. E. McCoy, *Biochemistry* **9**, 533 (1970).

exchange chromatography, is then reacted with ATP in the presence of rat liver pyridoxal kinase to form the final product. Rat liver pyridoxal kinase is used instead of the enzyme isolated from *E. coli* because the latter is susceptible to strong inhibition by pyridoxal. The use of the rat liver enzyme explains the yield of greater than 75% of the radiolabeled PLP in the present method as compared to the 9–20% yield reported by Eger and Rifkin.[2] After synthesis, [^{14}C]PLP is purified by Sephadex G-10 chromatography.

Preparation of MnO_2

A saturated solution of $KMnO_4$ is prepared by dissolving 19.5 g of $KMnO_4$ in 300 ml of H_2O and filtering to remove excess undissolved $KMnO_4$. This solution is added dropwise, with constant stirring, to 200 ml of 0.44 M $MnSO_4$ which has been heated and maintained at 90°. As $KMnO_4$ is added, the clear $MnSO_4$ solution gradually turns black and MnO_2 precipitates out of solution. Addition of $KMnO_4$ is continued until the supernatant of the suspension becomes purple tinged. The suspension is then kept at 90° with stirring for another 15 min before it is filtered through a Büchner funnel. The MnO_2 retained in the funnel is washed with hot water until the filtrate is colorless. This is followed by washing with 300–400 ml of methanol. The washed MnO_2 is dried in a vacuum oven at 50° overnight and stored under nitrogen until use. MnO_2 prepared in this manner is much better than commercially available MnO_2 in the oxidation of pyridoxine to pyridoxal.

Oxidation of [^{14}C]Pyridoxine to [^{14}C]Pyridoxal by MnO_2

Reagents

Manganese dioxide, prepared as described above
Hydrochloric acid, 0.1 N
Ammonium formate, 10 mM, pH 3.2
Sodium hydroxide, 1.0 N
[4,5-^{14}C]Pyridoxine

Procedure. MnO_2, 80 mg, is added to 20 ml of 0.1 N HCl containing 15 μmol of [^{14}C]pyridoxine. After the mixture has been shaken vigorously for 30 min at room temperature, it is centrifuged at 500 g for 10 min. The supernatant is titrated to pH 3.2 with 1 N NaOH and brought to a volume of 25 ml with 10 mM NH_4 formate (pH 3.2) before it is applied to the cation-exchange chromatography column.

AG 50W-X8 Cation-Exchange Chromatography to Purify [^{14}C]Pyridoxal

Preparation of the Column. The strongly acidic cation-exchange resin (Bio-Rad AG 50W-X8, hydrogen form, 200–400 mesh), 500 g, is washed four times, each for 10 min, with 500 ml of 1 N HCl to remove fluorescent contaminants. Excess acid is then removed by washing with deionized, distilled water. The resin is then washed with 500 ml of 1 N NaOH twice and rinsed with water until the pH is 7.4. The resin is next rinsed with 100 mM formic acid, followed by 10 mM formic acid, until the pH is 3.2 or lower. The resin is finally washed with 10 mM ammonium formate buffer (pH 3.2) to bring the pH and conductivity of the effluent to those of the formate buffer. The resin is packed in chromatographic columns 0.7 × 30 cm (Bio-Rad Econo-Columns, No. 737-1252) to a height of 10 cm. The columns are connected by plastic tubings to a proportioning pump which maintains the flow of the effluent at a rate of 1 ml/min.

Reagents

Ammonium formate, 10 mM, pH 3.2
Ammonium formate, 100 mM, pH 3.2
Sodium acetate, 95 mM, pH 4.2
Sodium acetate, 100 mM, pH 5.2
Sodium acetate, 120 mM, pH 5.6
Sodium phosphate, 100 mM, pH 6.5.

Procedure. The sample (25 ml) containing [^{14}C]pyridoxal (and a small amount of unreacted [^{14}C]pyridoxine) in 10 mM NH$_4$ formate (pH 3.2) is applied to the AG 50W-X8 column. The column is washed with 20 ml of 10 mM NH$_4$ formate (pH 3.2) and this is followed sequentially with the addition of 60 ml of 100 mM NH$_4$ formate (pH 3.2), 75 ml of 95 mM Na acetate (pH 4.2), 30 ml of 100 mM Na acetate (pH 5.2), 75 ml of 120 mM Na acetate (pH 5.6), and 60 ml of 100 mM Na phosphate (pH 6.5). All the eluate which precedes the addition of the 120 mM Na acetate (pH 5.6) buffer is discarded because pyridoxal elutes with the pH 5.6, 120 mM Na acetate buffer and pyridoxine elutes with the pH 6.5, 100 mM Na phosphate buffer. The effluent containing [^{14}C]pyridoxal and [^{14}C]pyridoxine is collected in 5-ml fractions and the radioactivity measured. Typically, greater than 90% of the radioactivity in the sample applied to the column is found to be [^{14}C]pyridoxal while the balance of the radioactivity is recovered as [^{14}C]pyridoxine.

In order to ensure that the [^{14}C]pyridoxal and [^{14}C]pyridoxine are eluted appropriately from the column, a sample blank (25 ml of 10 mM NH$_4$ formate, pH 3.2) and a standard solution containing 500 ng each of pyridoxal and pyridoxine in 25 ml of 10 mM NH$_4$ formate (pH 3.2) are chromatographed simultaneously on separate columns. Pyridoxal and pyridoxine which elute into the appropriate fractions are then assayed fluorometrically.

Pyridoxal is assayed by the cyanide method.[6] A 1-ml aliquot of each of the appropriate fractions which elute from the blank and the standard column is added to 1.0 ml of 1.0 M Na phosphate (pH 7.5) and 0.05 ml of 0.5 M KCN. As a control, another 1-ml aliquot from the same fraction is treated identically except 0.05 ml of H$_2$O is added in place of the 0.5 M KCN. These mixtures are incubated at 50° for 2 hr. After cooling to room temperature, 0.5 ml of 1.5 N NH$_4$OH is added. The fluorescence of these samples is measured with an excitation wavelength of 350 nm and an emission wavelength of 440 nm.

Pyridoxine, in the appropriate fractions, is also assayed by the cyanide method after it has been oxidized to pyridoxal with MnO$_2$.[7] For the oxidation step, 4.0 ml of each fraction eluted from the blank and the standard column is acidified by the addition of 0.5 ml of 1 N HCl. The acidified sample, 4.0 ml, is then added to 20 mg of MnO$_2$ in a plastic vial and the suspension is agitated vigorously at room temperature for 30 min. After centrifugation at 500 g for 10 min, 1.0 ml of the supernatant is added to 1.0 ml of 1 M Na phosphate (pH 7.5) and 0.05 ml of 0.5 M KCN (the test sample), and another 1.0 ml is added to 1.0 ml of 1 M Na phosphate (pH 7.5) plus 0.05 ml of H$_2$O (the control sample). The rest of the procedure is the same as described above for pyridoxal assay.

Preparation of Rat Liver Pyridoxal Kinase

Pyridoxal kinase is purified from the liver of male Sprague–Dawley rats according to the method of McCormick and Snell[8] through the second ammonium sulfate precipitation and dialysis step. This partially purified enzyme is stable when stored at −20° for up to 2 months. The enzyme is assayed at 37° in a mixture containing 0.3 mM pyridoxal, 0.5 mM ATP, 0.01 mM ZnCl$_2$, and 80 mM potassium phosphate (pH 6.0). The reaction is terminated after 1 hr by heating in a boiling water bath for 3 min. An aliquot of the supernatant fraction is assayed enzymatically for pyridoxal-P by the tyrosine apodecarboxylase method.[9,10] One unit of enzyme is that

[6] L. Tamura and S. Takanashi, this series, Vol. 18, Part A, p. 471.

[7] S. F. Contractor and B. Shane, *Clin. Chim. Acta* **21,** 71 (1968).

[8] D. B. McCormick and E. E. Snell, this series, Vol. 18, Part A, p. 611.

[9] L. Lumeng and T.-K. Li, this series, Vol. 62, p. 574.

[10] L. Lumeng, A. Lui, and T.-K. Li, *in* "Methods in Vitamin B-6 Nutrition" (J. E. Leklem, and R. D. Reynolds, eds.), p. 57. Plenum, New York, 1981.

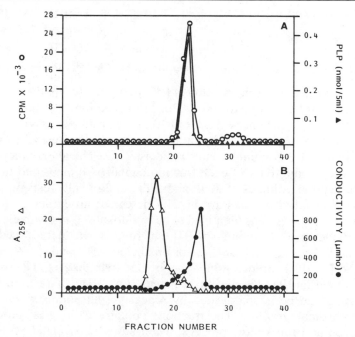

FIG. 1. Sephadex G-10 chromatography of a 3-ml reaction mixture, containing initially 0.33 μmol of [¹⁴C]pyridoxal, 80 mM potassium phosphate (pH 6), 10 mM ATP, 0.2 mM $ZnCl_2$, and 300 units of pyridoxal kinase, after 8 hr of incubation. (A) An aliquot of 0.01 ml from each fraction (5 ml) was assayed for radioactivity (○), and pyridoxal-P (PLP) was assayed by L-tyrosine apodecarboxylase (▲). (B) Adenine nucleotides and salt elutions were detected by absorbancy at 259 nm (△) and conductivity (●). (Reproduced from Lui et al.,[11] with permission of the publisher.)

amount which catalyzes the synthesis of 1 nmol of PLP in 1 hr at 37° in the above assay condition.

Enzymatic Phosphorylation of [¹⁴C]Pyridoxal

The incubation mixture for carrying out the enzymatic phosphorylation contains 0.33 mM of [¹⁴C]pyridoxal, 80 mM potassium phosphate (pH 6.0), 10 mM ATP, 0.2 mM $ZnCl_2$, and 300 units of pyridoxal kinase in a final volume of 3 ml. To ensure that ATP, which is unstable under the condition of incubation, does not become limiting in the reaction, 30 times as much of it as pyridoxal is used in the reaction. $ZnCl_2$ concentration is also increased proportionately to maintain an optimal ATP to Zn^{2+} ratio of 50:1. The mixture is incubated for 8 hr at 37° in Parafilm-sealed test tube wrapped in aluminum foil. The reaction is usually complete in 6–8 hr.

Sephadex G-10 Column Chromatography to Purify [^{14}C]Pyridoxal-P

The enzymatic phosphorylation of [^{14}C]pyridoxal to [^{14}C]PLP is terminated after 8 hr by heating to 100°. After centrifugation, the supernatant is applied to a 1.5 × 100-cm Sephadex G-10 column and eluted with distilled water. An aliquot from each effluent fraction (5 ml) is assayed for radioactivity and for PLP content by enzymatic assay.[9,10] The absorbancy at 259 nm and the conductivity of each fraction can also be measured to determine the elution volumes of ADP and ATP, and of salt, respectively.

Two radioactivity peaks are usually detected: an earlier peak (fractions 20–25, Fig. 1A) was identified to be pyridoxal-P by enzymatic assay, and a smaller peak (fractions 30–34) that follows was determined to be pyridoxal.[11] Only a small amount of unlabeled ADP and ATP contaminates the [^{14}C]PLP preparation since greater than 90% of the A_{259} peak comes off earlier than the pyridoxal-P peak (Fig. 1B).[11] If desired, further purification can be accomplished by repeat gel filtration (or by ion-exchange chromatography). The fractions containing the final product is lyophilized and the yellow powder, when stored desiccated at −20°, is stable for at least a year. The yield of [^{14}C]PLP from [^{14}C]pyridoxal is usually greater than 85%. The radiochemical purity of [^{14}C]PLP is determined by rechromatography on AG 50W-X8 cation-exchange column. Typically, greater than 95% of the radioactivity elutes in the first four fractions as [^{14}C]PLP.

[11] A. Lui, R. Minter, L. Lumeng, and T.-K. Li, *Anal. Biochem.* **112,** 17 (1981).

[17] Cation-Exchange High-Performance Liquid Chromatographic Analysis of Vitamin B6

By STEPHEN P. COBURN and J. DENNIS MAHUREN

The fact that vitamin B6 occurs in multiple forms ranging from very acidic to very basic presents a major analytical challenge. Also, concentrations may vary from several micrograms per gram in liver down to several nanograms per milliliter in plasma. Microbiological methods can be useful for measuring total vitamin B$_6$ content.[1] Enzymatic methods are available for specific measurement of coenzyme forms, especially pyridoxal phosphate.[2] However, a convenient method of individual quantitation of all seven major vitamin B$_6$ metabolites has remained elusive. A number of HPLC procedures have been proposed. Because of the wide range of chemical properties of the vitamers it is quite easy to find chromatographic conditions which will resolve mixtures of standard compounds. In Vol. 62 of this series, Williams[3] described methods for separating pyridoxal, pyridoxine, and pyridoxamine in simple mixtures such as pharmaceutical products. However, it is more difficult to find conditions which are applicable to biological samples, as indicated by Williams' comment, "Our limited experience in clean-up of vitamin B$_6$ from foodstuffs has met with variable success because of many interfering compounds present in the samples."[3] One problem is that in most systems at least one of the vitamers will elute very near the solvent front. Since biological samples often have large interfering peaks eluting with the solvent front it is essential to be able to delay the first vitamin B$_6$ compound long enough to avoid that initial interference.

One method of minimizing interference at any point in the chromatogram is selective detection. Because of a lack of specificity and sensitivity, ultraviolet detection has limited utility in vitamin B$_6$ analysis. We have found electrochemical detection to have good sensitivity but poor specificity; the multiple-electrode detectors which are now available may provide improved specificity. Fluorescence detection generally has good specificity and sensitivity. Unfortunately, pyridoxal phosphate exhibits poor fluorescence because the phosphate group increases the interaction between the aldehyde and phenol groups, thus altering the ionization of the phenol groups. Reagents which will react with the aldehyde group generally enhance fluorescence. Semicarbazide[4] and cyanide[5] have been common choices. In our experience the cyanide reaction produced the best sensitivity but also enhanced the fluorescence of several interfering substances. Gregory and Kirk[6] also noted interference with the cyanide procedure. While the cyanide procedure is difficult to incorporate directly into the chromatographic system, the semicarbazide reaction has been incorporated into a system.[4] We selected the bisulfite reaction because it provided sufficient additional selectivity to offset the slight reduction in

[1] T. R. Guilarte and P. A. McIntyre, *J. Nutr.* **111**, 1861 (1981).
[2] B. Chabner and D. Livingston, *Anal. Biochem.* **34**, 413 (1970).
[3] A. K. Williams, this series, Vol. 62, p. 415.
[4] J. T. Vanderslice and C. E. Maire, *J. Chromatogr.* **196**, 176 (1980).
[5] M. S. Chauhan and K. Dakshinamurti, this series, Vol. 62, p. 405.
[6] J. F. Gregory and J. R. Kirk, *J. Food Sci.* **42**, 1073 (1977).

fluorescence compared with the cyanide derivative and because it gave better sensitivity than the semicarbazone under acidic conditions.[7]

A comprehensive review of HPLC procedures for vitamin B_6 is available.[8] We will mention here only a few. Kothari and Taylor[9] examined the behavior of the following enzyme cofactors and related compounds including some B_6 vitamers on a C_{18} reverse-phase column: UDP-glucose, UDP-galactose, nicotanamide mononucleotide, adenosine triphosphate, folic acid, pyridoxal phosphate, pyridoxamine phosphate, pyridoxine, thiamin pyrophosphate, nicotinamide adenine dinucleotide, nicotinamide adenine dinucleotide phosphate, adenosine diphosphoglucose, flavin mononucleotide, vitamin B_{12}, and riboflavin. The authors stated that their technique was applicable to the analysis of these cofactors in tissues. However they gave no actual examples. The reverse-phase method proposed by Pierotti et al.[10] has the disadvantage of requiring hydrolysis of the phosphorylated forms. Similarly, the reverse-phase method proposed by Gregory et al.[11] requires deamination for pyridoxamine and pyridoxamine phosphate. Our experience suggests that every additional reaction increases the risk of interference. Therefore, it is preferable to have a method which can detect the unaltered molecules. Gregory and Feldstein[12] accomplish this in their latest reverse-phase method, which appears to be applicable to food samples but has marginal sensitivity for human plasma. Vanderslice and Maire[4] have developed an anion-exchange procedure which has recently been updated (Vanderslice et al.[13]).

Our preliminary testing suggested that with the reverse-phase method it might be difficult to resolve the fastest eluting vitamers from the other material eluting with the solvent front in biological samples. We were unable to reproduce the anion-exchange method of Vanderslice and Maire,[4] apparently because modifications in the process used to manufacture the anion-exchange resin have altered its performance. Therefore, we returned to cation-exchange methodology. Our goal was to develop a method which could resolve the seven major metabolic forms of vitamin B_6 (pyridoxal, pyridoxal phosphate, pyridoxine, pyridoxine phosphate,

---

[7] S. P. Coburn and J. D. Mahuren, Anal. Biochem. **129**, 310 (1983).

[8] S. P. Coburn, in "B_6-Pyridoxal Phosphate: Chemical, Biological and Medical Aspects" (D. Dolphin et al., eds.). Wiley, New York (in press).

[9] R. M. Kothari and M. W. Taylor, J. Chromatogr. **247**, 187 (1982).

[10] J. A. Pierotti, A. G. Dickinson, J. K. Palmer, and J. A. Driskell, J. Chromatogr. **306**, 377 (1984).

[11] J. F. Gregory, D. B. Manley, and J. R. Kirk, J. Agric. Food Chem. **29**, 921 (1981).

[12] J. F. Gregory and D. Feldstein, J. Agric. Food Chem. **33**, 359 (1985).

[13] J. T. Vanderslice, M. E. Cortissoz, S. G. Brownlee, and G. R. Beecher, Fed. Proc. Fed. Am. Soc. Exp. Biol. **42**, 1065 (1983).

pyridoxamine, pyridoxamine phosphate, and pyridoxic acid) and would provide sensitivity in the nanogram range.

Procedure

The following description incorporates some slight modifications from the original report.[7]

Instrumentation. A Spectra Physics (San Jose, CA) Model 8700 solvent delivery system, capable of ternary gradient formation; a Rheodyne (Cotati, CA) Model 7010 injector with 500-μl loop; an ISCO (Lincoln, NE) Model 314 syringe-metering pump for postcolumn reagent addition; a Farrand (Valhalla, NY) Mark I spectrofluorometer with 300-μl flow cell with a 7-54 filter (90% transmission at 330 nm, 10% at 400 nm) at the entrance to the excitation monochromator, excitation wavelength set at 330 nm and emission wavelength at 400 nm; and a Hewlett-Packard (Palo Alto, CA) Model 3390A integrator were used.

Columns. A guard column (4.5 mm $\times$ 5 cm) was followed by a 4.6 mm $\times$ 30-cm column, both of which were slurry packed with Vydac 401TP-B (Separations Group, Hesperia, CA) cation-exchange material (10-μm size) using a Model 705 column packer from Micromeritics (Norcross, GA). Newly packed columns must be conditioned by running the gradient program about 10 times.

Solvents

A. 0.02 N HCl
B. 0.1 M NaH$_2$PO$_4$ mixed with sufficient 0.1 N H$_3$PO$_4$ to adjust the pH to 3.3
C. 0.5 M NaH$_2$PO$_4$ mixed with sufficient 0.5 M NaOH to adjust the pH to 5.9.

Postcolumn Reagent. K$_2$HPO$_4$ (1.0 M) was mixed with sufficient 1.0 N H$_3$PO$_4$ to adjust the pH to 7.6. Just before use 1 mg NaHSO$_3$/ml was added.

Since one of the most common contaminants elutes close to pyridoxine, it is prudent to verify the purity of the solvents by occasionally running the gradient with no sample injected. Solvents are made with deionized filtered water. Redistillation or other purification may also be necessary. It is important that solvents be prepared frequently and refrigerated when not in use. Contaminating substances in the solvents will collect on the column and elute as peaks when the gradient progresses.

At the end of the day all parts of the system were flushed with water followed by methanol. The following morning after the methanol was washed out, the system was flushed with the solvent C until the baseline stabilized and then was equilibrated with the starting solvent. Failure to wash the system with the solvent C may cause erratic results in the first run of the day.

Chromatographic Conditions. The flow rates for the primary pump and postcolumn reagent were 1.5 ml/min and 4.5 ml/hr, respectively. A typical program for separating all seven compounds in plasma or tissues was 100% A from 0 to 10 min, linear gradient from 100% A at 10 min to 50% A and 50% B at 14 min, linear gradient from 50% A and 50% B at 14 min to 90% B and 10% C at 23 min, linear gradient from 23 min to 100% C at 29 min and maintaining 100% C until 35 minutes. Allowing an additional 10 min to reequilibrate the column with solvent A, such a program permitted an injection every 45 min.

Sample Preparation. Plasma, tissue, and food products were treated with trichloroacetic acid (TCA) to remove protein and free any protein bound B_6 forms. Plasma was prepared by slowly adding 2 ml 10% TCA to 2 ml plasma while mixing vigorously. After centrifugation the supernatant was extracted with 10 ml redistilled ethyl ether. The ether was discarded and the aqueous portion was filtered through a 0.45-μm membrane. Immediately prior to injection 5 μl containing 10 ng of 2-amino-5-chlorobenzoic acid was mixed with 1.0 ml of sample to provide an internal standard. The internal standard should not be allowed to stand in the sample for an extended length of time. Most other types of samples were prepared in a similar manner, with dilution as necessary to achieve desired peak sizes in injected samples. Pyridoxic acid in urine was measured directly after about a 1 : 500 dilution.

Discussion

Separation of a standard mixture is shown in Fig. 1. The unlabeled peak between pyridoxic acid and pyridoxamine phosphate is due to the change from solvent A to solvent B. Since we inject 0.5-ml samples, the chromatogram in Fig. 1 represents a sample concentration of 4 ng/ml for each vitamer. This provides adequate sensitivity to quantitate the B_6 vitamers in human plasma, which is one of the most challenging samples due to the low vitamin B_6 concentrations (Fig. 2). Although the ultimate goal of the chromatographer is to find a satisfactory isocratic system, the variety of vitamin B_6 compounds combined with the complexity of biological samples make it unlikely that an isocratic system will be satisfactory when information is needed on all the vitamers. One of the advantages of the complex gradient system described here is that it an easily be adjusted to changes in sample composition, column performance, or the selection

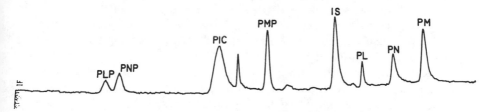

FIG. 1. Chromatogram produced by injecting 0.5 ml containing 2 ng of each of the following compounds: pyridoxal phosphate (PLP, 7.8 min), pyridoxine phosphate (PNP, 9.0 min), pyridoxic acid (PIC, 17.8 min), pyridoxamine phosphate monohydrochloride (PMP, 22.0 min), 2-amino-5-chlorobenzoic acid (IS, 28.0 min), pyridoxal hydrochloride (PL, 30.3 min), pyridoxine hydrochloride (PN, 33.2 min), and pyridoxamine dihydrochloride (PM, 35.9 min).[7]

of vitamers to be quantitated. The complete gradient for resolving all seven compounds requires about 35 min. However, if data are desired on only selected vitamers, the analyses can be conducted with short gradients or even isocratically. For example, pyridoxic acid in 5–10 diluted urine samples can be determined isocratically and consecutively in solvent A before cleaning the column with a stronger solvent. Also, pyridoxal, pyridoxamine, and pyridoxine in simple mixtures can be determined isocratically in solvent C in less than 10 min.

We selected Vydac 401TP for our initial trials because it was spherical, totally porous, and inexpensive. Since it gave satisfactory results, we have not tried other packings. Presumably, similar results could be developed for other cation exchangers with proper conditions. It appears that

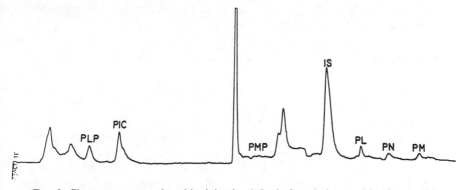

FIG. 2. Chromatogram produced by injecting 0.5 ml of a solution resulting from treating normal human serum with an equal volume of a 20% TCA solution. The sensitivity of the recorder is less than in Fig. 1 and the internal standard peak represents 5 ng rather than the 2 ng used in Fig. 1.[7]

the low cation-exchange capacity of the silica-based packing may contribute to the performance of the column. We have found it difficult to obtain comparable results with a 20-μm sulfonated styrene packing which has a much higher exchange capacity.

The method described here has been successfully applied to serum, brain, heart, lung, liver, pancreas, stomach, small intestine, large intestine, cecum, kidney, adrenal gland, bladder, penis, testes, skeletal muscle, bone marrow, and milk. Its application to normal urine is limited to quantitation of pyridoxic acid and pyridoxamine. The usual concentrations of the other vitamers are too low to quantitate accurately since urine usually produces some interfering peaks.

Good correlation was found between the HPLC results, pyridoxal phosphate measured enzymatically, and total vitamin B_6 determined by a radiometric microbiologic assay in serum from humans[14] and from cats and dogs.[15] However, severe interference was encountered in serum from guinea pigs, swine, horses, goats, and cattle eating natural diets. The interference was eliminated in guinea pigs and pigs fed a purified diet. Dietary modification was not tested in the other species. The problems encountered with serum from animals eating forage suggested that samples of plant origin might pose a problem. Preliminary tests with bread and green beans revealed interfering peaks. Gregory and Feldstein[12] reported that passing the sample through an anion-exchange column eliminated most interfering substances but also removed pyridoxic acid. We have not yet had an opportunity to test this procedure.

The elution times of substances such as pyridoxal phosphate and pyridoxic acid, which elute in solvent A, are very sensitive to both ionic strength and pH. If either the salt content or pH of the sample is too high, pyridoxal phosphate will elute faster and may be obscured by other fast-moving compounds.

The performance of the column does change with age. The gradient program suggested above is typical for columns which have been in use for about a week.

In routine use we expect precision to be within 10%. It is our impression that the limiting factor is the accuracy of the integration. We selected a 300-μl flow cell because of our available equipment and because we wanted maximum sensitivity. However, the large flow cell does broaden the peaks slightly and may increase the electronic error involved in identifying the start of a peak. Lui et al.[16] have obtained satisfactory results using this method with a smaller flow cell. However, he evaluated instru-

[14] M. P. Whyte, J. D. Mahuren, L. A. Vrabel, and S. P. Coburn, unpublished results (1984).
[15] S. P. Coburn, J. D. Mahuren, and T. R. Guilarte, J. Nutr. 114, 2269 (1984).
[16] A. Lui, L. Lumeng, and T-K. Li, Am. J. Clin. Nutr. 41, 1236 (1985).

ments from several manufacturers before finding one which gave acceptable performance. K. Dakshinamurti[17] also informed us that he readily duplicated our resolution but could not achieve comparable sensitivity with his detector. Some of these vitamers are not very fluorescent. In addition, the activation wavelength of 330 nm may not be a strong intensity wavelength in some types of lamps and is quite close to the emission wavelength of 400 nm. These factors combined pose a difficult challenge in using the fluorometer. Therefore, if possible, it is advisable before purchasing a fluorometer to verify its performance with this specific chromatographic system using the anticipated concentrations of the B₆ vitamers.

The postcolumn reagent serves two functions. The bisulfite ion reacts with the aldehyde groups to enhance the fluorescence of pyridoxal phosphate. The buffer raises the pH of the eluate to about 7, thus further enhancing the fluorescence of most of the vitamers. Pyridoxine phosphate which elutes in solvent A would not be detected at these wavelengths without increasing the pH. The increased pH also provides a convenient way to monitor that the postcolumn reagent pump is working properly. While the reactions between the bisulfite ion and the aldehyde group is essentially instantaneous, the length of tubing between the point of adding the postcolumn reagent and the detector must be sufficient to allow complete mixing of the postcolumn reagent stream with the column eluate. Incomplete mixing will cause a noisy baseline in the detector. This can usually be corrected simply by lengthening and/or coiling the tubing between the point of adding the reagent and the detector. Because the reagent pump is very close to the detector it is essential that it be pulse free. We use syringe pumps.

We periodically encounter an interfering peak corresponding to pyridoxine. We suspect that this is due to bacterial contamination of the buffers or water purification system. Briddon and Hunt[18] have encountered a similar problem in amino acid analysis. In both cases the protocols involve fairly lengthy gradients on ion-exchange columns; this may concentrate trace contaminants which are then eluted in the later stages of the gradient. Preparing solvents fresh in clean glassware each Monday usually controls this problem. Sometimes replacement of water purification cartridges may be needed.

We have not found pyridoxamine or pyridoxamine phosphate to be the major vitamer in fresh guinea pig liver, muscle, kidney, or adrenal tissues extracted with trichloroacetic acid.[7] We have occasionally encountered

---

[17] K. Dakshinamurti, personal communication (1983).
[18] A. Briddon and G. Hunt, Clin. Chem. (Winston-Salem, N.C.) 29, 723, (1983).

substances in plasma which eluted very close to pyridoxamine phosphate. In most cases spiking with pyridoxamine phosphate has demonstrated that the peak was not pyridoxamine phosphate. It does appear that the concentrations of the amine forms increase postmortem and as a result the amine forms may be the major forms in commercial meat products.[11] Heating may also increase the concentrations of the amine forms.[11] Therefore, we suspect that the high concentrations of amine forms reported in heated tissues[19] or using the cyanide reaction[20,21] may be artifacts. The ability of the method described here to detect the vitamers with relatively mild treatment may facilitate clarification of the vitamer distribution under physiological conditions.

[19] V. F. Thiele and M. Brin, *J. Nutr.* **90,** 347, (1966).
[20] M. S. Chauhan and K. Dakshinamurti, *Clin. Chim. Acta* **109,** 159 (1981).
[21] B. Shane, *in* "Human Vitamin B$_6$ Requirements," p. 111. Natl. Acad. Sci., Washington, D.C., 1978.

[18] Highly Sensitive Methods for Assaying the Enzymes of Vitamin B$_6$ Metabolism

By ALFRED H. MERRILL, JR. and ELAINE WANG

Pyridoxal (pyridoxine, pyridoxamine) + ATP + Zn^{2+}
$$\xrightleftharpoons{\text{kinase}} \text{pyridoxal (pyridoxine, pyridoxamine) 5'-phosphate + Zn}^{2+} \text{(+ ADP)}$$

Pyridoxine (pyridoxamine) 5'-phosphate + O$_2$
$$\xrightarrow{\text{PNP (PMP) oxidase}} \text{pyridoxal 5'-phosphate + H}_2\text{O}_2 \text{(+ NH}_4\text{)}$$

Pyridoxal (pyridoxine, pyridoxamine) 5'-phosphate + H$_2$O
$$\xrightleftharpoons{\text{phosphatase}} \text{pyridoxal (pyridoxine, pyridoxamine) + P}_i$$

$$\text{Pyridoxal (+ NAD)} \xrightleftharpoons{\text{oxidase (dehydrogenase)}} \text{pyridoxic acid (+ NADH)}$$

The B$_6$ vitamers absorbed by mammals from the diet are pyridoxine, pyridoxamine, and pyridoxal. All three are phosphorylated by a single kinase,[1,2] and pyridoxine and pyridoxamine 5'-phosphates are converted to pyridoxal 5'-phosphate (PLP) by a flavoprotein oxidase.[3] The 5'-phosphates are hydrolyzed by alkaline phosphatases[4] and pyridoxal is oxi-

[1] D. B. McCormick, M. E. Gregory, and E. E. Snell, *J. Biol. Chem.* **236,** 2076 (1961).
[2] J. T. Neary and W. F. Diven, *J. Biol. Chem.* **245,** 5585 (1970).
[3] M. N. Kazarinoff and B. B. McCormick, *J. Biol. Chem.* **250,** 3436 (1975).
[4] L. Lumeng and T.-K. Li, *J. Biol. Chem.* **250,** 8126 (1975).

dized to the dead-end catabolite pyridoxic acid by aldehyde oxidase[5] or pyridoxal dehydrogenase.[6] Since the liver is a major site of vitamin B_6 metabolism and contributes most of the pyridoxal 5'-phosphate found in plasma,[7] it has been used for most studies of these enzymes. The assays described in this chapter were optimized for small samples of human liver,[8] but can be applied to other tissues and animals. Since there is often substrate and product inhibition with species differences in the optimal conditions, references to other animals are given as a guide for more general application of these methods.

Assay Methods

All procedures involving pyridoxyl compounds are conducted in dim light to avoid photodegradation.

Pyridoxal Kinase (EC 2.7.1.35)

Principle. The most sensitive assay for this enzyme is to measure the formation of [³H]pyridoxine 5'-phosphate, which can be easily separated from [³H]pyridoxine using DEAE-cellulose columns[8] or paper disks.[9] The radiolabeled pyridoxine can be synthesized by reduction of pyridoxal with $NaBH_4$, or purchased from Amersham as a special order. It is also possible to use pyridoxal as the substrate and quantitate the PLP enzymatically (e.g., using apo-tyrosine decarboxylase)[10] or by preparation of fluorescent derivatives.[11] In all of the assays of crude enzyme preparations, it is important to ensure that phosphatases (by hydrolyzing the product) or aldehyde oxidase (by depleting the pyridoxal) are not interfering with the measurements. The former is usually controlled by using enzyme preparations free of membranes and buffers containing moderately high concentrations of phosphate. The latter is not pertinent when [³H]pyridoxine is used as the substrate. If the [³H]pyridoxine is synthesized by reduction of pyridoxal, loss of radiolabel can occur if the assays are conducted at neutral to basic pH, where pyridoxine 5'-phosphate oxidase is most active.

[5] R. Schwartz and N. O. Kjeldgaard, *Biochem. J.* **48**, 333 (1951).
[6] M. Stanulovic, V. Jeremic, V. Leskovac, and S. Chaykin, *Enzyme* **21**, 357 (1976).
[7] L. Lumeng, R. E. Brashear, and T.-K. Li, *J. Lab. Clin. Med.* **84**, 334 (1974).
[8] A. H. Merrill, Jr., J. M. Henderson, E. Wang, D. W. McDonald, and W. J. Millikan, *J. Nutr.* **114**, 1664 (1984).
[9] E. Karawya and M. L. Fonda, *Anal. Biochem.* **90**, 525 (1978).
[10] V. M. Camp, J. Chipponi, and B. A. Faraj, *Clin. Chem. (Winston-Salem, N.C.)* **29**, 642 (1983).
[11] V. Bonavita, *Arch. Biochem. Biophys.* **88**, 366 (1960).

Reagents

A. Potassium phosphate buffer, 0.2 M, pH 5.75
B. ATP, 10 mM (pH adjusted to 5.75 with KOH)
C. ZnCl$_2$, 0.8 mM
D. KCl, 0.6 mM
E. [^{3}H]pyridoxine, 0.3 mM (20 mCi/mmol). This compound is purified by chromatography on AG 50W-X8 (H$^+$ form) and diluted with unlabeled pyridoxine to yield the desired specific activity.
F. DEAE-cellulose (e.g., Whatman DE-52), equilibrated with 10 mM ammonium formate
G. Ammonium formate, 10 mM
H. KCl, 0.5 M

Procedure. Assay mixtures are prepared using 10 μl of each reagent A–E and brought to 100 μl with water and the enzyme sample (e.g., up to 200 μg of 100,000 g supernatant of liver). The enzyme is added last to initiate the reaction. After varying times (0–60 min) of incubation at 37°, 0.5 ml of ice-cold ammonium formate (10 mM) is added and the samples are transferred to small columns containing 0.2 ml (packed bed volume) of DEAE-cellulose. The columns are washed with 12 ml of cold 10 mM ammonium formate, 2 ml of water, and, after changing the column to a separate test tube, 2 ml of 0.5 M KCl. A 1-ml aliquot of this eluate is counted with 4 ml of an aqueous miscible scintillation cocktail (e.g., Aquasol-2, New England Nuclear). The enzyme units (in micromoles per minute) are calculated using the equation:

$$\frac{\text{Enzyme}}{\text{units}} = \frac{\text{cpm}_{\text{reaction}} - \text{cpm}_{\text{time 0}}}{\text{Specific activity} \times \text{min}} \times 2 \times \text{quench correction factor}$$

The correction factor adjusts for the quenching by KCl and is determined empirically for different scintillation cocktails.

Essentially no [^{3}H]pyridoxine and all of the 5′-phosphate are recovered by this procedure,[8] and pyridoxic acid, the other radiolabeled compound that might be formed indirectly in assays of crude mixtures, does not bind to the column under these conditions.

Other Tissues. Examples of other tissues that have been assayed for pyridoxal kinase using various methods include human blood cells,[12,13] rat,[1] beef,[2] and sheep[14] liver, pig brain,[15] and hepatomas.[16]

---

[12] L. R. Solomon and R. S. Hillman, *Biochem. Med.* **16**, 233 (1976).
[13] J. A. Kark, M. J. Haut, C. U. Hicks, C. T. McQuilkin, and R. D. Reynolds, *Biochem. Med.* **27**, 109 (1982).
[14] E. Karawya and M. L. Fonda, *Arch. Biochem. Biophys.* **216**, 170 (1982).

Pyridoxine (Pyridoxamine) 5'-Phosphate Oxidase (EC 1.4.3.5)

Principle. This enzyme is often assayed by quantitating, by a variety of methods,[17] the formation of PLP from either natural substrate. Alternatively, the oxidase will act on substrate analogs to release fluorescent[18] or radiolabeled[19,20] compounds that can be easily analyzed. An analog of proven usefulness is *N*-(5'-phosphopyridoxyl)[³H]tryptamine (NPP-tryptamine), from which the oxidase forms PLP and [³H]tryptamine. Tryptamine, unlike the starting compound, is efficiently extracted by toluene and can be added directly to scintillation cocktail. This method has been described for many tissues[19]; given below are the optimal conditions for human liver.[8]

Reagents

   A. Tris–HCl, 0.9 *M*, pH 8.0 at 37°
   B. FMN, 15 μ*M*
   C. Tryptamine–HCl, 0.25 *M*
   D. *N*-(5'-Phosphopyridoxyl)[³H]tryptamine, 20 mCi/mmol, prepared from PLP and [³H]tryptamine as described in ref. 19
   E. NaOH, 1 *M*
   F. Toluene

Procedure. Assay mixtures are prepared using 100 μl of each reagent A–C and 100 μl of the enzyme (e.g., 5–500 μg of 100,000 *g* supernatant from liver) plus water is added. After 5 min at room temperature, the NPP-tryptamine is added to initiate the reaction and the mixtures are incubated at 37° with gentle swirling (the surface area of the mixture should allow the oxygen concentration of the solution to be equivalent to that of air; with this volume this is accomplished, for example, by using test tubes with 13- to 16-mm diameters). After varying times (0–30 min), 1 ml of 1 *M* NaOH is added and the tubes are placed on ice for 15 min. The radiolabeled product is extracted with 2 ml of toluene, and the toluene layer clarified by brief centrifugation at approximately 500 *g*. A 1-ml aliquot of the toluene is counted in scintillation cocktail (e.g., Aquasol-2).

---

[15] F. Kwok and J. E. Churchich, *J. Biol. Chem.* **254**, 6489 (1979).

[16] J. W. Thanassi, L. M. Nutter, N. T. Meisler, P. Commers, and J.-F. Chiu, *J. Biol. Chem.* **256**, 3370 (1981).

[17] A. H. Merrill, Jr., M. N. Kazarinoff, H. Tsuge, K. Horiike, and D. B. McCormick, this series, Vol. 62, p. 568.

[18] M. E. Depecol and D. B. McCormick, *Anal. Biochem.* **101**, 435 (1980).

[19] L. Langham, B. M. Garber, D. A. Roe, and M. N. Kazarinoff, *Anal. Biochem.* **125**, 329 (1982).

[20] J.-D. Choi, D. M. Bowers-Komro, M. D. Davis, D. E. Edmondson, and D. B. McCormick, *J. Biol. Chem.* **258**, 840 (1983).

Activities can be expressed in picamoles [³H]tryptamine released per minute, which are calculated (as above) by subtracting the 0 time control from the samples with longer incubation, dividing by the specific activity of the substrate and the reaction time, and correcting for the size of the aliquot counted (i.e., half) and the counting efficiency of the scintillation cocktail. Since the substrate is used well below its K_m, it is important to adjust the protein concentration and time to use only a fraction of the substrate. Alternatively, the data can be expressed as fractional conversion[19] which is also proportional to enzyme amount.

Other Assays and Tissues. The most sensitive assays using a natural substrate are those that employ pyridoxine 5'-phosphate, since the oxidase is 6- to 7-fold more active with this substrate than with pyridoxamine 5'-phosphate.[20] Care must be taken in using this compound because it exhibits potent substrate inhibition of the oxidase.[20] Likewise, PLP is a strong inhibitor and the use of Tris buffer diminishes its effect.[21]

This enzyme has been studied most extensively in rabbit liver,[3,18,20] and in rat liver,[16,21] brain,[22] blood cells,[19,23] and other tissues.[19]

Pyridoxal 5'-Phosphate Phosphatase (Alkaline Phosphatase)

Principle. This enzyme can be assayed by numerous assays using either PLP or other phosphate-containing substrates (e.g., *p*-nitrophenyl phosphate). A sensitive assay using PLP is to incubate the enzyme with this substrate, remove the substrate from the product by ion-exchange chromatography, and quantitate the pyridoxal by fluorescence after its reaction with cyanide.[8,24]

Reagents

A. Triethanolamine-HCl buffer, 1.0 M, pH 9.0 at 25°
B. MgCl₂, 0.1 M
C. Pyridoxal 5'-phosphate, 2 mM
D. DEAE-cellulose (e.g., Whatman DE-52), equilibrated with 50 mM ammonium formate
E. Ammonium formate, 50 mM and 10 mM
F. KCN, 0.1 M in 0.1 M HEPES buffer, pH 8.3 at 23°
G. HEPES buffer, 1 M, pH 8.3 at 23°
H. Na₂CO₃, 0.4 M

Procedure. Assay mixtures are prepared using 25 μl of reagents A–C

[21] A. H. Merrill, K. Horiike, and D. B. McCormick, *Biochem. Biophys. Res. Commun.* **83**, 984 (1978).
[22] F. Kwok and J. E. Churchich, *J. Biol. Chem.* **255**, 882 (1980).
[23] J. E. Clements and B. B. Anderson, *Biochim. Biophys. Acta* **613**, 401 (1980).
[24] G. P. Smith, G. D. Smith, and T. J. Peters, *Clin. Chim. Acta* **114**, 257 (1981).

and brought to 100 μl with water and the enzyme (e.g., 5–80 μg of a crude homogenate or particulate fraction from liver). After varying periods of incubation at 37° with gentle swirling, 1 ml of ice-cold ammonium formate (50 mM) is added and the mixtures are transferred to small columns containing 0.2 ml (packed volume) of DEAE-cellulose. The column is washed with 1 ml of 10 mM ammonium formate and the two eluates are combined. A 1-ml aliquot is transferred to a screw-capped test tube, and 0.1 ml of 0.1 M KCN and 0.1 ml of 1 M HEPES buffer are added. After incubation at 50° for 1 hr, the samples are cooled to room temperature and 0.8 ml of 0.4 M Na$_2$CO$_3$ is added. The fluorescence of the samples and similarly treated pyridoxal standards solutions are measured with excitation and emission wavelengths of 350 and 434 nm, respectively. The nanomoles pyridoxal formed is calculated from the standard curve and corrected for the background fluorescence (from a time 0 control), aliquot size (half of the total), and the incubation time.

Other Tissues. Assays of this activity in other tissues have been conducted using similar incubation conditions.[4,24]

Pyridoxal (Aldehyde) Oxidase and Dehydrogenase

Principle. The oxidation of pyridoxal results in the formation of a strong fluorophor, pyridoxic acid, and in the case of the dehydrogenase reaction, a highly absorbant chromophor, NADH. Both can be used to follow these reactions and the methods, which are similar, are described below.

Reagents for Pyridoxal Oxidase

Potassium phosphate, 1 M, pH 7.0
Pyridoxal, 0.5 mM

Reagents for Pyridoxal Dehydrogenase

Sodium pyrophosphate buffer, 0.6 M, pH 9.5
Pyridoxal, 4 mM
NAD$^+$, 8 mM

Procedure. Both methods involve the preparation of assay stocks containing the reagents at one-tenth of their stock concentrations (above) and water (plus enzyme) to yield the volume required by the cuvettes being used. The cuvettes are placed in water-jacketed cuvette holders and allowed to equilibrate at 37° for several minutes. The background fluorescence or absorbance is recorded, the enzyme sample (e.g., 100–600 μg of 100,000 g supernatant) is added, and measurements are continued at 30-sec intervals until an initial velocity is obtained. Between measurements, the samples are protected from the light beam to minimize photodegradation of pyridoxal. Pyridoxic acid formation is measured at excitation and emission wavelengths of 312 and 430 nm, respectively, and the nanomoles of product is determined from a standard curve using varying amounts (0.05–0.75 μM) of pyridoxic acid (Sigma). The production of NADH is determined by the increase in absorbance at 340 nm using an extinction coefficient of 6200 M^{-1} cm^{-1}. To confirm that the change in absorbance or fluorescence is not due to artifacts such as other endogenous reductants or settling of fine particles in the enzyme solution, the assays are repeated with the addition of pyridoxal to initiate the reaction.

Other Tissues. Relatively few tissues have been examined for these activities,[5,6,8,25] and it is not known definitely which enzyme is responsible for the majority of the pyridoxal degradation *in vivo*. To assess the relative contribution of the relative oxidase versus dehydrogenase activities of a given sample, the rate of pyridoxic acid formation can be compared to the rate of NADH production, or the basal versus NAD$^+$-stimulated rate of pyridoxic acid formation can be determined.

Other Enzymes

Attempts to identify additional reactions that are important in the metabolism of the various forms of vitamin B$_6$ by mammalian tissues have not uncovered enzymes other than those described above. In considering the total ways in which pyridoxamine 5'-phosphate can be converted to PLP, however, the second half reaction of aminotransferases (enzyme–pyridoxamine 5'-phosphate + α-keto acid $\rightleftharpoons$ enzyme–PLP + amino acid) might be considered a member of this pathway.

[25] R. R. Bell, C. A. Blanchard, and B. E. Hakell, *Arch. Biochem. Biophys.* **147**, 602 (1971).

[19] Modified Purification of Pyridoxamine (Pyridoxine) 5'-Phosphate Oxidase from Rabbit Liver by 5'-Phosphopyridoxyl Affinity Chromatography

By DELORES M. BOWERS-KOMRO, TORY M. HAGEN, and DONALD B. McCORMICK

Pyridoxamine (pyridoxine) 5'-phosphate oxidase (EC 1.4.3.5) is an FMN-dependent enzyme responsible for the oxidation of the 5'-phosphates of pyridoxamine and pyridoxine as follows:

Pyridoxamine 5'-phosphate + O_2 + H_2O → pyridoxal 5'-phosphate + H_2O_2 + NH_3
Pyridoxine 5'-phosphate + O_2 → pyridoxal 5'-phosphate + H_2O_2

The enzyme was first purified to homogeneity by Kazarinoff and Mc-Cormick[1] who used rabbit liver as a good source of mammalian oxidase. The use of classic fractionation steps, e.g., acid, salt, solvent, and conventional columns, was presented in a preceding volume in this series.[2] Studies on the substrate specificity of the oxidase revealed that good binding could be achieved with sizeable secondary amine functions at the 4' position of the 5'-phosphopyridoxyl moiety.[1,3] This allowed development of N-(5'-phospho-4'-pyridoxal)amino-Sepharose as an affinity matrix for purification of the oxidase. By inclusion of such an affinity column in the purification of oxidase from rat brain, Cash et al.[4] were able to obtain good enrichment of this enzyme as well as pyridoxal kinase. A combination of steps from the original method used for purification of the liver oxidase with the affinity column was utilized by Churchich[5] for purification of oxidase from pig brain and has also been used by Nutter et al.[6] in their studies of oxidases from mammalian tissues.

The present description of oxidase purification from rabbit liver is a modification of our original procedure in which the temperature-sensitive step with ethanol and the final calcium phosphate column have been excluded, and an affinity column included. The result is a more facile overall purification which is suitable not only for less protein-containing extracts from tissues such as brain but also the more protein-rich extract from liver.

Assay Methods

The oxidase activity was measured by the formation of the pyridoxal 5'-phosphate, quantitated as the phenylhydrazone which absorbs light at 410 nm.[7] One enzyme unit catalyzes the formation of 1 nmol of pyridoxal 5'-phosphate per hour at 37° in 0.2 M Tris at pH 8. Protein was determined by the method of Lowry et al.[8]

[1] M. N. Kazarinoff and D. B. McCormick, J. Biol. Chem. 250, 3436 (1975).
[2] A. H. Merrill, M. N. Kazarinoff, H. Tsuge, K. Horiike, and D. B. McCormick, this series, Vol. 62 [83].
[3] M. N. Kazarinoff and D. B. McCormick, Biochem. Biophys. Res. Commun. 52, 440 (1973).
[4] C. D. Cash, M. Maitre, J. F. Rumigny, and F. Mandel, Biochem. Biophys. Res. Commun. 96, 1755 (1980).
[5] J. E. Churchich, Eur. J. Biochem. 138, 327 (1984).
[6] L. M. Nutter, N. T. Meisler, and J. W. Thanassi, Biochemistry 22, 1599 (1983).
[7] H. Wada, this series, Vol. 18 Part A [96].
[8] O. H. Lowry, N. J. Rosebrough, A. L. Farr, and R. J. Randall, J. Biol. Chem. 193, 265 (1951).

Phosphopyridoxyl-Affinose

A 5′-phosphopyridoxyl-Affinose was prepared according to Cash et al.[4] by reacting pyridoxal 5′-phosphate with N'-(ω-aminohexyl)-Sepharose (Sigma) and reducing the resulting imine with sodium borohydride. To quantitate the amount of B_6 moiety attached to the resin, a weighed portion of wet resin was digested in 1 N HCl for 30 min at 100°, and the solvent was removed under reduced pressure. A difference spectrum of acidic and neutral aliquots of the resuspended residue, when compared to pyridoxamine 5′-phosphate as standard, revealed that between 5 and 6 μmol of B_6 moiety is attacked per gram of wet resin.

Purification Procedure for Oxidase

The fractionation steps were done in the cold (0–4°) and in dim light. All buffers contained phenylmethylsulfonyl fluoride (0.1 mM) as a proteolysis inhibitor as well as 0.1 mM 2-mercaptoethanol for thiol protection. Steps 1, 2, 4, and 5 are essentially those described by us previously[1,2] and Steps 3 and 6 are adapted from Churchich[5] with some modification.

Step 1. Extraction. Frozen rabbit liver (250 g) is broken into pieces and is homogenized for 5 min in a Waring blender at medium setting with 1 liter of 0.02 M potassium phosphate (pH 7.0). The homogenate is then centrifuged at 18,000 g for 30 min and the resulting supernatant is filtered through four thicknesses of cheesecloth.

Step 2. Acid Treatment. The filtered supernatant is adjusted to pH 5 by dropwise addition of 2.0 N acetic acid (prechilled to 0–4°) and the solution allowed to stir for 10 min. The acid-treated extract is centrifuged at 18,000 g for 30 min and the precipitate discarded. The supernatant is adjusted to pH 6.8 by addition of prechilled 2 M KOH.

Step 3. Ammonium Sulfate Treatment and Dialysis. The supernatant is treated with ammonium sulfate to give a 40% saturation (243 g/liter). After the addition of the ammonium sulfate, the solution is stirred for 30 min and then centrifuged at 18,000 g for 30 min. The supernatant is treated with ammonium sulfate to give a 60% saturation (164 g/liter). Again the solution is stirred for 30 min and centrifuged at 18,000 g for 30 min; the supernatant is discarded. The precipitate is dissolved in 0.02 M potassium phosphate (pH 8.0) containing 2 μM FMN and dialyzed against this buffer overnight at 4°.

Step 4. DEAE-Sephadex A-50 Column. The dialysate is applied to a column (40 × 5 cm) of DEAE-Sephadex A-50 that is equilibrated with 0.02 M potassium phosphate (pH 8.0). The protein is washed onto this column with 50 ml of this buffer before a linear gradient between 0.1 and

PURIFICATION OF PYRIDOXAMINE (PYRIDOXINE) 5'-PHOSPHATE OXIDASE FROM
RABBIT LIVER

Step	Volume (ml)	Total protein (mg)	Total activity (units)	Specific activity (units/mg)	Yield (%)
1. Extraction	1050	54,100	339,200	6.3	100
2. Acid treatment	960	18,700	321,600	17.2	95
3. Ammonium sulfate treatment, dialysis	96	7,550	204,100	27.0	60
4. DEAE-Sephadex A-50	47	630	187,400	297	55
5. Sephadex G-100	19	112	85,230	761	25
6. Phosphopyridoxyl-Sepharose	5	1.1	19,400	17,600	5.7

0.2 M potassium phosphate (pH 8.0) totaling 1.5 liters is used to elute the enzyme. The fractions with oxidase activity (greater than 400 U/ml) elute at an ionic strength of about 0.16 M. These pooled fractions are concentrated by ultrafiltration (Amicon PM10 membrane) to ~40 ml.

Step 5. Sephadex G-100 Column. The DEAE concentrate is poured onto a Sephadex G-100 column (75 × 4 cm) that is equilibrated with 0.02 M potassium phosphate (pH 6.8). The column is eluted with this buffer, and the fractions containing enzyme activity (greater than 800 U/ml) are pooled. To protect the oxidase and ensure binding of coenzyme, 10^{-4} M FMN is gradually added so that the final concentration is 2 μM FMN in the pooled enzyme solution. The enzyme solution is then concentrated to ~25 ml.

Step 6. Phosphopyridoxyl-Sepharose Column. The G-100 concentrate is applied to the affinity column of phosphopyridoxyl-linked Sepharose (12 × 2.5 cm) equilibrated with 0.02 M potassium phosphate (pH 6.8) containing 2 μM FMN. A linear gradient from 125 ml of this buffer to 125 ml of the same buffer containing 0.4 M potassium chloride is used to elute most of the protein from the column (monitored by absorbance at 280 nm). The oxidase is then released from the affinity matrix by washing with 50 ml of 0.02 M potassium phosphate (pH 5.0) containing 1 mM pyridoxal 5'-phosphate and 0.4 M potassium chloride. This collected fraction is concentrated by ultrafiltration (or by using dialysis tubing and Carbowax) to ~6 ml, dialyzed against 0.02 M potassium phosphate (pH 6.8) for 12 hr with two or three changes of buffer, and assayed for activity and protein. The enzyme obtained by release from the affinity matrix is nearly homogeneous and can be stored shell-frozen at −80°. Typical overall purification results are as given in the table.

Higher yield can be obtained by collecting the oxidase which undergoes some nonspecific binding and is eluted at ionic strength of ~0.16 M (~60% along the gradient). This fraction often represents overloading of the capacity of the affinity column. Before a second pass over the affinity matrix is attempted, this material must be concentrated and dialyzed against 0.02 M potassium phosphate (pH 6.8) containing 2 μM FMN.

[20] Monoclonal Antibodies to Vitamin B6

By Jason M. Kittler, Dace Viceps-Madore, John A. Cidlowski, Natalie T. Meisler, and John W. Thanassi

Antisera containing polyclonal antibodies to vitamin B6 and its derivatives have been prepared by several groups of investigators.[1-4] One can elicit such antiera by immunizing rabbits with carrier molecules containing the haptenic 5'-phosphopyridoxyl (PPxy) group which is formed by derivatizing ε-amino groups of lysine residues in carrier molecules with pyridoxal 5'-phosphate (PLP) and sodium borohydride.[5,6] The applications of polyclonal antibodies to the study of vitamin B6 have been rather limited in scope.[7]

Methods for the production of monoclonal antibodies were reported first in 1975 by Köhler and Milstein.[8] The singular advantage of monoclonal antibodies stems from the fact that they are essentially reagents which are not subject to biological variability associated with antibodies prepared by methods involving conventional immunization of animals. Since their introduction, monoclonal antibodies have rapidly become commonly used research tools.

Because of the advantages of monoclonal antibody technology, we became interested in preparing stable hybridomas which produced mono-

[1] F. Córdoba, C. Gonzalez, and P. Rivera, Biochim. Biophys. Acta 127, 151 (1966).

[2] D. Ungar-Waron and M. Sela, this series, Vol. 18, Part A, p. 147.

[3] V. Raso and B. Stollar, Biochemistry 14, 584 (1975).

[4] J. W. Thanassi and J. A. Cidlowski, J. Immunol. Methods 33, 261 (1980).

[5] P. H. Strausbauch, A. B. Kent, J. L. Hedrick, and E. H. Fischer, this series, Vol. 11, p. 671.

[6] J. E. Churchich, Biochim. Biophys. Acta 102, 280 (1965).

[7] J. M. Kittler and J. W. Thanassi, in "Coenzymes and Cofactors: Pyridoxal Phosphate and Derivatives" (D. Dolphin, R. Poulson, and O. Avramovic, eds.). Wiley, New York (in press).

[8] G. Köhler and C. Milstein, Nature (London) 256, 495 (1975).

clonal antibodies to vitamin B$_6$. We wished to explore the applications of such antibodies to the study of PLP, the coenzymatically active form of the six B$_6$ vitamer forms. In addition to its unique versatility as a coenzyme, PLP has proven to be an extremely useful reagent in the derivatization of a large number of diverse proteins.[7,9]

Preparation of PPxy-Proteins[1]

The only report dealing with the preparation of hybridomas which produce antibodies to vitamin B$_6$ appeared in a 1983 publication by Viceps-Madore et al.[10] The carrier proteins containing the haptenic PPxy group happened to be a mixture of proteins that were partially purified from human placentas according to procedures described by Wrange et al.[11] It must be emphasized that no particular significance should be attached to the use of human placental proteins as carriers for the PPxy group. Any protein or proteins which can be reductively and covalently derivatized at lysine residues with PLP and sodium borohydride will probably serve the purpose. PPxy-proteins are easily prepared. No special precautions or conditions are required except as noted below.

Protein (5–15 mg/ml) is dissolved in ice-cold 10 mM PLP, previously adjusted to pH 7–7.5. In order to avoid photodecomposition of PLP or PPxy-proteins, solutions should be kept in subdued light or, preferably, in a work area illuminated by GE F40G0 yellow fluorescent bulbs. The solution is stirred for approximately 15 min to allow for completion of Schiff base formation between PLP and ε-amino groups of lysine residues. Small amounts of sodium borohydride solid or solution are added with continuous stirring. The PLP acts as an indicator and the intense yellow color of the vitamin bleaches to a pale straw yellow when the reduction is complete. Stirring is continued for another 15–30 min after the color is bleached. Excess sodium borohydride can be removed by acidification to pH 5 or less with acetic acid, followed by dialysis. Caution should be exercised in the acidification step as there is generation of hydrogen gas and the potential for considerable foaming. Alternatively, one can remove excess sodium borohydride and undesired small molecular weight contaminants, such as pyridoxine 5′-phosphate, by dialysis alone, using a neutral or slightly basic buffer of choice. The extent of derivatization by PLP can be most readily determined by measuring the absorption at 325

[9] J. M. Kittler, N. T. Meisler, D. Viceps-Madore, J. A. Cidlowski, and J. W. Thanassi, Anal. Biochem. **137,** 210 (1984).
[10] D. Viceps-Madore, J. A. Cidlowski, J. M. Kittler, and J. W. Thanassi, J. Biol. Chem. **258,** 2689 (1983).
[11] O. Wrange, J. Carlstedt-Duke, and J.-Å. Gustafsson, J. Biol. Chem. **254,** 9284 (1979).

nm, using a molar absorption coefficient of 8300 for pyridoxamine 5'-phosphate,[1,6] or by analysis of the derivatized protein for its phosphate content. Covalent linkage of PLP to the protein is established by precipitation of the protein with trichloroacetic acid and the concomitant loss of 325 nm absorption in the supernatant.

Preparation of Monoclonal Antibodies to Vitamin B_6

Procedures for immunization of mice, preparation of hybridomas, selection of clones, and large-scale production of monoclonal antibodies have been described in considerable detail.[12,13] What follows are the specific procedures used in the preparation of monoclonal antibodies to vitamin B_6.[10]

Immunization, Fusion, and Cloning of Hybridomas. Female BALB/c mice are injected intraperitoneally with 115 μg of PPxy-proteins in Freund's complete adjuvant. A second intraperitoneal injection (115 μg) is given 7–8 weeks later in incomplete adjuvant. Two months later, mice receive an intravenous injection with 230 μg of antigen. On the third day postintravenous injection, mice are killed and the spleens removed aseptically. Fusion of spleen cells and myeloma cells (X63-Ag8.653) is carried out essentially as described by Gefter *et al.*[14] Spleen cells are mixed with murine myeloma cells at a ratio of 4 : 1 and exposed to 35% polyethylene glycol 1000 for 8 min at 25°. The cells are then resuspended in Dulbecco's modified Eagle's medium supplemented with 100 μM hypoxanthine, 10 μM aminopterin, 30 μM thymidine, 20% calf serum, 0.1 mM nonessential amino acids, 10% National Cancer Tissue Culture Medium 109 (M.A. Bioproducts, Walkersville, MD), 100 units of penicillin/ml, 100 μg streptomycin/ml and dispensed (2.5 × 10^4 myeloma cells/well) in 96-well tissue culture plates. The cultures are incubated at 37° in a humidified 10% CO_2 incubator. Visible colonies appear within 1–2 weeks. The supernatants are removed from these colonies and tested for desired monoclonal antibody production by an enzyme-linked immunosorbent assay (ELISA) described below. Desired colonies are recloned by limiting dilution on Sprague–Dawley thymus feeder layers (10^7 cells/well of 96-well plates).[15] After initial selection of desired clones, it is necessary to supplement the growth medium with 4 mM pyridoxal and 0.5 mM pyridoxamine 5'-phos-

[12] G. Galfré and C. Milstein, this series, Vol. 73, Part B, p. 3.
[13] R. H. Kennett, T. J. McKearn and K. B. Bechtol, eds., "Monoclonal Antibodies." Plenum, New York, 1980.
[14] M. L. Gefter, D. H. Margulies, and M. D. Scharff, *Somatic Cell Genet.* **3,** 231 (1977).
[15] T. J. McKearn, *in* "Monoclonal Antibodies" (R. H. Kennett, T. J. McKearn, and K. B. Bechtol, eds.), p. 374, Plenum, New York, 1980.

phate, presumably to counteract a hybridoma-caused vitamin B$_6$ deficiency.

Hybridoma Selection by ELISA. Identification of wells which contain hybridomas secreting antibodies to the haptenic PPxy group is made by an ELISA procedure. For this purpose, PLP is reductively and covalently linked (as described above) to a carrier protein which had not been used previously in the immunization of mice. Thus, PPxy-bovine serum albumin (PPxy-BSA)[1] contains the haptenic PPxy group but no other antigenic determinants. The antigen preparation of PPxy-BSA at a concentration of 10 ng/100 μl in 50 mM carbonate/bicarbonate (coating) buffer (pH 9.6) is applied into the wells of 96-well poly(vinyl chloride) microtiter plates; maximum binding of PPxy-BSA occurs at an applied concentration of 50 ng/well. After 2 hr at 25°, the wells are filled with 1% BSA in coating buffer and kept overnight at 4°. Unbound protein is washed off with 10 mM phosphate-buffered saline (pH 7.4) containing 0.05% Tween 20 (PBS-T) (three washes, 2 min each). Antibody-containing hybridoma test supernatants are diluted 1 : 10 into PBS-T containing 1% BSA (PBS-BSA-T); 100-μl aliquots are added to the antigen-coated wells. Following an overnight incubation at 4°, unbound antibody is removed and the wells are washed three times with PBS-T (2 min each time). To the wells are then added goat anti-mouse Fab$_2$–horseradish peroxidase conjugate (100 μl) that has been previously diluted 1 : 10,000 with PBS-BSA-T. After 45 min at 37°, unbound second antibody is removed and the wells are washed with PBS-T as described above. Substrate for horseradish peroxidase is added to each well (150 μl of a 0.006% solution of H$_2$O$_2$ in 0.1 M sodium citrate buffer, pH 5.0), containing 1 mg o-phenylenediamine/ml). The enzymatic reaction is stopped after 1 hr at 25° by the addition of 75 μl of 4 M sulfuric acid. Absorbances in the wells are measured at 490 nm on a Dynatech microtiter plate ELISA reader.

Large-Scale Production of Monoclonal Antibodies to Vitamin B$_6$. The preparation of large amounts of monoclonal antibodies via ascites tumors production is a commonly used procedure[12,13] which most often involves priming recipient mice by intraperitoneal injection of the immunosuppressant pristane (tetramethylpentadecane). Several weeks later the animals are inoculated with about 10[7] hybridoma cells of interest. The ascites fluid which accumulates contains very high concentrations (10 mg/ml) of antibody. The fluid can be tapped off using an 18-gauge syringe needle, and the monoclonal antibody purified by a variety of procedures. The details of the experimental protocol for the large-scale production of a monoclonal antibody to the PPxy group have been published[7] and do not differ from the general procedure described elsewhere in this series.[12]

Characteristics of Monoclonal Antibodies to the PPxy Group[10]. Of 23 hybridoma clones which produced monoclonal antibodies directed against PPxy-human placental proteins, 18 (78%) produced antibodies that bound to PPxy-BSA but not BSA. Thus, the PPxy group is a potent haptenic antigen.

The anti-PPxy monoclonal antibodies obtained belong to the IgG class. Only one of them, designated E6(2)2, has received any significant use; it has been shown to be a member of the γ_1 subclass.[16] Seventeen of the 18 PPxy-positive hybridoma supernatants were tested for their abilities to bind various B_6 vitamer forms. In general, the phosphorylated vitamer forms were much preferred, with pyridoxamine 5'-phosphate and pyridoxine 5'-phosphate being more tightly bound than PLP. It is apparent that the monoclonal anti-PPxy antibodies can make fine discriminations based on the presence or absence of the 5'-phosphate group and the substitution and configuration about C-4' of the vitamin.

Applications of Monoclonal Anti-PPxy Antibodies

The applications to date of monoclonal antibodies to the PPxy group have centered around detection of PLP-binding proteins using immunoblot techniques. Blotting is "the process of transferring macromolecules from gels to an immobilizing matrix" such as nitrocellulose (NC).[17] Protein immunoblotting is the detection of proteins on blots using immunochemical techniques. Monoclonal antibody E6(2)2 has been shown to be extremely useful in the specific detection of PLP-binding or PLP-derivatized proteins on blots. The following describes what we consider to be the method of choice for the detection of such proteins.

Detection of PPxy-Proteins with Monoclonal Antibody E6(2)2. Protein-containing solutions, such as cytosolic tissue extracts, are incubated for 30 min with a 3% ice-cold aqueous sodium borohydride solution (0.3 ml/ml of extract) to reductively and covalently link endogenous or added PLP to PLP-binding proteins. Following dialysis against PBS, the proteins are separated on SDS–PAGE gels and transferred to NC by conventional procedures.[10]

Protein-containing NC blots are washed in PBS (pH 7.3) and then incubated in a blocking solution (37° for 1 hr, followed by several hours at 4°). The blocking solution that we have found most effective contains 1% BSA and $2\frac{1}{2}$% human plasma in PBS. The human plasma is outdated plasma obtained from a local blood bank. It is dialyzed against 5 mM hydroxylamine in 10 mM PBS, followed by dialysis against PBS alone.

[16] J. M. Kittler and J. W. Thanassi, unpublished observation (1984).
[17] J. M. Gershoni and G. E. Palade, *Anal. Biochem.* **124**, 396 (1982).

This treatment removes approximately 95% of the PLP found in plasma.[10,18] The NC blot is briefly washed in PBS and incubated for 30 min at 4° in 0.3% hydrogen peroxide in PBS. This step suppresses nonspecific peroxidase-like activity. The blot is then washed in PBS; washing consists of three 10-min incubations at either room temperature or in a cold room on a shaking device using a polypropylene food container fitted with a top. The washed blot is incubated in E6(2)2-containing ascites fluid, diluted 1 : 100,000 in 1% BSA in PBS, for 1½ hr at 37° and at 4° overnight. If ammonium sulfate-precipitated, lyophilized E6(2)2 is used, it is reconstituted at an equivalent concentration in 1% BSA in PBS.

Following washing in PBS-T, the NC blot is incubated with biotin-conjugated goat anti-mouse IgG (H and L chains) (Cappel, Malvern, PA.), diluted 1 : 10,000 in 1% BSA in PBS. The incubation time consists of 1½ hr at 37° and overnight at 4°. The blot is washed in PBS-T and incubated (37°, 1 hr; 4°, overnight) with horseradish peroxidase-conjugated avidin (Cappel, Malvern, PA.) diluted 1 : 10,000 in 1% BSA in PBS. After washing, the blot is developed in a staining solution which is composed of a freshly prepared, nitrogen-gassed, filtered solution composed of 40 mg of 3,3′-diaminobenzidine and 40 μl of 30% hydrogen peroxide in 100 ml of 50 mM Tris buffer (pH 7.6). Incubation with staining solution is allowed to proceed at room temperature with gentle shaking until the desired intensity of color is obtained. This usually occurs within minutes. The developed immunoblot is washed and stored in water until it can be photographed. The dried blot can be stored permanently.

Typical results obtained with this procedure are shown in Fig. 1 which compares a Coomassie blue-stained SDS–PAGE gel with the corresponding immunoblot. In comparing the lanes containing the molecular weight standards, it is apparent that E6(2)2 monoclonal antibodies to the PPxy group do not react with the molecular weight standard proteins, as expected. Phosphorylase b, which contains 1 mol of PLP per mole of enzyme, was treated with sodium borohydride according to the procedure of Strausbauch et al.[5]; it provides a reference PPxy-protein. The amount of PPxy-phosphorylase b applied to the gel was barely detectable by Coomassie blue stain. In contrast, the immunoblot procedure not only heavily stained PPxy-phosphorylase b but easily detected smaller amounts of faster moving contaminants. The third lanes contained 150 μg of liver cytosolic proteins. It is evident from these lanes that the immunoblot detection of PPxy-proteins using E6(2)2 has a remarkable discriminating ability at a concentration equivalent to a 1 : 100,000 dilution of

[18] M. H. Lipson, J. P. Kraus, L. R. Solomon, and L. E. Rosenberg, Arch. Biochem. Biophys. 204, 486 (1980).

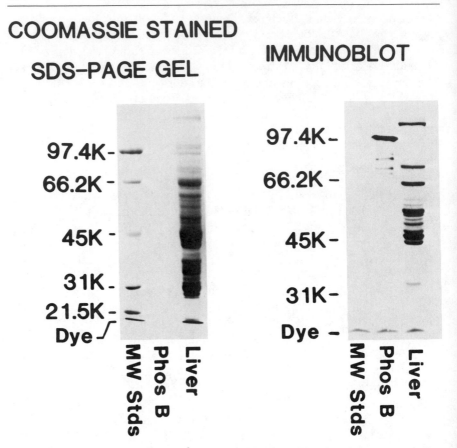

FIG. 1. Comparison of Coomassie blue-stained SDS–PAGE gel and the corresponding immunoblot visualized with monoclonal antibody E6(2)2 to the 5′-PPxy group.

the ascites fluid. The inclusion of 1 mM pyridoxamine 5′-phosphate to diluted ascites fluid blocks the interaction of E6(2)2 with PPxy-proteins, whereas 1 mM pyridoxamine is without effect. Thus, this particular monoclonal antibody is highly selective for phosphorylated vitamer forms. As expected, no staining takes place if horseradish peroxidase-conjugated avidin is left out. On the other hand, one must run controls in which the monoclonal anti-PPxy antibody and biotin-conjugated anti-mouse IgG (H and L chains) are left out as there are avidin-binding proteins in liver which can lead to the appearance of bands. The intense band having an apparent molecular weight greater than 100,000 is one such band.

Comments

The utility of monoclonal antibodies to the vitamin B$_6$ family is potentially very great. Their preparation was reported in 1983.[10] As of this writing, there has not been sufficient time to explore and exploit their applications. Monoclonal antibody E6(2)2 has been used to compare PLP-binding proteins in liver and hepatoma cytosolic extracts.[19,20] In addition, the reagent properties of PLP have been used to develop a general immunochemical method for detecting proteins on blots.[9] In this procedure, blotted proteins are derivatized *in situ* with PLP and sodium borohydride. The PPxy-proteins thus formed can be detected as described above. This method takes advantage of the chemical reactivity of PLP with the ε-amino groups of lysine residues in proteins, the ease of reduction of the PLP-lysine Schiff base with sodium borohydride, and the selectivity of E6(2)2 for the PPxy-group. One can easily detect nanogram quantities of proteins on NC blots by this technique.

Other applications under study involve the properties of steroid hormone receptors and their interactions with PLP. Immunoaffinity methods for the isolation of PPxy-derivatized peptides and proteins are also in progress. Immunohistochemical applications are yet another possibility. The differing selectivities of the various monoclonal antibodies to B$_6$ vitamer forms can, in principle, be applied to the analysis of tissues and fluids for their B$_6$ vitamer compositions.

In summary, the extremely large number of PLP-dependent enzymes and the outstanding chemical properties of PLP as a protein-derivatizing reagent make monoclonal antibodies to the vitamin B$_6$ family very promising research tools with numerous and far-ranging applications.

Acknowledgments

The methods described in this chapter were developed with the support of USPHS Grants AM25316, AM 20892, AM 20762, AM 25490, and CA-35878 and BRSG Grant 2-32908 from the University of Vermont College of Medicine. The authors wish to acknowledge Mr. Robert Shaw for his expert typing of the manuscript.

[19] J. M. Kittler, D. Viceps-Madore, J. A. Cidlowski, and J. W. Thanassi, *Biochem. Biophys. Res. Commun.* **112**, 61 (1983).
[20] J. W. Thanassi, *in* "Chemical and Biological Aspects of Vitamin B-6 Catalysis" (A. E. Evangelopoulos, ed.) p. 11. Alan R. Liss, Inc., New York, 1984.

[21] Pyruvoyl-Dependent Histidine Decarboxylase from *Lactobacillus* 30a: Purification and Properties

By ESMOND E. SNELL

A number of lactic acid bacteria and clostridia produce inducible histidine decarboxylases that contain a pyruvoyl group at the active center, rather than pyridoxal 5'-phosphate.[1] Of these, the enzyme from *Lactobacillus* 30a has been studied most thoroughly. Its assay, purification, and properties are described below.

Assay Method

Principle

Release of the carboxyl group of histidine occurs according to Eq. (1).

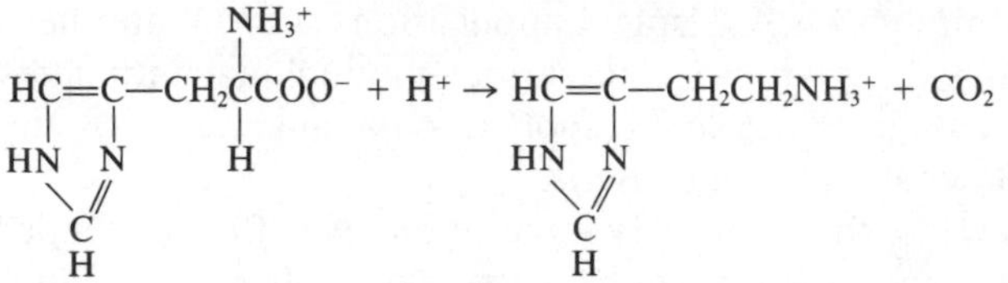

$$HC{=}C{-}CH_2\overset{\overset{\displaystyle NH_3^+}{|}}{C}COO^- + H^+ \rightarrow HC{=}C{-}CH_2CH_2NH_3^+ + CO_2 \qquad (1)$$

The CO_2 released is measured either manometrically or as $^{14}CO_2$ from L-[1-^{14}C]histidine.

Procedure

Manometric Assay.[2] The side arm of the Warburg vessel contains 50 μg of bovine serum albumin (which protects the enzyme from partial inactivation during thermal equilibration and assay) in 0.05–0.5 ml of 0.2 M ammonium acetate buffer (pH 4.8). The main compartment of the vessel contains 5 mg (24 μmol) of L-histidine monohydrochloride monohydrate in enough ammonium acetate buffer to bring the total liquid contents of the flask to 3.0 ml. After 15 min equilibration at 37°, the contents of the side arm and main compartment are mixed and the evolution of CO_2 is recorded at 2.5-min intervals for 5–10 min, depending upon its rate, care being taken to avoid significant substrate depletion during the assay

---

[1] P. A. Recsei and E. E. Snell, *Annu. Rev. Biochem.* **53**, 357 (1984).
[2] G. Y. Chang and E. E. Snell, *Biochemistry* **7**, 2005 (1968).

METHODS IN ENZYMOLOGY, VOL. 122

period. Activities are expressed in terms of micromoles of CO_2 released per milligram of protein per minute.

Radiometric Assay. This assay measures release of $^{14}CO_2$ from L-[1-^{14}C]histidine at $37°$. It is a slight modification of previously described[3-5] methods. The assay is carried out in 10×75-mm test tubes that contain 0.2 ml of reaction mixture consisting of 0.2 M ammonium acetate and 20 mM L-[1-^{14}C]histidine at pH 4.8. After addition of the enzyme (about 0.1-1 μg of the homogenous protein or its equivalent in impure fractions), the reaction tube is immediately connected by a rubber tube to a scintillation vial containing a strip (1.5×8 cm) of Whatman No. 1 filter paper moistened with 0.2 ml of 25% 2-phenethylamine in methanol. After incubation for 10 min at $37°$ with gentle shaking, 0.2 ml of 10% perchloric acid is injected to stop the reaction. The acidified solution in incubated for an additional hour at $37°$. Scintillator cocktail (10 ml) consisting of 6 g of Omnifluor in 1 liter of toluene and Triton X-100 mixture ($2:1$) is then added to each vial, and radioactivity is determined in a suitable liquid scintillation spectrometer. Counting efficiency, determined by CO_2 release from NaH$^{14}CO_3$ (in an Isocap/300 spectrometer), is about 89%. This system can measure up to 20 μmol of released CO_2 with linear correlation. One unit of carboxylase activity is defined as the amount of enzyme that catalyzes decarboxylation of 1 μmol of L-histidine per minute under these conditions. Specific activity represents units of enzyme per milligram of protein.

Protein Concentrations. Protein concentrations in impure enzyme preparations are determined by the Lowry method[6] using bovine serum albumin as a standard; the purified enzyme is determined by its absorbance at 280 nm, using a value for $E_{1\%}^{1\,cm}$ of 16.2 for the pure protein.

Growth of *Lactobacillus* 30a

Stock Cultures. Lactobacillus 30a (American Type Culture Collection 33222) is carried as stab cultures by monthly transfer in the crude medium of Table I, solidified by addition of 2% agar. After 24 hr of growth at $37°$, cultures are held at $4°$ until use or until the next transfer.

Growth of Cells for Enzyme Purification. Wholly defined,[7] semidefined,[7] and crude[2] media for growing cells of *Lactobacillus* 30a with high

[3] S. Tanase, B. M. Guirard, and E. E. Snell, *J. Biol. Chem.* **260**, 6738 (1985).
[4] P. A. Recsei and E. E. Snell, *Biochemistry* **9**, 1492 (1970).
[5] A. Ichiyama, S. Hakamura, Y. Nishizuki, and O. Hayaishi, *J. Biol. Chem.* **245**, 1699 (1970).
[6] O. H. Lowry, N. J. Rosebrough, A. L. Farr, and R. R. Randall, *J. Biol. Chem.* **193**, 265 (1951).
[7] B. M. Guirard and E. E. Snell, *J. Bacteriol.* **87**, 370 (1964).

TABLE I
CRUDE MEDIUM FOR *Lactobacillus* 30a

Component	Amount per liter of double-strength medium[a]
Yeast extract[b]	20 g
Sucrose	20 g
Casein hydrolyzate[b]	15 g
Histidine hydrochloride · H_2O	2 g
Potassium acetate	3 g
Ascorbic acid	1 g
Pyridoxamine hydrochloride	33 μg
Salts A[c]	20 ml
Salts B[c]	5 ml

[a] Adjust to pH 5.4 with acetic acid.
[b] Commercial yeast extract available from Difco Laboratories, Anheuser-Busch, or other suppliers is satisfactory; commercial casein hydrolysates (low salt), e.g., from Difco Laboratories, also are satisfactory.
[c] Salts A contain 165 g of $KH_2PO_4 \cdot H_2O$ and 165 g of $K_2HPO_4 \cdot 3H_2O$ per liter of solution. Salts B contain 80 g of $MgSO_4 \cdot 7H_2O$, 4 g of NaCl, 4 g of $FeSO_4 \cdot 7H_2O$, 4 g of $MnSO_4 \cdot H_2O$, and 1 ml of concentrated HCl per liter of solution.

histidine decarboxylase activity have been described. *Lactobacillus* 30a requires each of the amino acids for growth; when enzyme labeled specifically in a single amino acid residue is desired, defined media containing that label must be used. For purification of unlabeled enzyme, the crude medium of Table I is used, since it is cheaper and easier to prepare than other media. Quantities (15 liter) of medium in 20-liter bottles are inoculated with 500 ml of a culture of *Lactobacillus* 30a grown in the same medium, then incubated for 16–24 hr at 37° with no aeration. The cells are then harvested in a Sharples centrifuge. Two to three grams of wet cells per liter of medium are obtained.

Because of the acidity of the medium, the large inocula used, and the rapid growth of the bacterium, it is not necessary to sterilize the medium in the 20-liter bottles if they are inoculated immediately. However, it is convenient to sterilize the medium for 15 min at 121° so it can be kept for a few days prior to inoculation without fear of contamination.

Purification Procedure[2]

1. Preparation of Acetone-Dried Cells. Harvested cells (300–500 g wet weight) of *Lactobacillus* 30a are mixed with a volume of water equal to the wet cell volume and stirred to produce a uniform slurry. This slurry is added slowly with rapid stirring to 5–10 times its volume of cold (−20°) acetone. The cells are collected in a large Büchner funnel and washed with a volume of cold acetone twice that of the original cell slurry, and then with a similar quantity of cold, anhydrous, peroxide-free ether. They are then spread on sheets of filter paper to day at room temperature. The acetone-dried cells can be stored at least 1 year at 5° with no loss in decarboxylase activity.

2. *Preparation of Cell Extracts.* Acetone-dried cells (200–300 g) from Step 1 are stirred with 1 liter of ammonium acetate buffer (pH 4.8) at 37° for 5 hr. Cell debris is removed by centrifugation and reextracted with 700-ml portions of the same buffer until the last extract contains only about 5% of the total activity extracted from the cells.

3. *Ammonium Sulfate Fractionation of the Cell Extract.* The pooled cell extract from Step 2 is cooled to 0° and finely divided ammonium sulfate is added in small portions with stirring to 45% of saturation. The small precipitate is discarded. Additional ammonium sulfate is then added to 70% saturation. The active precipitate is collected by centrifugation, suspended in 200 ml of water, and dialyzed at room temperature against several changes (1 liter each) of 0.025 N potassium chloride for a total of at least 3 hr.

4. *Heat Treatment.* The dialyzed protein from Step 3 is heated by swirling 200-ml portions in a 500-ml Erlenmeyer flask in a water bath maintained at 75° until the temperature reaches 70°. It is maintained at 70° for 2 min, then cooled in running water. The precipitate is centrifuged out and discarded.

5. *Acetone Fractionation.* The supernatant solution from Step 4 is cooled to 0° in an ice–acetone bath and nine-tenths volume of cold (−20°) acetone per volume of supernatant is added slowly. Precipitated material is centrifuged out and discarded. Additional acetone (five-tenths volume per volume of supernatant) is added and the precipitate is collected by centrifugation. Finally, more acetone (five-tenths volume per volume of supernatant) is added and the precipitate is again collected by centrifugation. The enzyme is largely precipitated in either the second or third acetone precipitate (or both). The final acetone supernatant is discarded.

6. *Sephadex Filtration and Crystallization.* The active protein fraction from Step 5 is dissolved in 75 ml of ammonium acetate buffer (pH 4.8) and passed at room temperature over a 4 × 50-cm column of Sephadex G-200 equilibrated with the same buffer. The flow rate is about 50 ml/hr. Fractions are collected for 15 min each, and the active fractions are pooled and chilled to 0°. Aqueous ammonia is added to pH 6–8; then finely ground ammonium sulfate is added to bring the solution to 50% saturation at 0°. If the solution becomes turbid it is centrifuged and the insoluble matter is

TABLE II

ACTIVITY OF HISTIDINE DECARBOXYLASE AT VARIOUS STAGES OF THE
PURIFICATION PROCEDURE

Active fraction	Volume (ml)	Total protein (mg)	Specific activity (μmol/min·mg)	Overall yield of activity (%)
Dried cells (260 g)			0.164[a]	
Extract	1900	26,400	2.07	(100)
Ammonium sulfate fractionation	200	9,200	4.92	91
Heat treatment	191	1,530	29.6	91
Acetone fractionation	80	895	43.4	77
Sephadex filtration and crystallization	230	422	69.5	58

[a] Expressed in terms of dried cells instead of protein.

discarded. The enzyme crystallizes from the supernatant solution when additional finely ground ammonium sulfate is added to 70% of saturation at 0°. The crystals obtained after a few hours at 0° are harvested by centrifugation, dissolved in 75 ml of water, and rerun over the Sephadex column. If the elution profile shows a poor separation between the enzyme peak and a smaller peak of inactive protein with a lower molecular weight, the active fractions are pooled, crystallized, and again passed over the column. The active fractions from this final Sephadex column are pooled, crystallized, dissolved in a small amount of water, and recrystallized a number of times. The crystals are normally stored in a 50% saturated ammonium sulfate solution at −10° and are stable indefinitely. A sample protocol of the purification procedure is given in Table II.[8]

Structure and Properties

Physicochemical Properties. Electrophoresis of histidine decarboxylase denatured in sodium dodecyl sulfate shows the presence of two dissimilar subunits, a small β subunit ($M_r \cong 8000$, pI 5.55) and a larger ($M_r \cong 28,000$, pI 5.45) α subunit that contains the pyruvoyl group blocking its NH$_2$ terminus.[9,10] Crystallographic and ultracentrifugal studies show that

[8] J. Rosenthaler, B. M. Guirard, G. W. Chang, and E. E. Snell, *Proc. Natl. Acad. Sci. U.S.A.* **54**, 152 (1965).
[9] W. D. Riley and E. E. Snell, *Biochemistry* **7**, 3520 (1968).
[10] W. D. Riley and E. E. Snell, *Biochemistry* **9**, 1485 (1970).

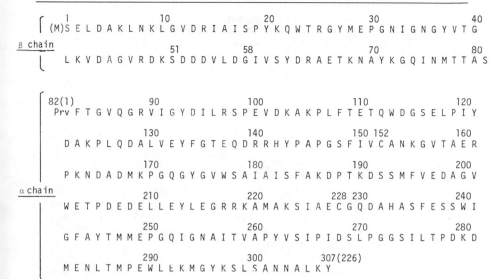

FIG. 1. Amino acid sequence of the β chain and α chain of histidine decarboxylase from *Lactobacillus* 30a.

the native enzyme ($M_r = 208,000$; $s_{20,w} = 9.4$ S, pI 4.4) has the subunit composition $(\alpha\beta)_6$ at pH 4.8 and both low and high ionic strength; at pH 7.0 a dissociation reaction $(\alpha\beta)_6 \rightleftharpoons 2(\alpha\beta)_3$, is observed ($M_r = 104,000$; $s_{20,w} = 6.9$) and is essentially complete at low ionic strength.[11] The subunits are readily separated after denaturation in 5 M guanidine · HCl by chromatography over diethylaminoethyl cellulose (DE-52).[12] Reassociation of the separated subunits, neither of which is catalytically active alone, occurs in yields up to 40% under proper conditions with regeneration of enzymatic activity.[13]

The complete amino acid sequence of the β chain (81 residues, $M_r = 8856$)[12] and the α chain (226 residues, $M_r = 24,892$),[14] and hence of the entire enzyme,[14] is shown in Fig. 1. The enzyme derived from mutant 3 of *Lactobacillus* 30a (a nitrosoguanidine mutant[15]) has two amino acid replacements, both in the β chain: Ser-51 is replaced by Ala and Gly-58 by Asp. In addition, about 15% of the mutant β chains contain Met-Ser at the

[11] M. L. Hackert, W. E. Meador, R. M. Oliver, J. B. Salmon, P. A. Recsei, and E. E. Snell, *J. Biol. Chem.* **256**, 687 (1981).
[12] G. L. Vaaler, P. A. Recsei, J. L. Fox, and E. E. Snell, *J. Biol. Chem.* **257**, 12770 (1982).
[13] S. Yamagata and E. E. Snell, *Biochemistry* **18**, 2964 (1979).
[14] Q. K. Huynh, P. A. Recsei, G. L. Vaaler, and E. E. Snell, *J. Biol. Chem.* **259**, 2833 (1984).
[15] P. A. Recsei and E. E. Snell, *J. Bacteriol.* **112**, 624 (1972).

TABLE III
SUBSTRATES AND INHIBITORS OF HISTIDINE DECARBOXYLASE FROM *Lactobacillus* 30a[a]

Substrates[b]	Activity (μmol/min·mg)	K_m (mM)	Inhibitors (competitive)[c]	K_I (mM)
L-Histidine	67	0.90	Imidazolepropionic acid	1.8[d]
N^π-Me-L-histidine	0.78	>200	Urocanic acid	2.1[d]
(1,2,4-Triazole-3-)-DL-alanine	0.71	>200	Imidazole	3.2[d]
β-(2-Pyridyl)-DL-alanine	0.52	>200	*N*-Me-imidazole	7.2[d]
(Thiazole-2-)-DL-alanine	0.10		Pyridine	1.4
N^τ-Me-L-histidine	0.025		Histamine	11[e]
2-Thiol-L-histidine	0.026		Imidazoleacetic acid	130[d]
			Imidazolecarboxylic acid	970[d]

[a] All comparisons are at pH 4.8.
[b] From Chang and Snell.[2] Me designates methyl.
[c] Carbonyl reagents (e.g., KCN, hydroxylamine) and -SH reagents [e.g., *p*-chloromercuribenzoate, dithiobis(2-nitrobenzoic acid), iodoacetate] are noncompetitive inhibitors of the decarboxylase. α-Methyl-L-histidine, α-fluoromethylhistidine, and D-histidine did not inhibit this enzyme at the concentrations tested.
[d] From Rosenthaler *et al.*[8]
[e] From Recsei and Snell.[4]

NH_2 terminus rather than Ser.[12] These replacements decrease stability of the enzyme and change its pH activity profile (the mutant 3 histidine decarboxylase is inactive at pH 7.0, where the wild-type enzyme retains substantial activity), but do not decrease its activity at pH 4.8, its optimum.[16]

One of the two -SH groups of the enzyme titrates with dithiobis(2-nitrobenzoic acid) (DTNB) in the native enzyme at pH 7.0 with complete loss of enzymatic activity;[17] the second -SH group titrates only after denaturation of the enzyme. The reactive -SH group has been identified as belonging to Cys-147 (residue 228 in the diagram), which lies in a hydrophilic portion of the α chain.[18] When the TNB group is replaced by -CN, however, the enzyme regains activity but shows altered kinetic parameters.[16] This -SH group is therefore not a catalytic residue, but activity is sensitively dependent upon conformation of this part of the protein chain.

Catalytic Properties. The substrate and inhibitor specificity of this histidine decarboxylase are shown in Table III. The enzyme is highly specific for histidine; ornithine, arginine or lysine are neither substrates

[16] P. A. Recsei and E. E. Snell, *J. Biol. Chem.* **257**, 7196 (1982).
[17] R. S. Lane and E. E. Snell, *Biochemistry* **15**, 4175 (1976).
[18] Q. K. Huynh and E. E. Snell, *J. Biol. Chem.* **260** (in press).

nor inhibitors. V_{max} of the wild-type enzyme (about 80 μmol/min·mg between pH 4.8 and 5.8)[4] is reduced only by about 40% even at pH 2.8 and 7.8. K_m values, however, are greatly increased below pH 4.0 or above pH 6.5; V_{max}/K_m is maximal and almost constant between pH 4.8 and 6.0.[4] Schiff-base adducts formed between substrate histidine or product histamine and the carbonyl group of the pyruvoyl group of the enzyme [Prv = residue 82(1) of the α subunit] are trapped by reduction to the corresponding secondary amines when $NaBH_4$ is added to reaction mixtures.[4] Formation of such Schiff bases is viewed as the primary event resulting in labilization and loss of the carboxyl group during catalysis. Enzymatic decarboxylation of L-histidine in ^2H_2O proceeds stereospecifically with incorporation of one deuterium into the α-methylene group[2] with retention of the overall configuration.[2,19]

[19] A. R. Battersby, M. Nicoletti, J. Staunton, and R. Vlegaar, *J. Chem. Soc., Perkin Trans.* 1980 (1) p. 43.

[22] Prohistidine Decarboxylase from *Lactobacillus* 30a

By ESMOND E. SNELL and QUANG KHAI HUYNH

Preparation, assay, and properties of the histidine decarboxylase (HisDCase) from *Lactobacillus* 30a have been descibed elsewhere in this volume.[1] HisDCase arises by an unusual, intramolecular, nonhydrolytic cleavage of an inactive prohistidine decarboxylase (proHisDCase)[2] during which an internal serine residue of each proenzyme subunit is converted to a pyruvoyl residue at the NH_2 terminus of the larger of two daughter subunits in the active enzyme.[3-5] Purification and properties of this proenzyme are described here.

Assay Method and Growth of Bacteria

Principle. ProHisDCase is first converted to active HisDCase, which is then assayed by manometric or radiometric methods already described.[1]

[1] This volume [21].
[2] P. A. Recsei and E. E. Snell, *Biochemistry* **12**, 365 (1973).
[3] P. A. Recsei and E. E. Snell, *Annu. Rev. Biochem.* **53**, 357 (1984).
[4] W. D. Riley and E. E. Snell, *Biochemistry* **9**, 1485 (1970).
[5] P. A. Recsei, Q. K. Huynh, and E. E. Snell, *Proc. Natl. Acad. Sci. U.S.A.* **80**, 973 (1983).

Conversion of ProHisDCase to HisDCase. Proenzyme is converted quantitatively to active enzyme by incubating up to 10 mg of protein in 1 ml of 1 M potassium phosphate buffer (pH 7.6) for 24 hr at 37°.[2]

Bacterial Culture. The source of the proenzyme is mutant 3 of *Lactobacillus* 30a.[2] This organism is carried either as stab cultures in agar as described for wild-type *Lactobacillus* 30a,[1] or as lyophilized cultures for long-term storage. For production of proHisDCase, 5–10% inocula are used. For example, for 10 liters of medium, serial transfers from the stab culture to 5 ml, 100 ml, 1 liter, and finally 10 liters of liquid crude medium[1] are made at 12-hr intervals with incubation at 37°. Cells are then collected in a Sharples centrifuge, and acetone-dried as described.[1]

Purification Procedure

1. Extraction of Cells. Acetone-dried cells (10 g) of *Lactobacilllus* 30a mutant 3 are stirred with 100 ml of 0.2 M ammonium acetate (pH 4.8) for 12 hr at 37°. Insoluble materials are removed by centrifugation and reextracted with 50 ml of the same buffer for 2 hr at 37°.

2. Ammonium Sulfate Fractionation. The pooled extracts from Step 1 are cooled to 0° and ammonium sulfate is added with stirring to 45% of saturation. The small precipitate is discarded and additional ammonium sulfate is added to 70% of saturation. The active precipitate is collected by centrifugation.

3.Heat Treatment. The precipitate from step 2 is suspended in 100 ml of distilled water (the pH should be about 4.8) and heated by swirling in a 75° water bath until the temperature reaches 70°. After 2 min at 70°, the solution is cooled in running water, the precipitate removed by centrifugation, and the supernatant solution dialyzed against 0.2 M ammonium acetate, pH 4.8 at 4°. The small precipitate which forms during dialysis is centrifuged out and discarded; the supernatant is stored in 0.2 M ammonium acetate, pH 4.8 at 4°. SDS–PAGE electrophoresis at this point shows the single π band of the proenzyme, admixed with small amounts (usually <5%) of the α and β subunits of the activated proenzyme. A protocol of the purification procedure is shown in the table.

Comment. The success of this simple procedure results from induction of high levels of the proenzyme during cell growth and its stability to acetone and to heat at a fairly low pH. Proenzyme can also be purified by the same procedures used for the active enzyme[1,2] if the final step, crystallization above pH 7.0, during which the proenzyme is converted to active enzyme,[2] is eliminated. ProHisDCase and HisDCase copurify at all steps; the purified proenzyme therefore contains variable amounts of active en-

PURIFICATION OF PROHISTIDINE DECARBOXYLASE FROM ACETONE-DRIED CELLS (10 g)
OF *Lactobacillus* 30a MUTANT 3

Step	Volume (ml)	Total activity[a] (units)	Total protein (mg)	Specific activity (units/mg protein)	Yield (%)
1. Crude extract	135	411	1907	0.2	(100)
2. Ammonium sulfate	100	390	618	0.6	95
3. Heat treatment	98	324	107	3.0[b]	79

[a] Before activation. One unit of activity is the amount of proenzyme which produced 1 μmol CO_2/min at 37°. This activity results from traces of activated proenzyme that copurify with proenzyme (see text). Specific activity is expressed as units per milligram protein.

[b] Following activation, this specific activity rises to 74.5.

zyme depending on prior history of the preparation. Formation of active enzyme is minimized by restricting the cell growth period, by maintaining the pH below 5.0 during fractionation and storage. Activation can be avoided at higher pH values if monovalent cations (especially K^+ or NH_4^+ ions) are absent.[6,7]

Properties

Structure of ProHisDCase. ProHisDCase contains six identical pyruvate-free π subunits; on activation each π subunit gives rise to two nonidentical subunits (α and β) of active HisDCase according to Eq. (1).

$$(\pi)_6 + 6H_2O \longrightarrow (\alpha\beta)_6 + 6NH_3 \tag{1}$$

At the subunit level, this conversion occurs as shown in Eq. (2) (Prv = pyruvoyl group).[8]

$$
\begin{array}{c}
\overset{\pi \text{ subunit}}{} \quad \overset{81}{} \quad \overset{82}{} \\
H_2NSer \text{------------} Ser \cdot Ser \text{------------} TyrOH + H_2O
\end{array}
$$

$$
\begin{array}{c}
\overset{81}{} \quad \overset{82}{} \\
\longrightarrow H_2NSer \text{------------} SerOH + Prv \text{------------} TyrOH + NH_3 \tag{2} \\
\underset{\beta \text{ subunit}}{} \qquad \qquad \underset{\alpha \text{ subunit}}{}
\end{array}
$$

The amino acid sequence of the wild-type proHisDCase is that shown for HisDCase,[1] but with Prv-82 replaced by Ser-82 and with an amide bond

[6] P. A. Recsei and E. E. Snell, *in* "Metabolic Interconversion of Enzymes 1980" (H. Holzer, ed.), p. 335. Springer-Verlag, Berlin and New York, 1981.

[7] E. E. Snell and P. A. Recsei, *Colloq. Ges. Biol. Chem.* **32**, 177 (1981).

[8] P. A. Recsei, Q. K. Huynh, and E. E. Snell, *Proc. Natl. Acad. Sci. U.S.A.* **80**, 973 (1983).

between Ser-81 and Ser-82.[9] In the mutant proHisDCase described here, Ser-51 and Gly-58 of the wild-type proenzyme are replaced by Ala-51 and Asp-58,[10] respectively. The M_r of the π subunit of proHisDCase, calculated from its amino acid sequence.[9] is 33,731 and for the native proenzyme (π_6) is 202,386. Under conditions of high pH and low ionic strength, proHisDCase appears to occur also as a π_3 oligomer.[11]

Activation of ProHisDCase. Reaction (2) is unusual in that chain cleavage per se is nonhydrolytic; the oxygen of the newly formed carboxyl terminus of the β chain, Ser-81, is transferred from Ser-82.[8] The conversion is first order with respect to protein and proceeds optimally at pH 7.6. Rates of activation vs pH show that acidic titration of a group with a pK near 6.8 and basic titration of a group with a pK near 8.2 decrease the activation rate; this rate is essentially zero below pH 5.0.[6] Activation requires monovalent cations; at pH 7.6 their order of effectiveness is $K^+ > NH_4^+ > Rb^+ > Na^+ \cong Cs^+ > Li^+$, and again is unusual in that the activation rate is half-order with respect to K^+ concentration.[6] An activation energy of 15.2 kcal/mol at pH 7.6 was calculated from an Arrhenius plot. Wild-type proHisDCase can be isolated admixed with large amounts of active HisDCase by application of the same procedures used for isolation of the mutant 3 proHisDCase to cells of *Lactobacillus* 30a (wild-type). For this purpose, the culture is grown initially in low-histidine medium, and histidine is added as inducer 2 hr before harvesting the cells. Activity of the contaminating HisDCase can be destroyed by treatment with $NaBH_4$; after dialysis the activation characteristics of the wild-type and mutant enzymes can be compared. In this way it was shown that activation of the wild-type proHisDCase also required monovalent cations, and proceeded about three times as fast as that of mutant proHisDCase ($t_{1/2} = 1.4$ vs 4.3 hr, both in 0.9 M potassium phosphate buffer (pH 7.6) at 37°).[12]

[9] Q. K. Huynh, P. A. Recsei, G. L. Vaaer, and E. E. Snell, *J. Biol. Chem.* **259**, 2833 (1984).
[10] G. L. Vaaler, P. A. Recsei, J. L. Fox, and E. E. Snell, *J. Biol. Chem.* **257**, 12770 (1982).
[11] M. L. Hackert, W. E. Meador, R. M. Oliver, J. B. Salmon, P. A. Recsei, and E. E. Snell, *J. Biol. Chem.* **256**, 687 (1981).
[12] P. A. Recsei and E. E. Snell, *J. Biol. Chem.* **257**, 7196 (1982).

[23] Pyridoxal Phosphate-Dependent Histidine Decarboxylase from *Morganella* AM-15

By ESMOND E. SNELL and BEVERLY M. GUIRARD

Unlike the inducible pyruvoyl-dependent histidine decarboxylases (HisDCases) from four different gram-positive bacteria,[1,2] this same enzyme from gram-negative bacteria so far studied[3,4] requires pyridoxal 5'-phosphate (PLP) as coenzyme. In this respect the latter enzyme resembles mammalian HisDCase.[5,6] The currently best studied representative of the bacterial PLP-dependent HisDCases is that from *Morganella* AM-15.[4,7] Its purification and properties are described here.

Assay Method

Either manometric or radiometric determination of CO_2 released during decarboxylation can be used. Both methods are described elsewhere in this volume.[8] However, because the *Morganella* HisDCase has a higher pH optimum and is more labile than HisDCase from *Lactobacillus* 30a, the 0.2 M ammonium acetate buffer (pH 4.8) used in the described assay[8] is replaced by 0.1 M potassium phosphate, 0.1 mM EDTA, 0.01 mM PLP, and 0.02 mM dithiothreitol, pH 6.5.[7] With this modification, the radiometric assay is used without further change. If the manometric assay is used, the higher pH also requires that two-armed Warburg flasks be used. An acid tip (0.1 ml of 10 N H_2SO_4) from the second side arm at the end of the reaction period is made to release dissolved CO_2. One unit of activity corresponds to release of 1 μmol of CO_2 per minute at 37°. Specific activity represents the number of units per milligram of protein.

[1] P. A. Recsei and E. E. Snell, *Annu. Rev. Biochem.* **53,** 357 (1984).

[2] P. A. Recsei, W. M. Moore, and E. E. Snell, *J. Biol. Chem.* **258,** 439 (1982).

[3] B. M. Guirard, S. Tanase, and E. E. Snell, *in* "Chemical and Biological Aspects of Vitamin B-6 Catalysis" (A. E. Evangelopolis, ed.), Part A, p. 235. Alan R. Liss, Inc., New York.

[4] S. Tanase, B. M. Guirard, and E. E. Snell, *Fed. Proc., Fed. Am. Soc. Exp. Biol.* **43,** 1998 (1984).

[5] L. Hammer and S. Hjerten, *Agents Actions* **10,** 93 (1980).

[6] Y. Taguchi, T. Watanabe, H. Kubota, H. Hayashi, and H. Wada, *J. Biol. Chem.* **259,** 5214 (1984).

[7] S. Tanase, B. M. Guirard, and E. E. Snell, *J. Biol. Chem.* **260,** 6738 (1985).

[8] This volume [21].

Protein Concentrations. In impure preparations, protein is determined by the Lowry method[9] following dialysis to remove dithiothreitol, if present. Purified preparations are determined from their absorbance at 278 nm, $E_{1\,cm}^{1\%} = 14.5$.

Growth of Morganella AM-15

Morganella AM-15 is obtained from the American Type Culture Association (35200) as a lyophilized culture. This is dispersed in 10 ml of sterile Brain Heart Infusion medium (Difco) and incubated overnight at 30° without aeration. Portions of this culture are lyophilized or carried as agar deeps for day-to-day laboratory use, with transfer every second week. For large-scale growth, a medium (CYHG) containing 3% casein hydrolysate (Difco), 0.2% yeast extract (Difco), and 1% L-histidine hydrochloride is adjusted to pH 6.5. After sterilization and cooling, separately sterilized glucose solution is added to a final concentration of 2%. A 10% inoculum is used, initially from 10 ml of Brain Heart Infusion medium into 100 ml of the CYHG medium, then to 1 liter, 10 liters, etc., as desired. An incubation period of 18 hr at 27–30° between transfers permits maximum growth (about 2 g of wet cells per liter). The collected cells are washed with buffer A (0.05 *M* potassium phosphate, 0.05 *M* potassium succinate, 2 m*M* NaEDTA, 1 m*M* dithiothreitol, and 0.01 m*M* pyridoxal-P, pH 6.0) and stored at −20° until used.

Purification of HisDCase[7]

All purification steps are carried out at 4°. Because this enzyme loses activity in the absence of stabilizing agents, all buffers used during purification contain 0.1 m*M* NaEDTA, 0.1 m*M* dithiothreitol (DTT), and 5 μM PLP unless other concentrations are stated. Results of the purification procedure are summarized in the table.

Step 1. Cell Breakage. Washed cells (200 g, wet weight) are suspended in 500 ml of buffer A and divided into two 250-ml portions. Each portion is treated for a total of 30 min in a Heat Systems Ultrasonic Sonicator with cooling after each 3 min of treatment. The lysate is centrifuged at 12,000 g for 30 min. The precipitate is resuspended in 300 ml of Buffer A and again sonicated for a total of 30 min as before. After centrifuging at 12,000 g for 30 min, all supernatant solutions are combined and centrifuged at 105,000 g for 30 min. The residue is discarded.

[9] O. H. Lowry, N. J. Rosebrough, A. L. Farr, and R. R. Randall, *J. Biol. Chem.* **193**, 265 (1951).

PURIFICATION OF HISTIDINE DECARBOXYLASE FROM *Morganella* AM-15

Step	Volume (ml)	Total protein (mg)	Specific activity (units/mg)	Yield (%)	Purification (-fold)
1. Extraction[a]	1000	19,000	0.46	100	1
2. Polyethyleneimine treatment	950	14,350	0.59	96	1.3
3. Ammonium sulfate fractionation	150	8,060	1.01	92	2.2
4. Dialysis at pH 5.4	200	5,080	1.51	87	3.3
5. CM-Sephadex	285	333	21.3	81	46.3
6. Hexyl-Sepharose	70	95	60.5	65	131
7. DEAE-Sephadex	70	72	65.3	53	142
8. Second Hexyl-Sepharose	78	60	66.5	45	144
9. Sephacryl S-300	130	48[b]	72.0[b]	40	156

[a] From 200 g of cells.
[b] Calculated using a value of $A_{278\,nm}^{1\%} = 14.5$ for the homogenous enzyme.

Step 2. Polyethyleneimine Treatment. Aqueous 5% (w/w) Polymin P, adjusted to pH 6.0 with HCl, is added to the supernatant solution from Step 1 to provide 12 g of Polymin P per 100 g of protein. After 30 min the precipitate is centrifuged out and discarded. The absorbance ratio, A_{260}/A_{280}, at this point is about 1.0.

Step 3. Ammonium Sulfate Fractionation. Solid ammonium sulfate is added to the supernatant from Step 2 to 35% of saturation. The solution is stirred for 30 min, and then centrifuged at 12,000 g for 20 min. Insoluble material is discarded. To the supernatant solution, additional ammonium sulfate is added to 70% saturation. After 30 min, the active precipitate is collected by centrifugation.

Step 4. Dialysis at pH 5.4. The precipitate from Step 3 is dissolved in buffer A, then dialyzed for 40 hr at 4° against 10 mM potassium phosphate buffer/5 mM potassium succinate/1 mM DTT/EDTA/PLP buffer (pH 5.4). The precipitate that forms is discarded.

Step 5. Carboxymethyl-Sephadex Column Chromatography. The dialyzed solution from Step 4 is passed through a CM-Sephadex C-50 column (5.0 × 75 cm) equilibrated with 10 mM potassium phosphate/5 mM potassium succinate/DTT/EDTA/PLP buffer (pH 5.4), and washed out with the same buffer. Fractions having HisDCase activity are collected and adjusted to pH 6.0, then reduced in volume to about 100 ml by ultrafiltration using a Diaflo membrane YM10 (Amicon).

Step 6. Hexyl-Sepharose Column Chromatography. The solution from Step 5 is applied to a hexyl-Sepharose column (2.6×55 cm) preequilibrated with 20 mM potassium phosphate/DTT/EDTA/PLP buffer (pH 6.0). The column is developed in a linear gradient of KCl formed from 500 ml of this buffer and 500 ml of 0.3 M KCl in the same buffer. HisDCase was eluted at a KCl concentration of 0.1 M.

Step 7. DEAE-Sephadex Column Chromatography. Combined active fractions from the hexyl-Sepharose column are reduced in volume by ultrafiltration and diluted into 50 mM potassium phosphate buffer (pH 6.4). The sample is chromatographed on a DEAE-Sephadex A-50 column (2.2×40 cm) equilibrated with 50 mM potassium phosphate buffer (pH 6.4). The enzyme elutes in a KCl linear gradient (0–0.4 M) in the same buffer at a concentration of 0.25 M KCl.

Step 8. Second Hexyl-Sepharose Column Chromatography. The active fractions from Step 7 are combined and reduced in volume by ultrafiltration on a YM10 membrane. The concentrated enzyme solution is then refractionated on a hexyl-Sepharose column as described in Step 6.

Step 9. Sephacryl S-300 column chromatography: The combined active fractions from Step 8 are filtered through a Sephacryl S-300 (4×40 cm) column previously equilibrated with 50 mM potassium phosphate/DTT/EDTA/PLP buffer (pH 6.0). The active fractions are combined, reduced to 5 ml by ultrafiltration, and stored at 4° in buffer A.

Comment on Purification Procedure. The last few steps of this procedure do not increase activity of the preparation greatly, but are required to remove small amounts of a persistent electrophoretic impurity. Furthermore, this procedure has not been widely used and still yields somewhat variable results with different investigators. Several steps of an alternative purification procedure[3] are available if needed.

Properties

Stability. Activity of this HisDCase is readily lost during purification and storage for reasons not yet clear. Buffers containing succinate, PLP, EDTA, and DTT stabilize, but do not completely prevent such loss in our hands.

Purity, Subunit Structure, and Molecular Weight.[7] Both the native and denatured protein from Step 9 of the purification procedure give single bands on gel electrophoresis, corresponding to M_r values of 170,000 and 43,000, respectively. Threonine is the only NH$_2$-terminal amino acid, and these findings together indicate the presence of four identical subunits. pI values of 3.9 and 4.6, respectively, were found for the native and denatured protein.

PLP Dependency, Resolution, and Spectral Properties.[7] In the absence of sulfhydryl compounds, this HisDCase exhibits absorption bands at 416, 333, and 278 nm, with $A_{416}/A_{278} = 0.108$ and $A_{333}/A_{413} = 0.36$. It is inactivated by reduction with cyanoborohydride, and ε-N-pyridoxyllysine appears in acid hydrolysates of the reduced enzyme. The enzyme is resolved (and inactivated) by dialysis against cysteine, and is reconstituted by combination with one PLP per subunit.

Catalytic Parameters.[7] V_{max} for decarboxylation of histidine by this HisDCase decreases from about 100 μmol/min·mg at pH 5.0 to 56 μmol/min·mg at pH 8.0. K_m is lowest (1.2 mM) and V_{max}/K_m is highest at pH 6.5–6.7. The activation of energy for histidine decarboxylation is 10.5 kcal/mol.

Substrate Specificity and Inhibitors.[3,7] N^τ-Methyl-L-histidine and β-(2-pyridyl)-DL-alanine at concentrations of 25 mM are decarboxylated at rates 60 and 50% those of L-histidine; N^π-methyl-L-histidine, L-arginine, L-lysine, and L-ornithine are neither substrates nor inhibitors of this enzyme at concentrations up to 100 mM.

As expected for a PLP-enzyme, HisDCase is inhibited by many carbonyl reagents, of which the hydrazine analog of histidine, 2-hydrazino-3-(4-imidazolyl)propionic acid, a competitive inhibitor, is most potent (K_I 0.002 mM). A large number of other substrate analogs also are competitive inhibitors, including L-histidine methyl (K_I 0.5 mM) or ethyl (K_I 0.65 mM) ester, carnosine (K_I 1.2 mM), α-methyl-DL-histidine (K_I 3.0 mM), urocanic acid (K_I 20 mM), and imidazoleacetic acid (K_I 30 mM). α-Fluoromethyl-L-histidine is a potent mechanism-based inhibitor that irreversibly inactivates the holoenzyme, but not the apoenzyme of HisDCase[10]; the mechanism of its action is not yet known.

Interactions of Inhibitors with HisDCase.[7] Several inhibitors listed above interact with the PLP of holoHisDCase as shown by spectral changes. These include histidine methyl ester and also DTT and other thiols, which protect the enzyme against oxidative inactivation at low concentrations, but act as mixed type inhibitors at high concentrations. α-Methylhistidine gradually inactivates HisDCase apparently by undergoing a decarboxylation-dependent transamination reaction: PLP of the holoenzyme is released as pyridoxamine 5'-phosphate (PMP) with formation of the inactive apoenzyme. Activity is maintained if excess PLP is added, with gradual accumulation of PMP. Other products of the reaction have not been determined.

---

[10] H. Hayashi, S. Tanase, and E. E. Snell, unpublished data.

Section VI

Nicotinic Acid: Analogs and Coenzymes

[24] Simultaneous Extraction and Combined Bioluminescent Assay of Oxidized and Reduced Nicotinamide Adenine Dinucleotide

By Matti T. Karp and Pauli I. Vuorinen

A bacterial bioluminescence-based method for measuring NADH was introduced by Stanley in 1971.[1] In this assay light is emitted when bacterial luciferase and NAD(P)H dehydrogenase (FMN) [NAD(P)H : FMN oxidoreductase] catalyze the oxidation of NAD(P)H, $FMNH_2$, and a long-chain aliphatic aldehyde. The method does not involve the laborious and unreliable enzymatic cycling as in some older methods.[2,3] However, it is sensitive enough for the biological specimens in the milligram range.[4] In the last couple of years, rather specific luminescence reagents for both NADH and NADPH has been developed. This is possible because in bacteria there exists specific oxidoreductases for both forms of nucleotides.[5] The concentrations of bacterial luciferase, NADH : FMN oxidoreductase (NADH-OR), FMN, and an aldehyde can be fixed so that the resulting enzyme reactions exhibit reaction kinetics suitable for analytical purposes.[6] This fact is made use of in the following combined assay system for NAD⁺ and NADH.

Principle

When nicotinamide adenine dinucleotides are extracted and thereafter measured with enzymatic cycling, either the oxidized (acid extraction) or reduced (alkaline extraction) form have to be destroyed before measuring one or the other.[2] However, both nucleotide forms are efficiently extracted from biological material with buffered 70% ethanol.[7] In this procedure the extracted nucleotides are measured in a single cuvette by cou-

[1] P. E. Stanley, *Anal. Biochem.* **39**, 441 (1971).
[2] O. H. Lowry, J. V. Passoneau, D. W. Schulz, and M. K. Rock, *J. Biol. Chem.* **236**, 2746 (1961).
[3] F. M. Matschinsky, this series, Vol. 18, p. 3.
[4] A. Ågren, S. E. Brolin, and S. Hjerten, *Biochem. Biophys. Acta* **500**, 103 (1977).
[5] J. T. Lavi, T. N. E. Lövgren, and R. P. Raunio, *FEMS Microbiol. Lett.* **11**, 197 (1981).
[6] T. Lövgren, A. Thore, J. Lavi, I. Styrelius, and R. Raunio, *J. Appl. Biochem.* **4**, 103 (1982).
[7] M. T. Karp, R. P. Raunio, and T. N. E. Lövgren, *Anal. Biochem.* **128**, 175 (1983).

pling the alcohol dehydrogenase (ADH) reaction to bioluminescent reaction. The three coupled enzyme reactions which form the basis of the bioluminescent assay of NAD^+ and $NADH$ are as follows:

$$NAD^+ + C_2H_5OH \xrightarrow{ADH} NADH + CH_3CHO \qquad (1)$$

$$NADH + H^+ + FMN \xrightarrow{NADH-OR} FMNH_2 + NAD^+ \qquad (2)$$

$$FMNH_2 + RCHO + O_2 \xrightarrow{luciferase} FMN + RCOOH + H_2O + light \qquad (3)$$

The assay system initially contains all the reactants of the above three enzyme reactions except for ADH. The purpose of the assay is to achieve constant light emission in two steps which reflect, first, the NADH content and, second, the NAD^+ content of the sample. In between the two steps, the system is calibrated with standards. The light emitted in the first step [reactions (2) and (3)] is directly proportional to the NADH content in the test sample. The assay is calibrated with known amounts of NADH after which ADH is added to convert NAD^+ to NADH [reaction (1)]. The resulting light emission is directly proportional to the NAD^+ content in the test sample. Finally, the assay system can be calibrated by adding a known quantity of NAD^+ (see Fig. 1).

Reagents

NADH Monitoring Kit (LKB-Wallac, Turku Finland, No. 1243-103) containing NADH Monitoring Reagent, NADH standard, and decanal. Reconstitute according to manufacturer's instructions.

Nicotinamide adenine dinucleotide standard, NAD^+ (Boehringer No. 127302, or Sigma No. N-7004). Dissolve 3.32 mg NAD^+ in 0.9 ml distilled water and then make up to 1.0 ml (5 mM), dilute further 1 : 10,000 to give a 5×10^{-7} M standard solution.

NADH standard. Reconstitute according to manufacturer's instructions. Take 1 ml of the standard and add 9 ml distilled water to give a 5×10^{-7} M standard solution.

Potassium phosphate buffers, 0.1 and 0.01 M, pH 7.0.

Extraction solution. Prepare a 70% ethanol solution (v/v) in 0.01 M phosphate buffer, pH 7.0.

Alcohol dehydrogenase (Sigma No. A-3263). Prepare a solution of 1 mg/ml in 0.1 M phosphate buffer, pH 7.0.

Equipment. Any manual or automated luminometer which can continuously monitor an enzymatic reaction, such as LKB-Wallac 1250 or 1251 Luminometers, can be used. A potentiometric chart recorder is connected to the luminometer. Also needed are a machine for powdering

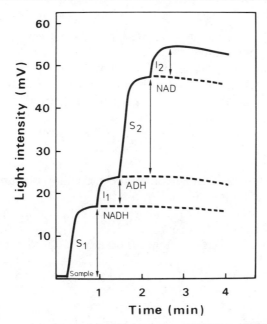

FIG. 1. Assay procedure for combined bioluminescent NAD⁺ and NAD determination. A sample is injected to the reaction mixture, which contains all the necessary compounds for the bacterial bioluminescent NADH measurement. At the first step of the assay, a constant light emission (S_1) is obtained which reflects the NADH content of the sample. The system is internally calibrated with 5 pmol NADH (I_1), after which in the second step of the assay, ADH is added to convert NAD⁺ to NADH (S_2). At the end of the assay, the system can be once more calibrated with 5 pmol NAD⁺ (I_2).

tissue samples at the temperature of liquid nitrogen, such as B. Braun Microdismembrator (Melsungen, West Germany), and a Vortex mixer.

Extraction Procedure

Tissue samples are freeze-clamped immediately after removal from the test animal, with tongs cooled to the temperature of liquid nitrogen. The tissues are then powdered either manually or with a microdismembrator. About 1 mg of this powder is transferred to 0.5 ml buffered 70% ethanol solution and mixed thoroughly at 22° with a vortex mixer at regular intervals over a period of 20 min. The mixture is then centrifuged at 3000 g for 20 min, and the NAD⁺ and NADH content of the supernatant is measured as described below. The protein content of the precipitate is measured with the Coomassie brilliant blue method.[8]

[8] M. M. Bradford, *Anal. Biochem.* **72,** 248 (1976).

Assay Procedure

The following additions in normal luminometer polystyrene cuvettes should be performed rapidly when using a manually operated luminometer. An automated luminometer should be adjusted so that the reactants are added directly in the cuvette standing in front of the photomultiplier tube. Phosphate buffer and the measuring chamber of the luminometer should be temperature controlled to 25°.

Pipet in the assay cuvette 380 μl of 0.1 M phosphate buffer, pH 7.0, and 100 μl of NADH Monitoring Reagent. Mix and record the background, B (mV).

Add 10 μl of the test sample in 70% (v/v) ethanol. Mix and measure the constant level of light emission, S_1 (mV).

Add 10 μl of the NADH standard, 5×10^{-7} M. Mix and measure the increase in light emitted, I_1 (mV).

Add 2 μl of the ADH solution. Mix and measure the increase in light emitted, S_2 (mV).

Add 10 μl of the the NAD$^+$ standard, 5×10^{-7} M. Mix and measure the light emitted, I_2 (mV).

Calculation of the Results

$$\text{NADH (mol/mg protein)} = \frac{(S_1 - B) \text{ (mV)}}{I_1 \text{ (mV)}} \times \frac{\text{conc. of NADH standard } (5 \times 10^{-7} M)}{\text{protein (mg/ml)}} \times 10^{-3}$$

$$\text{NAD}^+ \text{ (mol/mg protein)} = \frac{S_2 \text{ (mV)}}{I_2 \text{ (mV)}} \times \frac{\text{conc. of NAD}^+ \text{ standard } (5 \times 10^{-7} M)}{\text{protein (mg/ml)}} \times 10^{-3}$$

Discussion

A standard curve for NADH measurement with the NADH Monitoring Reagent is shown in Fig. 2. The figure contains also a curve for nonspecific reaction of NADPH as measured with the same reagent. The light response to NADPH is less than 1% of that of NADH. A similar curve as shown for NADH could be obtained for NAD$^+$. Thus the measurement range for both forms of nucleotides (NAD$^+$ and NADH) is 5×10^{-14} to 10^{-9} mol/assay tube. As shown by Karp et al.,[7] NAD$^+$ and NADH concentrations can vary in the sample so that biological specimens can be measured for their NAD$^+$ and NADH content with the present method. This is further illustrated in the table, where the concen-

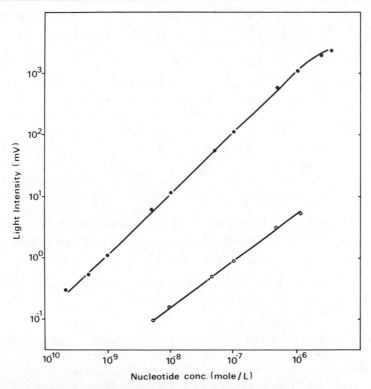

FIG. 2. Standard curve for NADH measurement with the NADH Monitoring Reagent. Specific response for NADH (●). Unspecific response for NADPH (○). Each point represents the mean of three determinations.

NADH, NAD$^+$, AND NAD$^+$/NADH LEVELS OF SOME RAT TISSUES AS DETERMINED BY THE COMBINED BIOLUMINESCENT ASSAY[a]

Tissue	NADH (pmol/mg protein)	NAD$^+$ (nmol/mg protein)	NAD$^+$/ NADH
Brain	37 ± 7	2.17 ± 0.18	71.2 ± 8.2
Heart	52 ± 4	5.74 ± 0.33	94.7 ± 9.4
Kidney	33 ± 3	4.64 ± 0.15	149.2 ± 15.9
Liver	277 ± 34	6.72 ± 0.78	26.4 ± 3.5
Lung	22 ± 2	1.33 ± 0.08	72.5 ± 10.6
Skeletal muscle	109 ± 13	3.64 ± 0.25	36.6 ± 3.6

[a] Male Spraque–Dawley rats (250–300 g) were killed by dislocation of the neck. The tissues were quickly removed and assayed as described in the extraction and assay procedure. Results are expressed in pmol and nmol/mg protein ± SEM. The number of rats was 10 and duplicate measurements were made.

trations of NAD$^+$ and NADH of some rat tissues are measured with the present bioluminescent method.

Troubleshooting

Each lot of ADH should be tested for the presence of NAD$^+$ and NADH contaminants. In liver samples, ADH activities can be so high that a change in the actual NAD$^+$/NADH ratio can occur. Therefore it is advisable to test NADH recovery by adding the NADH standard to the extraction solution. The amount of tissue sample should not be raised above 5 mg of protein.

[25] Purification of Reduced Nicotinamide Adenine Dinucleotide by Ion-Exchange and High-Performance Liquid Chromatography

By DEXTER B. NORTHROP, COLIN J. NEWTON, and STEVEN M. FAYNOR

The presence of impurities in NADH preparations was identified some time ago by Dalziel[1] and Fawcett et al.[2] These impurities plus others found more recently are known to effect the results of enzymatic kinetic experiments.[3–6] Chromotography on DEAE-cellulose has been the traditional method of purification,[7] but other anion-exchange resins have also been used, including DEAE-Sephadex,[8,9] QAE-Sephadex,[10] and benzoylated DEAE-cellulose.[11] However, Loshon et al.[12] have shown that while anion-exchange chromotography separates NAD and NADH well,

[1] K. J. Dalziel, Nature (London) 191, 1098 (1961).
[2] C. P. Fawcett, M. M. Ciotti, and N. O. Kaplan, Biochem. Biophys. Acta 54, 210 (1961).
[3] K. J. Dalziel, Biochem. J. 84, 240 (1962).
[4] P. E. Strandjord and K. J. Clayson, J. Lab. Clin. Med. 67, 144 (1966).
[5] B. F. Howell, S. Margolis, and R. Schaffer, Clin. Chem. (Winston-Salem, N.C.) 22, 1648 (1976).
[6] B. F. Howell, this series, Vol. 66, p. 55.
[7] E. J. Pastore and M. Friedkin, J. Biol. Chem. 236, 2314 (1961).
[8] I. Wenz, W. Loesche, U. Till, H. Petermann, and A. Horn, J. Chromatogr. 120, 187 (1976).
[9] W. Loesche, I. Wenz, U. Till, H. Petermann, and A. Horn, this series, Vol. 66, p. 11.
[10] E. Haid, P. Lehmann, and J. Ziegenhorn, Clin. Chem. (Winston-Salem, N.C.) 21, 884 (1976).
[11] L. Kurz and C. Frieden, Biochemistry 16, 5207 (1977).
[12] C. A. Loshon, R. B. McComb, L. Bond, G. N. Bowers, W. H. Coleman, and R. H. Gwynn, Clin. Chem. (Winston-Salem, N.C.) 23, 1576

at least one of the dehydrogenase inhibitors cannot be separated from NADH on DEAE-cellulose or DEAE-Sephadex. This was demonstrated using the reversed-phase high-performance liquid chromatographic (HPLC) procedure of Margolis *et al.*,[13] now the method of choice for analyzing samples of NADH.[14] Because the reversed-phase partition chromotographic system provided the only means of detecting all of the impurities, a method was derived to adapt it to their removal as well.[18] The final product is greater than 99% pure as judged by the HPLC procedure of Margolis.

Chromatography on AG MP-1

Forty milligrams of NADH are dissolved in 0.5 M ammonium bicarbonate (pH 9.0) and applied to a 1.1 × 30-cm AG MP-1 (100–200 mesh) anion-exchange column, equilibrated previously with this same buffer at 4° in the dark. The column is eluted in the dark at 4° with starting buffer until the absorbance at 340 nm returns to baseline levels (~125 ml). The eluant is then changed to 1 M ammonium bicarbonate (pH 9.0) to elute NADH. The 260/340 absorbance ratio shows impurities eluting on the leading edge of the NADH peak; these are discarded. Fractions with absorbance ratios of 2.26 ± .05 are combined and saved.

High-Performance Liquid Chromotography

The product from the AG MP-1 column if pumped onto a semipreparative Bondapak C_{18}/Porasil B (0.78 cm × 61 cm × 2; 37–50 μm particle diameter) reversed-phase column, equilibrated previously with 1 M ammonium bicarbonate at pH 9.0 and 25°. The column is washed with 500 ml of 1 M ammonium bicarbonate and a broad, low peak of impurities is eluted. The eluant is then changed to 50 ml of 0.1 M ammonium bicarbonate (pH 9.0), to decrease the ionic strength. Finally, NADH is eluted in a sharp peak (~20 ml) with 95% ethanol, made basic with a small amount of concentrated ammonium hydroxide.

[13] S. Margolis, B. F. Howell, R. Schaffer, *Clin. Chem. (Winston-Salem, N.C.)* **22**, (1976).

[14] The purity of NADH has previously been assessed by measuring its absorbance in solution at 260 and 340 nm and calculating the ratio of the absorbances.[15] A value of less than 2.3 has been accepted as evidence of reasonable purity, assuming that impurities lack absorbance at 340 nm. However, impurities have been isolated which have 340-nm absorbance and spectral ratios less than 2.3[9,16,17]; hence this criterion can no longer be defended.

[15] A. L. Lehninger, *Biochem. Prep.* **2**, 92 (1952).

[16] S. Margolis, B. F. Howell, R. Schaffer, *Clin. Chem. (Winston-Salem, N.C.)* **23**, 1581 (1977).

[17] J.-F. Bielman, C. Ladinte, and G. Weimann, *Biochemistry* **18**, 1212 (1979).

[18] C. J. Newton, S. M. Faynor, and D. B. Northrop, *Anal. Biochem.* **132**, 50 (1983).

Dry Storage of Purified NADH

Because moisture accelerates the decomposition of NADH[4,19,20] the purified NADH is stored in 1,2-propanediol over molecular sieves to maintain dryness.[21,22] Water is removed from the 95% ethanol fraction by stirring with anhydrous sodium sulfate. The fraction is filtered, its volume reduced by rotary evaporation, and the resultant slurry suspended in 1,2-propanediol which has been dried over sodium sulfate and distilled. Sonication assists dissolution of NADH (solubility ≃ 15 mg/ml; small amounts of material failing to dissolve are removed by decantation). Several activated molecular sieves (type 4A, Grade 514, 8–12 mesh) are added to each milliliter of NADH solution and the preparation is stored at $-20°$.[23]

Removal of Propanediol

For many purposes, microliter volumes of the purified NADH in 1,2-propanediol may be conveniently pipetted and used directly. When necessary, the diol can be quickly removed by combining an equal volume of the NADH preparation and 1 M potassium phosphate or 1M ammonium bicarbonate (pH 9), and pumping the mixture onto the Porasil B reversed-phase chromatographic column.[24] The column is eluted with dilute ammonium hydroxide (pH 10). Buffer and diol elute prior to the NADH in an overlapping peak, so the first 10% of the purified nucleotide is discarded.

[19] V. H. Gallati, *J. Clin. Chem. Clin. Biochem.* **21,** 884 (1976).
[20] S. E. Godtfredsen and M. Ottesen, *Carlsberg Res. Commun.* **43,** 171 (1978).
[21] I. E. Modrovich, U.S. Patent 4, 153, 511. (1979).
[22] P. Carter and M. E. Rose, *Clin. Biochem. (Ottawa)* **13,** 38 (1980).
[23] Attempts to obtain a dry solid form of purified NADH suitable for enzymatic assay have not been successful. Lyophilization and precipitation always introduce nucleotide impurities derived from NADH, with the only exception being barium precipitation. However, barium is deleterious to many enzymes, and a method to remove it without introducing significant breakdown of NADH has not been devised.
[24] D. B. Northrop and C. J. Newton, *Anal. Biochem.* **103,** 359 (1983).

[26] Desalting of Nucleotides by Reverse-Phase High-Performance Liquid Chromatography

By DEXTER B. NORTHROP and COLIN J. NEWTON

Difficulties in desalting nucleotides by gel filtration have been well-documented.[1-3] During attempts to scale up reverse-phase analytical high-performance liquid chromotography (HPLC) to remove nucleotide contaminants of NADH,[4] an alternative and effective desalting method was found.[5]

Principle. When chromatographed on reverse-phase columns, charged amphiphilic compounds form a tight association with the nonpolar stationary phase in the presence of high salt, but favor the polar mobile phase in the absence of salt. Hence, salt passes through the column while charged amphiphilic compounds are retarded until they are salt free.

[1] C. Bernofsky, *Anal. Biochem.* **68,** 311 (1975).
[2] P. C. Engel, *Anal. Biochem.* **82,** 512 (1977).
[3] F. M. Dickinson, *Anal. Biochem.* **92** 523 (1977).
[4] C. J. Newton, S. M. Faynor, and D. B. Northrop, *Anal. Biochem.* **132,** 50 (1983).
[5] D. B. Northrop and C. J. Newton, *Anal. Biochem.* **103** 359 (1980).

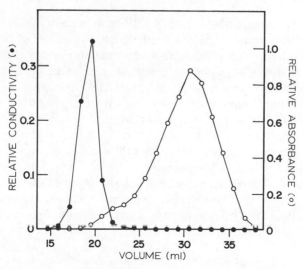

FIG. 1. Reverse-phase desalting of NADH in potassium phosphate. (●) Relative conductivity; (○) relative absorbance.

Example. A 7.8 mm i.d. × 61-cm steel column was packed with C_{18} Porasil B resin (Waters Associates), washed with 70% methanol, and equilibrated with deionized water. Ten milligrams of NADH dissolved in 1 ml of 1 M potassium phosphate (pH 9) was pumped onto the column and eluted with dilute ammonium hydroxide (pH 10) at a flow rate of 1 ml/min. The conductivity of collected fractions was measured with Lab-line Bionometer, to follow salt elution. The absorbance of collected fractions was measured on a Spectronic 710 at a nonmaximal wavelength (387.5 nm) to follow nucleotide elution. Figure 1 shows the elution profile so obtained. Desalting is effected by frontal elution with a low ionic strength eluant, producing some overlap between salt and nucleotide. Barely detectable conductivities in the nucleotide peak (<0.005) approximated the molarity of the NADH. Fractions of nucleotide with conductivities less than 0.005 constituted 89% of the total absorbance measured.

[27] Preparation of 3-Aminopyridine 1,N^6-Ethenoadenine Dinucleotide and 3-Aminopyridine 1,N^6-Ethenoadenine Dinucleotide Phosphate

By BRUCE M. ANDERSON and CONSTANCE D. ANDERSON

Structural analogs of NAD and NADP have been of major importance in studies of reactions catalyzed by dehydrogenases, ADP-ribosyltransferases, and other pyridine nucleotide-dependent enzymes. Most of these coenzyme analogs have been prepared enzymatically through the transglycosidation reaction catalyzed by NAD^+ glycohydrolases (NADases).[1] In a limited number of cases, the direct chemical conversion of NAD to a desired analog has been successful. The conversion of NAD to 3-aminopyridine adenine dinucleotide (AAD) through the hypobromite reaction (Hofmann reaction) exemplifies this type of synthesis.[2] The AAD formed in this fashion can subsequently be converted to a site-labeling reagent through diazotization.[3] Chemical alteration of NAD and NADP has been somewhat limited by most chemical reactions. However, having alternate chemical methods available is an advantage since the enzymatic formation of site-labeling derivatives can be complicated by having de-

[1] L. J. Zatman, N. O. Kaplan, and S. P. Colowick, *J. Biol. Chem.* **200**, 197 (1953).
[2] T. L. Fisher, S. V. Vercellotti, and B. M. Anderson, *J. Biol. Chem.* **248**, 4293 (1973).
[3] J. K. Chan and B. M. Anderson, *J. Biol. Chem.* **250**, 67 (1975).

sired products effectively inactivate enzymes used for synthesis. Although the hypobromite reaction was successfully used for the conversion of NAD to AAD, this reaction was not applicable for the conversion of NADP to 3-aminopyridine adenine dinucleotide phosphate (AADP) since the 2'-phosphoryl group of NADP was rapidly hydrolyzed under conditions of the hypobromite reaction.[4] It was therefore necessary to prepare the 3-aminopyridine analog of NADP enzymatically through the transglycosidation reaction catalyzed by NADases.[4]

Since the diazotized 3-aminopyridine analogs of NAD and NADP were demonstrated to be effective site-labeling reagents for selected dehydrogenases,[3,5,6] it was felt that the incorporation of a fluorescing adenine derivative into these analog structures would greatly increase the sensitivity of analyzing enzymes modified through site-labeling processes. This prompted the synthesis of dinucleotides containing both the diazotizable 3-aminopyridine and the fluorescing 1,N^6-ethenoadenine moieties. Both NAD and NADP analogs containing these structural alterations have been successfully prepared.[7,8]

Preparation of 3-Aminopyridine 1,N^6-Ethenoadenine Dinucleotide

3-Aminopyridine 1,N^6-ethenoadenine dinucleotide (ε-AAD) was prepared from nicotinamide 1,N^6-ethenoadenine dinucleotide (ε-NAD) and 3-aminopyridine through the transglycosidation reaction catalyzed by *Bungarus fasciatus* snake venom NADase. ε-NAD was prepared by the method of Barrio *et al.*[9] with some modifications. NAD (100 mg) was dissolved in 100 ml of 0.1 M sodium citrate buffer (pH 4.5) containing 1.7 ml chloroacetaldehyde, and the resulting mixture was incubated at 40° for 27 hr. The reaction was essentially complete in 24 hr as determined by diluted samples assayed by high-performance liquid chromatography (HPLC). HPLC assays were performed on a Spectra-Physics SP8000 HPLC with an Alltech RSIL AN (4.6 × 250-mm) 5-μm anion-exchange column. The mobile phase used was 150 mM potassium phosphate buffer (pH 3.3) with a flow rate of 1 ml/min.

[4] B. M. Anderson, J. H. Yuan, and S. V. Vercellotti, *Mol. Cell. Biochem.* **8,** 89 (1975).

[5] N. K. Amy, R. H. Garrett, and B. M. Anderson, *Biochim. Biophys. Acta* **480,** 83 (1977).

[6] B. M. Anderson, S. T. Kohler, and C. D. Anderson, *Arch. Biochem. Biophys.* **188,** 214 (1978).

[7] D. A. Yost, M. L. Tanchoco, and B. M. Anderson, *Arch. Biochem. Biophys.* **217,** 155 (1982).

[8] B. M. Anderson and C. D. Anderson, *Anal. Biochem.* **134,** 50 (1983).

[9] J. R. Barrio, J. A. Secrist, and N. J. Leonard, *Proc. Natl. Acad. Sci. U.S.A.* **69,** 2039 (1972).

The ε-NAD reaction mixture was applied to a previously washed Dowex 1-X8, 200–400 mesh (formate) column (2 × 30 cm). The column was then washed with distilled water until the UV absorbance of the eluate was less than 0.02. Elution of ε-NAD was achieved by a linear ammonium formate salt gradient, 0–200 mM (2 × 1 liter). ε-NAD was eluted at an ammonium formate concentration of approximately 100 mM. Fractions having an absorbance ratio at 265/275 nm of 1.2 were pooled and lyophilized to dryness. The resulting material was dissolved in 15 ml of distilled water and desalted on a Sephadex G-10 column (3.5 × 110 cm). Desalted fractions containing ε-NAD were concentrated to dryness by lyophilization and represented an overall yield of 70%.

ε-NAD (200 mg), purified as above, was dissolved in 50 mM potassium phosphate buffer, pH 7.5 (20 ml), and to this solution were added 560 mg of 3-aminopyridine (recrystallized from diethyl ether). The transglycosidation reaction was initiated by the addition of 20 units of purified *B. fasciatus* NADase,[10] and incubated at room temperature for 18 hr. The formation of ε-AAD was monitored by HPLC as described above.

The ε-AAD reaction mixture was applied to a previously washed Dowex 1-X8, 200–400 mesh (formate) column (2 × 20 cm). The column was washed free of 260-nm absorbing material with distilled water. The desired product, ε-AAD, was eluted with a linear ammonium formate salt gradient, 0–200 mM (2 × 1 liter), and fractions having a 260/275 nm absorbance ratio averaging 1.76 were pooled and lyophilized to dryness. The resulting ε-AAD was dissolved in 30 ml of distilled water and desalted in 2 × 15-ml portions on a Sephadex G-10 column (3.5 × 110 cm). Desalted ε-AAD fractions were concentrated to dryness by lyophilization and represented an overall yield of 87%.

Preparation of 3-Aminopyridine 1,N^6-Ethenoadenine Dinucleotide Phosphate

The route of synthesis of 3-aminopyridine 1,N^6-ethenoadenine dinucleotide phosphate (ε-AADP) differs from that used for the synthesis of ε-AAD. Attempts to convert nicotinamide 1,N^6-ethenoadenine dinucleotide phosphate to ε-AADP through the transglycosidation reaction catalyzed by snake venom NADase resulted in lower than expected yields presumably due to specificity properties of the snake venom enzyme. A more effective preparation of ε-AADP was accomplished by converting AADP to ε-AADP through incubation with chloroacetaldehyde. AADP was first prepared enzymatically from NADP and 3-aminopyridine using

[10] D. A. Yost and B. M. Anderson, *J. Biol. Chem.* **256,** 3647 (1981).

the transglycosidation reaction described previously[11] with pig brain NAD-ase at pH 8.1, or substituting purified snake venom NADase at pH 7.5.

The subsequent reaction of AADP with chloroacetaldehyde was carried out using the method of Barrio et al.[9] with the following modifications. AADP (100 mg) was dissolved in 100 ml of 0.1 M sodium citrate buffer (pH 4.5) containing 1.7 ml chloroacetaldehyde. The resulting mixture was incubated at 40° for 24 hr. At timed intervals, diluted samples of the reaction mixture were assayed by HPLC, as previously described, with the mobile phase being 100 mM potassium phosphate (pH 4.0) and a flow rate of 1 ml/min. The three major components of the reaction mixture were readily resolved during the course of the reaction (Fig. 1).

At the completion of the reaction, the reaction mixture was applied to a Dowex 1-X8, 200–400 mesh (formate) column (2 × 30 cm) which had been washed for 20 hr with distilled water. The desired product, ε-AADP, was eluted from the column with a linear ammonium formate salt gradient, 0–2 M (2 × 1 liter), and fractions having a 260/275 nm absorbance ratio averaging 1.73 were pooled and lyophilized to dryness. The resulting ε-AADP was dissolved in 15 ml of distilled water and desalted on a Sephadex G-10 column (4.5 × 90 cm). Desalted ε-AADP fractions were concentrated to dryness by lyophilization. The purified ε-AADP showed only one absorbing peak at 260 nm in HPLC analysis and on repeated experiments, the average yield of ε-AADP was 50%.

Properties of ε-AAD and ε-AADP

The spectral properties of ε-AAD and ε-AADP are shown in the table. Each analog exhibited three major UV absorbance maxima, and millimolar extinction coefficients are listed for each of these wavelengths. Excitation of the coenzyme analogs at their respective UV absorbance maxima resulted in fluorescence emission at 423 nm for ε-AAD and 419 nm for ε-AADP. In each case, maximum fluorescence emission was obtained by excitation at the absorbance maximum in the 300-nm region. Both analogs exhibited intramolecular quenching of fluorescence at neutral pH and an enhancement of fluorescence upon protonation of the ethenoadenine moiety at lower pHs.

The aminopyridine moieties of the two analogs were readily diazotized under previously described conditions,[3] and the resulting diazonium derivatives were observed to azo-couple with N-(1-naphthyl)ethylenediamine dihydrochloride at pH 2.0 to form azo dyes absorbing maximally at 512 nm. In each case, azo dye formation decreased with increasing pH

---

[11] B. M. Anderson and T. L. Fisher, this series, Vol. 66, p. 81.

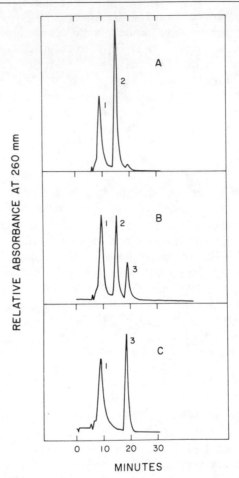

FIG. 1. HPLC analysis of the conversion of AADP to ε-AADP. Separations were performed on an Alltech RSIL anion-exchange column eluted with 0.1 M potassium phosphate buffer (pH 4.0) at a flow rate of 1 ml/min. Peak 1, chloroacetaldehyde; peak 2, AADP; peak 3, ε-AADP. (A) The reaction at 40 min; (B) 3 hr; and (C) 20 hr.

with essentially no azo dye formation above pH 3.5. The lack of azo-coupling at neutral pH is attributed to the conversion of the diazonium derivatives to their respective diazotates. Although unreactive in azo-coupling reactions, the diazotates react readily at neutral pH with sulfhy-dryl compounds. The second-order rate constants at pH 7.0 for the reactions of diazotized ε-AAD were 11 M^{-1} and 4.6 M^{-1} for cysteine and glutathione, respectively. In reactions of diazotized ε-AADP at pH 6.4, a

SPECTRAL PROPERTIES OF THE COENZYME ANALOGS

Compound	Absorbance maxima (nm)	Millimolar extinction coefficient	Fluorescence emission maxima (nm)
ε-AAD	226	36.2	423
	256	11.6	423
	315	4.6	423
ε-AADP	256	11.5	419
	275	6.4	419
	310	4.3	419

second-order rate constant of $3.5\ M^{-1}$ was obtained for both cysteine and glutathione. This reactivity of the diazotized analogs toward sulfhydryl groups supports their use as site-labeling reagents for pyridine nucleotide-dependent enzymes having essential sulfhydryl groups. This application has been documented by the site-labeling of yeast alcohol dehydrogenase by diazotized ε-AAD at neutral pH.[7] The fluorescence of the ethenoadenine moiety attached through this site-labeling sulfhydryl modification reaction provided greater sensitivity in analyzing the inactivated enzyme.

[28] Simultaneous Photometric Determination of Oxygen Consumption and NAD(P)H Formation or Disappearance Using a Continuous Flow Cuvette

By MANASE M. DÂNŞOREANU *and* MARIUS I. TELIA

The simultaneous determination of oxygen consumption and NAD(P)H formation or disappearance is important for the investigation of oxidative phosphorylation with intact mitochondria and mitochondrial or bacterial membranes, as well as for the study of the microsomal hydroxylation reactions involving drugs or metabolites. Taking advantage of the peculiarities of the absorption spectra of HbO₂/Hb (HbO₂, oxyhemoglobin; Hb, deoxygenated hemoglobin) and NAD(P)H/NAD(P), one can measure alternatively, in the same sample, both the formation or disappearance of NAD(P)H, at 334 nm (in the vicinity of an isosbestic point of the HbO_2/Hb spectra), and the oxygen consumption, at 436 nm.[1] As only small amounts of biological material are required, there are no disturbing effects due to turbidity.

The semiautomatic measurement of oxygen consumption and NAD(P)H formation or disappearance requires solutions to the following problems: (1) the provision of a discontinuous flux of reaction mixture from a glass reservoir; (2) the achievement of the appropriate oxygen partial pressure in the reservoir and cuvette; (3) the continuous stirring of the reaction medium in the cuvette in order to maintain the optical homogeneity; (4) the maintenance of constant temperature.

This chapter describes a cuvette and other devices, simple, precise, and easy to handle, which can be adapted to a commercially available instrument, such as the Eppendorf Spectralline photometer.

Semiautomatic Flow System for the Measurement of Oxygen Consumption and NAD(P)H Formation or Disappearance

The major components of the system are a thermostatted glass reservoir, a roller pump, and a thermostatted, stoppered glass cuvette with a continuous mixing device.[1-3] It is very important that all the tubes connecting the cuvette to the reservoir are as little permeable to oxygen as possible. Glass tubes of about 1.0-nm diameter joined with thick-walled natural rubber are suitable.

Glass Reservoir. This reservoir ensures both the primary thermoregulation (by water circulation) and the suitable degree of deoxygenation (by nitrogen gas bubbling) of the reaction mixture. It is stopped with a rubber plug pierced by two glass tubes which penetrate to the bottom of the reservoir. One is connected to the nitrogen gas source and the other to the inlet of the roller pump. In addition, the plug is provided with two orifices: one is the gas outlet and the other may be used for returning the reagents to the reservoir. The closed-circuit system is used to prevent wasting the reaction mixture in the period of preparation of the measurements. A magnetic stirrer ensures the mixing of the reactants during the period in which nitrogen is not bubbled into the reaction mixture.

Roller Pump. This pump accomplishes the transport of the reaction mixture from the reservoir into the measuring cuvette at a minimum rate

[1] M. Dânşoreanu, M. Telia, C. T. Ţărmure, M. Oargă, M. Markert, A. Ivanof, and O. Bârzu, *Anal. Biochem.* **111,** 321 (1981).

[2] L. Mureşan, M. Dânşoreanu, A. Ana, A. Bara, and O. Bârzu, *Anal. Biochem.* **104,** 44 (1980).

[3] M. Dânşoreanu, M. Markert, I. Lascu, C. T. Ţărmure, J. Frei, and O. Bârzu, *Int. J. Biochem.* **15,** 1191 (1983).

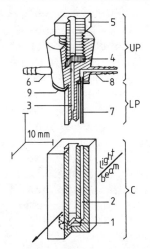

FIG. 1. The stoppered glass cuvette for the continuous flow of reagents. Two vertical sections show the details of the device. For the significance of the symbols see the text.

of 2 ml/min. It must be provided with natural rubber tubing because of the high permeability coefficient for oxygen of the silicone rubber.

Stoppered Glass Cuvette. As shown in Fig. 1 the square-shaped cuvette (C) (12 × 12 mm) is made of glass and has a height of 30 mm. The interior plane section is also square shaped (4 × 4 mm) but the lower part, housing the magnetic stirring bar (1), is cylinder shaped, with a diameter of 5.5 mm and a height of 3 mm. For the purpose of introducing the solution into the cuvette, a vertical channel (2) 1 mm in diameter is built into one of the side walls, which opens into the cylindrical portion of the cuvette. The stopper is made of Lucite. Its bottom part (LP), 4 × 4 mm in cross section, protrudes 14 mm into the cuvette and limits the reaction volume to about 0.2 ml. To help the evacuation of gas bubbles, the lower surface of the stopper is concave. This part of the stopper is pierced by an axial channel (3) which continues up to the top (UP). To prevent the mixing of the contents of the cuvette with the liquid from the rest of the system, the lower part of the channel is narrowed to 0.3 mm diameter over a portion 3 mm in height. The upper opening of the axial channel is covered with a silicon rubber disk (4) held in position by the top piece (5), which is screwed into the stopper body with an 8-mm-diameter 0.75-mm thread. The needle of a 10-μl Hamilton microsyringe may reach the inside of the cuvette by passing through the channel crossing the piece (5), piercing the disk (4) and through the axial channel of the stopper (3). The evacuation of the cuvette is made through the side branch (6), which opens into the central channel of the stopper. For the introduction of the liquid into the cuvette, there is a plastic tube (7), of 1 mm outside and 0.3 mm inside diameter, entering the channel (2). The tube is connected to the second side branch (8) of the stopper. The tightness of the assembly is ensured by the rubber seal (9), 0.3 mm thick, pressed upon the insertion of the stopper. A small amount of silicon grease may also be employed.

Other types of cuvettes, reported previously, are characterized by slightly larger working volumes (0.4–0.5 ml).[2,3]

The cuvette holder (not shown in Fig. 1) is made of metal and ensures the thermoregulation of the cuvette by water circulation. The glass cuvette and the thermostatted cell holder with the magnetic stirrer underneath (15–1800 rpm)[4] fit exactly into the Eppendorf Spectralline photometer (Model 1101 M, 6114 S, or 6118) as has been previously described.[5]

Experimental Procedure

The preparation of human red cell hemolysate and the purification and storage of hemoglobin have been described previously in this series.[6] However, good results can often be obtained with nonpurified hemolysates.[5] For special purposes, highly purified hemoglobin may be obtained following the method described by Porumb et al.[7]

A thermostatted glass reservoir of 50 ml is filled with reaction mixture containing HbO_2 at 0.075 mM final concentration (details are given separately). When the desired temperature has been achieved, the pump is turned on at a rate of 2 ml/min and the liquid is transported from the reservoir through the measuring cuvette using a closed-circuit system until no more bubbles appear. With the magnetic stirrer of the cuvette in motion, the recorder is switched on and it should draw, at 436 nm, a smooth line. Any rapid, low-amplitude oscillations are indicative of the formation of the air bubbles. They can be eliminated by turning the pump on for a few minutes more. The level marked by the recorder represents the baseline and should be set to the value of 0.0–0.1 absorbance units (E_0 in Fig. 2). If the reaction mixture contains 0.2 mM NAD(P)H, with the 334-nm filter instead of the 436-nm filter, the level of the absorbance will be greater by 1.275 units (E_0' in Fig. 2). After the pump is turned off, nitrogen is bubbled into the reservoir (arrow b in Fig. 2) to achieve a 30–

[4] All components are now available from Eppendorf Gerätebau, Hamburg, Federal Republic of Germany.

[5] O. Bârzu, M. Dânşoreanu, R. Munteanu, A. Ana, and A. Bara, Anal. Biochem. 101, 138 (1980).

[6] O. Bârzu, this series, Vol. 54, p. 485.

[7] H. Porumb, I. Lascu, D. Matinca, M. Oargă, V. Borza, M. Telia, O. Popescu, G. Jebeleanu, and O. Bârzu, FEBS Lett. 139, 41 (1982).

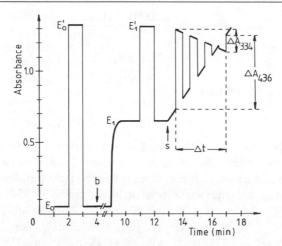

FIG. 2. Simultaneous photometric determination of oxygen consumption and NADH oxidation. For the significance of the symbols, see Experimental Procedure.

40% deoxygenation of HbO_2. This requires about 5 min, and its accomplishment is indicated by an obvious change of color toward violet. This is accompanied by an increase of the absorbance by about 0.6–0.8 unit at 436 nm (E_1 in Fig. 2), and by a decrease of about 0.01–0.015 unit at 334 nm (E_1' in Fig. 2), measurable after running the pump for 2–3 min.

The pump is once again turned off, the outlet of the cuvette is connected to a waste vessel, and a sample of microsomes, mitochondria, or bacterial membranes is injected into the cuvette. The reaction is triggered by injecting either the biological material or substrate (arrows in Fig. 2). The increase of absorbance at 436 due to oxygen consumption is recorded for about 20–30 sec, after which the 334 nm filter is placed in position and the recording of the absorbance is continued for another 20–30 sec. This procedure is repeated two to four times (Fig. 2).

The assay of a new sample can be performed after the removal of the content of the cuvette by allowing the pump to work for 45–60 sec at a rate of 2 ml/min.

Calculation of the Oxygen Consumption and NAD(P) Formation
   or Disappearance

The correspondence between the experimentally measured variations in absorbance and the rates of oxygen consumption (Q_O) or NAD(P)H formation/disappearance ($Q_{NAD(P)H}$) for each sample is described by the following algebraic equations:

$$Q_O \text{ (ng} \cdot \text{atoms/min)} = - \frac{2000V \, \Delta A_{436} f}{\Delta \varepsilon_1 d \, \Delta t}$$

$$= -23.1 \, \Delta A_{436}/\Delta t \qquad (1)$$

$$Q_{\text{NAD(P)H}} \text{ (nmol/min)} = \frac{1000V}{\varepsilon_{\text{NAD(P)H } 334} d \, \Delta t} \left(\Delta A_{334} + \frac{\Delta \varepsilon_2}{\Delta \varepsilon_1} \, \Delta A_{436} \right)$$

$$= 84.6(\Delta A_{334} - 0.02 \, \Delta A_{436})/\Delta t \qquad (2)$$

where V is the reaction volume (0.203 ml), which may be estimated indirectly by determining the dilution of a colored solution of both known concentration and extinction coefficient; d is the light pathlength (0.4 cm); ΔA_{334} and ΔA_{436} are the variations of absorbance at 334 and 436 nm, corresponding to an interval of time Δt (Fig. 2). $\Delta \varepsilon_1 = \varepsilon_{\text{Hb } 436} - \varepsilon_{\text{HbO}_2 \, 436} = 65.0 \text{ m}M^{-1} \text{ cm}^{-1}$; $\Delta \varepsilon_2 = \varepsilon_{\text{Hb } 334} - \varepsilon_{\text{HbO}_2 \, 334} = -1.3 \text{ m}M^{-1} \text{ cm}^{-1}$; $\varepsilon_{\text{NAD(P)H } 334} = 6.0 \text{ m}M^{-1} \text{ cm}^{-1}$. f is a correction factor (1.48 in this case) which depends on the HbO_2 concentration and the affinity for oxygen of both hemoglobin and the oxygen-consuming system[8,9] and its value can be evaluated experimentally.[6,9] Equations (1) and (2) will yield negative or positive results depending on whether the reactions lead to consumption or formation of the components assayed.

Determination of the $Q_{\text{NAD(P)H}}/Q_O$ Ratio of Several Oxygen-Consuming Systems

Rat Liver Microsomes. Rat liver microsomes are capable of oxidizing NADPH both in the absence and in the presence of xenobiotics (RH) which become hydroxylated (ROH).[10,11] In the first case, of "free" oxidation of NADPH, the Q_{NADPH}/Q_O ratio is 1; in the second case, of "coupled" oxidation of NADPH, this ratio is 0.5. As the two processes run in parallel it is obvious that the experimentally obtained ratio will be between 0.5 and 1. If the lipid peroxidation is inhibited by adding EDTA to the reaction mixture[12] and one denotes by $Q_{\text{NADPH(f)}}$ and $Q_{\text{NADPH(c)}}$ the free and coupled oxidation rates of NADPH (the second being equal to the rate of hydroxylation of RH), one may write[1] $Q_{\text{NADPH(f)}} = 2Q_{\text{NADPH}} - Q_O$ and $Q_{\text{NADPH(c)}} = Q_O - Q_{\text{NADPH}}$.

The reaction mixture contained 50 mM phosphate buffer (pH 7.4), 0.5 mM EDTA, and 0.075 mM (on heme basis) HbO_2. After partial deoxygen-

[8] W. W. Kielley, this series, Vol. 6, p. 272.
[9] O. Bârzu, L. Mureşan, and G. Benga, *Anal. Biochem.* **46,** 374 (1972).
[10] R. W. Estabrook, this series, Vol. 52, p. 43.
[11] A. Y. H. Lu and S. B. West, *Pharmacol. Ther., Part A,* **2,** 337 (1978).
[12] A. H. Archakov, G. F. Zhirnov, and I. I. Karuzina, *Biochem. Pharmacol.* **26,** 389 (1977).

OXYGEN CONSUMPTION AND NADPH OXIDATION BY
RAT LIVER MICROSOMES[a]

Additions	Q_{NADPH} or Q_O (nmol/min or ng · atom/min · mg protein)	Control rats	Induced with phenobarbital[b]
No addition	Q_{NADPH}[c]	−21.8	−28.1
	Q_O	−20.8	−27.4
	Q_{NADPH}/Q_O	1.05	1.03
Hexobarbital	Q_{NADPH}	−25.8	−43.9
(1 mM)	Q_O	−31.4	−58.4
	Q_{NADPH}/Q_O	0.82	0.75
	$Q_{NADP(f)}$	−20.2	−29.4[d]
	$Q_{NADPH(c)}$	− 5.6	−14.5

[a] From Dânşoreanu et al.[1]
[b] The rats were treated with a single dose of phenobarbital (75 mg/kg) 48 hr before sacrifice.
[c] The significance of the notations is given in the text.
[d] Note that free NADPH oxidation is not affected by the presence of hexobarbital.

ation of HbO_2, the mixture was pumped into the cuvette and 1–4 μl microsomes (20–40 mg of protein/ml) was injected. After optical density equilibration, the reaction was started with 20–40 nmol NADPH (1–2 μl of a 20 mM solution).

The results of the measurements, preformed at 30°, are shown in the table.

Mitochondria and Submitochondrial Particles. ATP synthesis can be measured indirectly via NADPH formation with an excess of glucose, $NADP^+$, hexokinase and glucose-6-phosphate dehydrogenase. In this case the Q_{NADPH}/Q_O ratio will be negative because the oxygen consumption is accompanied by $NADP^+$ reduction to NADPH.

The reaction mixture contained 180 mM sucrose, 50 mM KCl, 15 mM phosphate buffer (pH 7.4), 0.5 mM EDTA, 5 mM $MgCl_2$, 10 mM glucose, 0.075 mM HbO_2, 0.25 mM $NADPH^+$, 40 mg bovine serum albumin, and 50 units of both hexokinase and glucose-6-phosphate dehydrogenase (for 25 ml of solution). After suitable deoxygenation of HbO_2, 1 μl of succinate or glutamate (0.5 μmol), 1–2 μl of mitochondrial suspension (20–60 μg of protein), 1 μl (10 nmol) of Ap^5A [P^1, P^5-di(adenosine 5')-pentaphosphate], and 1 μl of ADP (40 nmol) were successively injected into the cuvette. In such an experiment performed at 30° with rat liver mitochondria, with succinate as substrate, after addition of ADP the Q_O is −115 ± 8 ng · atom/min · mg protein and the Q_{NADPH}/Q_O ratio is −1.8 ± 0.07. With glutamate these values are -48 ± 6 ng atom/min $\cdot$ mg protein and -2.6 ± 0.11, respectively.

Bacterial Membranes. The bacterial membranes have high specific NADH-oxidase activity, the Q_{NADH}/Q_O ratio being equal to unity. This ratio was used for the control of the calibration in the measurement of oxygen concentration with the Clark electrode,[13] but it can be used in the case of any other method having the same purpose.

The reaction mixture contained 50 mM phosphate buffer (pH 7.4), 0.2 mM NADH, and 0.075 mM HbO$_2$. Samples of 1–4 μl containing bacterial membranes (2.1 protein/ml) were injected into the cuvette. For 10 different measurements using various amounts of bacterial preparation the average value of Q_{NADH}/Q_O was -0.98 ± 0.02.

Comments

The method described above combines the advantages of the spectrophotometric method for the determination of oxygen consumption based on the use of oxyhemoglobin as oxygen donor and indicator, namely a sensitivity 10–100 times higher than the oxygraphic or manometric methods[9,14] and fast response,[6] with those of a flow system in connection with a cuvette of small volume. Thus, using quantities of protein of 0.1 mg at the most, accurate estimations of the NAD(P)H/O ratio can be obtained within a time interval of 3–4 min/sample. The errors of the method in the estimation of this ratio are less than 5%, being usually a consequence of the lack in accuracy of the different additions to the cuvette content. These additions must be of such concentration that their total volume does not exceed 8 μl (4% maximal dilution of the sample).[5] Using this method one must bear in mind that hemoglobin in solution is both the donor and indicator of the oxygen content. Thus, all agents affecting the optical and functional properties of HbO$_2$ can lead to erroneous results. In this respect, the correction factor f in Eq. (1) depends on temperature, pH, and the affinity for oxygen of the oxidizing system. It is necessary to check it, as previously described by Bârzu,[6] for every individual case. The H$_2$O$_2$-generating systems cannot be assayed without special precautions because H$_2$O$_2$ may oxidize the hemoglobin to methemoglobin.[2,5,6] On the other hand, those agents affecting the free SH groups or interacting with the heme are to be avoided. Otherwise, one should use an alternative method,[3] in which the oxygen concentration is measured with a Clark electrode.

[13] R. W. Estrabrook, this series, Vol. 10, p. 41.
[14] O. Bârzu and V. Borza, *Anal. Biochem.* **21,** 344 (1967).

Acknowledgments

We wish to thank Eppendorf Gerätebau (Hamburg, Federal Republic of Germany), for making available facilities for development of the methodology described in this work.

[29] Spectrophotometric Assay of NADase-Catalyzed Reactions

By BRUCE M. ANDERSON and DAVID A. YOST

NAD glycohydrolases (NADases) catalyze the hydrolysis of the nicotinamide–ribosidic bond in the pyridine nucleotide coenzymes, NAD and NADP. A number of assay methods have been used to monitor these reactions and to study kinetic parameters of these enzymes. The spectrophotometric measurement at 327 nm of the cyano adducts of NAD or NADP formed in the cyanide addition reaction[1] is frequently used to study less purified preparations of NADases. Disappearance of the pyridine nucleotides can also be monitored enzymatically using the appropriate assay mixtures for yeast alcohol or yeast glucose-6-phosphate dehydrogenases. With more highly purified preparations of NADases, continuous titrimetric methods can be employed as exemplified by the pH-stat assay used to study snake venom NADase.[2] For greater sensitivity, a fluorimetric assay using nicotinamide $1,N^6$-ethenoadenine dinucleotide as substrate has been employed.[2,3-5] The release through hydrolysis of the intramolecular interactions responsible for the quenching of fluorescence of this dinucleotide serves as the basis for the fluorimetric assay. Although a variety of methods is available for studies of NADases, difficulties can arise in the application of any of these methods due to factors such as the presence of interfering substances, the use of toxic chemicals, and the unavailability of proper instrumentation. In addition to the well-established oxidation–reduction roles for NAD and NADP, the more recent studies of the functioning of NAD as a substrate for ADP-ribosylation reactions support further the need for alternate methods of assay.

[1] S. P. Colowick, N. O. Kaplan, and M. M. Ciotti, J. Biol. Chem. 191, 447 (1951).
[2] D. A. Yost and B. M. Anderson, J. Biol. Chem. 256, 3647 (1981).
[3] J. R. Barrio, J. A. Secrist, and N. J. Leonard, Proc. Natl. Acad. Sci. U.S.A. 69, 2039 (1972).
[4] C.-Y. Lee and J. Everse, Arch. Biochem. Biophys. 157, 83 (1973).
[5] P. H. Pekala and B. M. Anderson, J. Biol. Chem. 253, 7453 (1978).

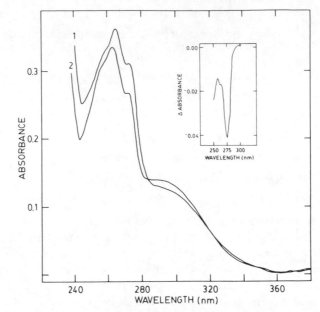

FIG. 1. Difference spectrum due to ε-NAD hydrolysis. Spectra were as follows: spectrum 1, before the addition of NADase, 35 μM ε-NAD, and 50 mM potassium phosphate buffer (pH 7.5) in 3 ml of mixture in a 1-cm cuvette; spectrum 2, 10 min after the addition of 1 unit of purified snake venom NADase. Inset is the difference spectrum between spectrum 1 and 2.

The direct spectrophotometric assay[6] described herein provides an additional method for monitoring these reactions.

Assay Method

A spectrophotometric method for assaying NADase-catalyzed reactions was developed using nicotinamide 1,N^6-ethenoadenine dinucleotide (ε-NAD) as substrate. The minor structural alteration in the purine moiety of this coenzyme analog does not greatly affect the functioning of this dinucleotide as a substrate for NADases and as a coenzyme for many dehydrogenases. Spectral changes accompanying the hydrolysis of the nicotinamide–ribose linkage in this dinucleotide serve as the basis of the spectrophotometric assay. The hydrolysis of ε-NAD to yield nicotinamide and ε-ADP-ribose was catalyzed by snake venom NADase previously purified to electrophoretic homogeneity according to Yost and Anderson.[2] Spectral changes observed upon complete hydrolysis of the dinucleotide at pH 7.5 are shown in Fig. 1. The UV absorbance of the

[6] D. A. Yost and B. M. Anderson, *Anal. Biochem.* **116,** 374 (1981).

dinucleotide decreases during hydrolysis with a maximum decrease occurring at 275 nm. The hydrolysis of ε-NAD was also monitored on a Spectra-Physics SP8000 HPLC with separations performed on an Alltech RSIL AN (4.6 × 250-mm) 5-μm column using 150 mM potassium phosphate buffer (pH 3.5) as the mobile phase. The maximum decrease measured spectrophotometrically at 275 nm at eight concentrations of ε-NAD was correlated with complete hydrolysis documented by HPLC assay. A linear relationship between the absorbance change at 275 nm and the concentration of ε-NAD hydrolyzed was observed over a concentration range of 5–50 μM. From these data, a millimolar extinction coefficient for ε-NAD hydrolysis of 0.89 was calculated. The change in absorbance at 275 nm served as a direct spectrophotometric assay for determining initial velocities of ε-NAD hydrolysis as catalyzed by purified snake venom NADase. Reactions were carried out in 50 mM potassium phosphate buffer, pH 7.5 at 25°, in 1-cm cuvettes containing 3 ml of reaction mixture. The micromoles of ε-NAD hydrolyzed per given time period were calculated using the established extinction coefficient.

Materials and Methods

The substrate, ε-NAD, was prepared from NAD by the method of Barrio *et al.*[3] with some modifications. NAD (100 mg) was dissolved in 100 ml of 0.1 M sodium citrate buffer (pH 4.5) containing 1.7 ml chloroacetaldehyde, and the resulting mixture was incubated at 40° for 27 hr. This ε-NAD reaction mixture was applied to a previously washed Dowex 1-X8, 200–400 mesh (formate) column (3 × 40 cm). The column was washed with distilled water until the UV absorbance of the eluate was less than 0.02. Elution of ε-NAD was achieved by a linear ammonium formate salt gradient, 0–200 mM (2 × 1 liter). ε-NAD eluted from the Dowex column when the ammonium formate concentration was approximately 100 mM. Fractions having a 265/275 nm absorbance ratio of 1.2 were pooled and lyophilized to dryness. The resulting material was dissolved in 15 ml of distilled water and desalted on a Sephadex G-10 column (3.5 × 110 cm). Desalted ε-NAD-containing fractions were again concentrated by lyophilization. This procedure routinely yielded greater than 70% recovery of an off-white powder characterized as ε-NAD. The purity of ε-NAD was established through HPLC and by demonstration of the accepted 265/275 nm absorbance ratio.

Bungarus fasciatus snake venom NADase was purified according to the method of Yost and Anderson.[2] Bovine seminal fluid NADase was purified according to Anderson and Yuan.[7] Spectrophotometric measurements were made on a Beckman Acta M VI recording spectrophotometer.

[7] B. M. Anderson and J. H. Yuan, this series, Vol. 66, p. 144.

COMPARISON OF SNAKE VENOM NADase
ASSAY PROCEDURES

Assay method	K_m (μM)	Sensitivity[a]
KCN addition	25	>0.08
Titrimetric	20	0.02 to 0.2
Fluorometric	18	0.0005 to 0.02
Spectrophotometric	21	0.007 to 0.08

[a] Expressed as enzymatic hydrolysis units, where 1 unit equals 1 μmol ε-NAD cleaved/min.

Fluorimetric measurements were obtained using a Perkin-Elmer 650-40 spectrophotofluorometer. Titrimetric measurements were made using a Radiometer type TTT80 titrator, PHM82 pH meter, type ABU80 autoburette assembly equipped with a 0.25-ml burette and a GK 2320C combination electrode. Titrimetric, fluorimetric, and cyanide addition assays were performed as described previously.[2]

Applications

Initial velocities of snake venom NADase-catalyzed hydrolysis of ε-NAD were determined at five concentrations of ε-NAD varying from 10 to 50 μM. From double-reciprocal plots of these data a K_m of 21 μM for ε-NAD and a V_{max} of 650 μmol/min · mg protein were determined in these reactions. The K_m value for ε-NAD is in excellent agreement with the 20 μM value obtained previously[2] using a titrimetric assay. At saturating substrate concentration, initial velocities of ε-NAD hydrolysis were proportional to enzyme concentration in the range of 5–60 ng of purified enzyme. Similar results were obtained when bovine seminal fluid NADase was used in place of the snake venom enzyme. Again kinetic parameters agreed well with values previously established using other assay methods.

The snake venom NADase-catalyzed hydrolysis of ε-NAD was studied using four different assay methods. Due to differences in the sensitivities of the methods employed, different substrate concentration ranges were necessary. The K_m value for ε-NAD obtained in the spectrophotometric method agreed well with the values observed using the other three methods (see table).

Comments

Of the four different methods of assay of ε-NAD hydrolysis shown in the table, the cyanide addition assay requires an additional sampling step and is the least sensitive but has been successfully applied in studies of crude preparations of enzyme. In comparison, the continuous spectrophotometric assay is 10-fold more sensitive than the cyanide addition assay and overlaps the sensitivity of the titrimetric assay. The presence of side reactions involving the production, utilization, or buffering of hydrogen ion can limit the application of the titrimetric assay method. The fluorimetric assay is the most sensitive method and of obvious advantage in reactions catalyzed by low concentrations of enzyme. However, interference from fluorescing contaminants or reactants can create problems with this assay. For example, inactivation of snake venom NADase by pentanedione could not be monitored by this assay method due to the interfering fluorescence of this reagent.[8] The spectrophotometric assay also has limitations in that crude enzyme preparations and other UV-absorbing reaction components can interfere with spectral measurements at 275 nm. However, the simplicity of the spectrophotometric method was a distinct advantage in monitoring enzyme activity in fractions obtained in enzyme purification procedures. The availability of the spectrophotometric method provides an alternate means of studying NADase-catalyzed reactions and an additional choice of assay for varying reaction conditions. The spectrophotometric method was also used to study the NADase purified from bovine seminal fluid, and due to the general acceptance of ε-NAD as a NADase substrate, should serve as an assay method for NADases from many different sources.

[8] D. A. Yost and B. M. Anderson, *J. Biol. Chem.* **257,** 767 (1982).

[30] Snake Venom NAD Glycohydrolase: Purification, Immobilization, and Transglycosidation

By BRUCE M. ANDERSON, DAVID A. YOST, and CONSTANCE D. ANDERSON

$$NAD + H_2O \rightarrow nicotinamide + ADP \ ribose + H^+$$

Assay Methods

Principle. the enzyme activity can be assayed titrimetrically by measuring the production of hydrogen ion, spectrophotometrically by converting NAD in samples of the reaction mixture to the NAD-cyano adduct absorbing at 327 nm, or spectrophotometrically by monitoring the 275-nm absorbance decrease accompanying the hydrolysis of the alternate substrate, nicotinamide $1,N^6$-ethenoadenine dinucleotide (ε-NAD). The discontinuous cyanide addition assay is used predominantly for assay of crude preparations of enzyme.

Assay Based on Cyanide Addition Reaction

Reagents

Potassium cyanide, 1 M
Potassium phosphate buffer (pH 7.5), 100 mM
Potassium phosphate buffer (pH 7.5), 50 mM
NAD, 3 mM (neutralized)

Procedure. The standard reaction mixture contains 1.5 ml of 100 mM potassium phosphate buffer (pH 7.5), 1 ml of 3 mM NAD, enzyme, and water to a final volume of 3.0 ml. The reaction mixture is incubated at 38° and 0.2-ml aliquots are transferred to 2.8 ml of 1 M potassium cyanide at timed intervals. Absorbance at 327 nm is read against a blank of 0.2 ml of 50 mM potassium phosphate buffer (pH 7.5) plus 2.8 ml of 1 M potassium cyanide. The number of micromoles of NAD remaining is calculated using the millimolar extinction coefficient of 6.2 for the NAD-CN adduct.[1]

Definition of Unit and Specific Activity. One unit of enzyme activity is defined as that amount which catalyzes the hydrolysis of 1 μmol of NAD per minute. Specific activity is expressed as units per milligram protein, determined by the method of Bradford.[2]

Assay Based on Titrimetry

Reagents

NAD, 3 mM (neutralized
NaCl, 1 M
NaOH, 1 mM

Procedure. The reaction mixture equilibrated and incubated at 38° contains 0.1 ml 1 M NaCl, 0.4 ml of 3 mM NAD, enzyme, and water to a final volume of 4.0 ml. All solutions are prepared with water degassed by boiling. Reactions are initiated by the addition of enzyme, and the rate of consumption of 1 mM NaOH required to maintain pH 7.5 is recorded. Automatic titration (pH-stat) is performed on a Radiometer type TTT80

---

[1] S. P. Colowick, N. O. Kaplan, and M. M. Ciotti, *J. Biol. Chem.* **191**, 447 (1951).
[2] M. M. Bradford, *Anal. Biochem.* **72**, 248 (1976).

titrator, PHM82 pH meter, type ABU80 autoburette assembly equipped with a 0.25-ml burette, and a GK 2320C combination electrode. Thermostatted reaction vessels are used to maintain the desired temperature. Rates determined titrimetrically are expressed as microequivalents H^+ released/min which equals micromoles NAD hydrolyzed per minute.

Assay Based on Hydrolysis of Alternate Substrate

Reagents

ε-NAD, 1 mM (neutralized)
Potassium phosphate buffer (pH 7.5), 50 mM

Procedure. The standard reaction mixture contains 0.15 ml of 1 mM ε-NAD, and 2.85 ml of 50 mM potassium phosphate buffer (pH 7.5). Reactions are initiated by addition of enzyme and the decrease in absorbance at 275 nm read against a buffer blank is recorded. Activity is expressed as micromoles ε-NAD hydrolyzed per minute, calculated using the established millimolar extinction coefficient of 0.89.[3] In those cases in which greater sensitivity is required, the hydrolysis of ε-NAD is monitored fluorometrically through excitation at 300 nm and fluorescence emission at 410 nm.

Purification Procedure

All purification steps are performed at 4°. Crude *Bungarus fasciatus* (banded krait) venom was purchased from Sigma. Column fractions are assayed spectrophotometrically with ε-NAD as substrate.

Step 1. Phosphocellulose Chromatography. Whatman phosphocellulose P-11 was activated as described below prior to use in column chromatography. Phosphocellulose P-11 was stirred in a 3-fold volume of 0.5 N NaOH for 15 min; after settling, the fines were removed by decanting. The suspension was then washed with 10 volumes of water through a fritted glass funnel; the pH after this step was 8.0–8.5. The resin was then stirred in a 3-fold volume of 0.5 N HCl for 15 min and brought to pH 6.0 by washing with water as before. The phosphocellulose resin was then stirred in a 5-fold volume of 0.5 M Tris–HCl buffer (pH 8.2). The resin was titrated to pH 8.2 with 6 N NaOH, and prior to use, the phosphocellulose columns were equilibrated with 0.05 M Tris–HCl buffer (pH 8.2).

Lyophilized crude *B. fasciatus* venom (200 mg) was dissolved in 10 ml 0.05 M Tris–HCl buffer, pH 8.2 (fraction 1), and applied to a phosphocellulose column (1.5 × 40 cm) previously equilibrated as above. The column

[3] D. A. Yost and B. M. Anderson, *Anal. Biochem.* **116,** 374 (1981).

TABLE I
PURIFICATION OF NADASE FROM *Bungarus fasciatus* SNAKE VENOM

Fraction	Total protein (mg)	Total activity (units)	Specific activity (units/mg)	Yield (%)	Purification (-fold)
1. Crude	200	260	1.3	100	1
2. Phosphocellulose	8.1	280	35	108	27
3. Sephadex G-100	0.65	260	400	100	308
4. Amicon Matrex Gel Blue A	0.19	250	1320	96	1030

was washed with buffer until absorbance at 280 nm was essentially zero. A linear gradient of 0.05 M Tris–HCl buffer (pH 8.2), and 0.05 M Tris–HCl buffer (pH 8.2), 0.2 M potassium chloride (600 ml each) was then applied to the phosphocellulose column. Fractions of 5 ml were collected and those containing NADase activity were pooled and concentrated by ultrafiltration (fraction 2).

Step 2. Gel Filtration Chromatography. A column was packed with Sephadex G-100 (4.0 × 100 cm) and equilibrated with 0.05 M sodium pyrophosphate buffer (pH 7.0). Fraction 2 was applied to the column and eluted with the same buffer at a flow rate of 0.6 ml/min. Fractions of 5 ml were collected and those containing NADase activity were pooled and concentrated by ultrafiltration (fraction 3).

Step 3. Affinity Chromatography. A column was packed with Amicon Matrex Gel Blue A (0.9 × 5 cm) and equilibrated with 0.05 M sodium pyrophosphate buffer (pH 7.0). Fraction 3 was applied and the column washed with the same buffer. After all the 280-nm-absorbing material had been removed, the column was eluted with 0.05 M sodium pyrophosphate buffer (pH 7.0), 0.5 M KCl. After 20–30 column volumes of this buffer wash, 0.005 M potassium phosphate buffer (pH 8.0), 2 M KCl was applied to the column. Fractions of 1 ml were collected and those containing NADase activity were pooled, concentrated, and washed free of salt using ultrafiltration (fraction 4). A summary of the purification procedure is shown in Table I.

Properties

Purified snake venom NADase is a glycoprotein composed of two subunits of M_r 62,000 each. Catalysis of NAD hydrolysis occurs through an ordered uni-bi mechanism with nicotinamide the first product released

TABLE II
KINETIC PARAMETERS OF THE PYRIDINE BASE
EXCHANGE REACTION

Pyridine base	K_m (mM)	V_{max} (μmol/min · mg)
3-Aminopyridine	3.2	18.2
3-Pyridylcarbinol	7.9	87.8
Pyridine	1.4	9.2
3-Acetylpyridine	20.0	93.2
3-Pyridylacetonitrile	6.1	24.3
3-Methylpyridine	6.5	35.1

and ADP-ribose the second product released.[4] The enzyme exhibits a pH optimum of 7.5 and is only slightly affected by changes in ionic strength. At pH 7.5 and 38°, the K_m for NAD is 14 μM with a V_{max} of 1380 μmol of NAD cleaved/min · mg protein. The activation energy for the enzymatic NAD hydrolysis is 15.7 kcal/mol. The enzyme is noncompetitively inhibited by a variety of pyridine bases and competitively by pyridine nucleotides, ADP-ribose, ADP, and AMP. The enzyme catalyzes a transglycosidation reaction with NAD and a number of pyridine derivatives to form pyridine-modified coenzyme analogs.[5] Analog formation is monitored by high-performance liquid chromatography (HPLC). In a typical transglycosidation reaction, the reaction mixture (1 ml) contains 1.5 mM NAD, 50 mM 3-aminopyridine, 1 hydrolytic unit of NADase (740 ng), and 50 mM potassium phosphate buffer (pH 7.5). Conversion to the major product, 3-aminopyridine adenine dinucleotide, is essentially complete in 2 hr.[5] Substrates and products are separated and assayed by HPLC using a Spectra-Physics SP8000 HPLC with an Alltech RSIL AN, 5-μm anion-exchange column (4.6 × 250 mm), and 100 mM potassium phosphate buffer (pH 3.5) employed as the mobile phase. Initial velocities of transglycosidation can be measured by HPLC assay, and the kinetic parameters for six such reactions are shown in Table II.

The enzyme also catalyzes an alcoholysis of NAD to form O-alkyl-ADP-ribosides and an ADP-ribosylation of imidazole derivatives.[6] Prod-

[4] D. A. Yost and B. M. Anderson, *J. Biol. Chem.* **256,** 3647 (1981).
[5] D. A. Yost and B. M. Anderson, *J. Biol. Chem.* **257,** 767 (1982).
[6] D. A. Yost and B. M. Anderson, *J. Biol. Chem.* **258,** 3075 (1983).

ucts in these reactions are separated and identified by HPLC analysis. Inactivation of the enzyme by pentanedione and Woodward's Reagent K indicates the importance of a lysyl residue and a carboxyl group in the reactions catalyzed. The enzyme can be immobilized on concanavalin A-Sepharose (Con A-Sepharose) with 70% retention of catalytic activity.

Immobilization and Transglycosidation

Principle. The NADase from *B. fasciatus* snake venom was adsorbed on Con A-Sepharose (Sigma) and demonstrated to retain both hydrolase and transglycosidase activities in the bound form. The matrix-bound enzyme was stable to repeated washings with buffer and storage at 4° and was used repeatedly for a preparative-scale synthesis of 3-acetylpyridine adenine dinucleotide. The matrix-bound NADase was 75% as active as the soluble enzyme form and exhibited identical kinetic parameters in the hydrolysis of ε-NAD. It was also demonstrated that the immobilized enzyme could be prepared directly from crude snake venom, thus avoiding the time required for purification. The application of immobilized snake venom NADase for the preparation of pyridine nucleotide coenzyme analogs has many advantages over procedures used previously for analog synthesis.

Immobilization of Purified Snake Venom NADase. NADase from crude *B. fasciatus* snake venom was purified as described previously. Con A-Sepharose containing 10 mg Con A per milliliter packed gel was prepared for incubation experiments by washing the gel three times with 50 mM potassium phosphate buffer (pH 7.5) to remove thimerosal and metals used as preservatives. Three milliliters of Con A-Sepharose suspension (1.5 ml packed gel) was washed three times with 10 ml buffer, centrifuged at low speed in a clinical centrifuge, and resuspended to a final volume of 8 ml with 50 mM potassium phosphate buffer (pH 7.5). To this gel suspension was added 1 ml containing 3.4 units of purified snake venom NADase. The mixture was stirred gently at 37° for 15 min and centrifuged at low speed, and the supernatant fraction was assayed fluorimetrically for residual (unbound) NADase activity, using ε-NAD as substrate. No enzyme activity was observed to remain in this supernatant fraction. The capacity of Con A-Sepharose to bind the purified snake venom NADase was investigated by adding increments of free enzyme to a given amount of affinity matrix until NADase activity was detected in the supernatant fraction after centrifugation. From these studies it was calculated that 21.6 units of purified NADase could be bound per milliliter of packed Con A-Sepharose gel.[7] The enzyme activity of the matrix-

[7] B. M. Anderson and C. D. Anderson, *Anal. Biochem.* **140,** 250 (1984).

bound NADase was assayed by incubating an aliquot of this gel suspension with NAD at 37° and, following the hydrolysis of NAD, using the cyanide assay with filtered reaction samples removed at timed intervals. NADase activity was observed to be 75% that of free NADase prior to immobilization. The fact that properties of the immobilized NADase are very similar to those of the soluble form of the enzyme is readily documented in studies of ε-NAD as substrate.

A fresh sample of NADase bound to Con A-Sepharose was used for studies of initial velocities of the hydrolysis of ε-NAD. The reaction mixtures at 37° contained 50 mM potassium phosphate buffer (pH 7.5), 0.017 unit of immobilized NADase, and ε-NAD varied from 2.5 to 50 μM in a total volume of 1.1 ml. Hydrolysis of ε-NAD was monitored fluorimetrically with a Perkin-Elmer 650-40 spectrophotofluorometer. The K_m value of 21 μM determined in these studies agreed well with the value of 20 μM determined previously with the soluble form of the enzyme.[4] Matrix-bound NADase suspensions can be washed after incubation reactions and stored in 50 mM potassium phosphate buffer (pH 7.5), 0.01% thimerosal at 4° with little loss in catalytic activity. Prior to additional use, the matrix-bound NADase suspension is first washed with buffer to remove the thimerosal preservative.

Immobilization of Crude Snake Venom NADase. In considering the application of matrix-bound NADase for the preparation of coenzyme analogs, it would be a distinct advantage if the enzyme could be adsorbed on the Con A-Sepharose gel from the crude snake venom preparation, thus avoiding the time required for the purification procedure. The major problem with this approach is the presence, in the snake venom, of a phosphodiesterase capable of catalyzing the hydrolysis of the pyrophosphate linkages of dinucleotides. A solution of 66 mg of crude *B. fasciatus* snake venom dissolved in 10 ml of 10 mM potassium phosphate buffer (pH 7.0) was prepared and assayed for both NADase and phosphodiesterase activities. NADase activity was measured with the cyanide assay and phosphodiesterase activity was determined using 9.1 mM bis-*p*-nitrophenyl phosphate in 0.1 M Tris–HCl buffer (pH 8.5), following the absorbance change at 405 nm. The crude snake venom solution was incubated at 37° with 1 mM EDTA, which was observed to irreversibly inactivate the phosphodiesterase. After a 60-min incubation with 1 mM EDTA there was no loss in NADase activity but an 85% loss of phosphodiesterase activity. After storage overnight at 4°, this enzyme preparation was observed to have only 6% of the original phosphodiesterase activity remaining.

This filtered NADase solution (10 ml) containing 42 units of NADase activity was incubated with 10 ml of packed Con A-Sepharose gel in a total of 50 ml of 50 mM potassium phosphate buffer (pH 7.5). After 30-min stirring at 37°, all of the NADase was bound to the gel and only a trace of phosphodiesterase activity remained. The Con A-Sepharose-bound NADase gel was washed three times with buffer and assayed for NADase and phosphodiesterase activities. The matrix-bound NADase was observed to be 70% as active as the free NADase prior to immobilization, and the gel appeared to have bound only 0.0012 unit of phosphodiesterase activity per milliliter of packed gel. When the matrix-bound crude snake venom NADase bound to Con A-Sepharose gel (unit for unit), the rate of analog synthesis was similar, and the analog yield was essentially the same. This indicates that the low phosphodiesterase activity adsorbed from the crude venom on the Con A-Sepharose gel had very little effect on analog synthesis. The immobilized NADase from crude snake venom was stable to repeated washings after incubation reactions, and was stored in 50 mM potassium phosphate buffer (pH 7.5), 0.01% thimerosal at 4°.

Transglycosidation Reactions Using Immobilized NADase. A preparative scale synthesis of 3-acetylpyridine adenine dinucleotide is described below. Five units of bound purified snake venom NADase suspended in 5 ml buffer were added to 45 ml of 50 mM potassium phosphate buffer (pH 7.5), containing 100 mg of NAD and 0.1 M 3-acetylpyridine. The reaction mixture was incubated at 37° with stirring and the progress of the reaction was monitored as follows. At timed intervals a filtered 0.1-ml aliquot of the reaction mixture was added to 2.8 ml of 90 mM unbuffered Tris containing 0.5 M ethanol. After the addition of 1 mg yeast alcohol dehydrogenase in 0.1 ml, the absorbance change was recorded at 340 and 365 nm. A plot of the 365/340 ratio as a function of time demonstrates the conversion of NAD to the 3-acetylpyridine analog, with essentially no NAD remaining after 2 hr. The yield of the coenzyme analog was 67%. The above reaction was repeated the same day with the same but washed matrix-bound NADase, and a 63% yield of analog was obtained.

Small-scale transglycosidation reactions as above were carried out daily over a 7-day period using the same immobilized NADase suspension throughout. Each day the transglycosidation reaction was repeated, with the immobilized NADase recovered by centrifugation, washed, and stored overnight at 4°. The matrix-bound NADase exhibited the same activity everyday for 5 days, with an approximate 10% loss of activity on the sixth day. The same enzyme preparation was then stored at 4° for 7 days and showed no further loss of transglycosidase activity on assay. The stability of the immobilized NADase as an effective, reusable preparation for analog synthesis was demonstrated. In addition, the ability to remove the immobilized enzyme cleanly from the synthetic reaction by filtration or centrifugation provides a distinct advantage for the ultimate purification of the coenzyme analog produced. The inherent properties of the snake venom NADase make it a more appropriate catalyst for transglycosidation reactions than other NADases previously used. The snake venom enzyme works equally well with NAD and NADP,[4] it catalyzes a variety of ADP-ribose transfer reactions,[5,6] and most importantly, it does not self-inactivate during catalysis as observed with other NADases.

Section VII

Flavins and Derivatives

[31] Luminometric Determination of Flavin Adenine Dinucleotide

By KARL DECKER and ARI HINKKANEN

FAD is the predominant biologically active form of riboflavin. It is an ubiquitous cofactor of redox enzymes, e.g., dehydrogenases, transhydrogenases, reductases, oxidases, and oxygenases. The isoalloxazine part of the molecule appears to be capable of different modes of reactions allowing various redox mechanisms.[1] FAD-dependent enzymes participate in anaerobic as well as aerobic electron and hydrogen transport. FAD exists within a cell in at least three distinguishable pools: (1) as free FAD; (2) in noncovalent attachment to proteins but with restricted exchangeability, e.g., in D-amino-acid oxidase (D-AAO, EC 1.4.3.3); (3) covalently bound to the polypeptide chain of an apoenzyme (for review see ref. 2). In bacteria, pool 2 was found to be the largest of the three.[3] These different forms of FAD can be separated preparatively[3] and analyzed individually. The biosynthesis of flavoproteins depends, at least in some cases, on the intracellular concentration of FAD.[3]

In most cells so far analyzed, the total FAD content is in the range of 15–70 nmol/g wet tissue. It is, therefore, desirable to have an assay method available that is both highly sensitive and specific with respect to other forms of flavins [riboflavin, flavin mononucleotide (FMN), 5-deazaflavin]. Several methods have been described that fulfill these prerequisites to various extents; they include a spectrophotometric assay based on the combination of the reactions catalyzed by D-AAO and lactate dehydrogenase (LDH, EC 1.1.1.27),[3,4] a bioluminometric method employing luciferase[5] and a procedure using fluorescence quenching during flavodoxin reconstitution.[6] The latter two methods necessitate the prior conversion of FAD to FMN. The procedure outlined below requires commercially available, cheap reagents and quite commonplace equipment only.

[1] V. Massey and P. Hemmerich, *Biochem. Soc. Trans.* **8**, 246 (1980).
[2] K. Decker, *in* "Flavins and Flavoproteins" (V. Massey and H. C. Williams, eds.), p. 465. Elsevier/North-Holland Biomedical Press, Amsterdam, 1982.
[3] A. Hinkkanen and K. Decker, *in* "Flavins and Flavoproteins" (V. Massey and H. C. Williams, eds.), p. 478. Elsevier/North-Holland Biomedical Press, Amsterdam, 1982.
[4] C. DeLuca, M. M. Weber, and N. O. Kaplan, *J. Biol. Chem.* **223**, 559 (1956).
[5] E. W. Chappelle and G. L. Picciolo, this series, Vol. 18, p. 381.
[6] S. G. Mayhew and J. H. Wassink, this series, Vol. 16, p. 217.

Assay Method

Principle

The specificity of the assay is provided by the apoprotein that recombines to fully active D-AAO [Eq. (1)] with FAD only. FMN, riboflavin, or 5-deazaflavin produce neither chemiluminescence nor are they inhibitory when added together with FAD to the chemiluminescence assay.[7] The superior sensitivity of the method is guaranteed by the high affinity of the apo-D-AAO for FAD as well as by the large photon yield accompanying the oxidation of luminol [6-amino-2,3-dihydro-1,4-phthalazinedione; (Eq. (3)] catalyzed by horseradish peroxidase (PO, EC 1.11.1.7).

$$\text{apo-D-AAO} + \text{FAD} \rightarrow \text{holo-D-AAO} \tag{1}$$
$$\downarrow$$
$$\text{D-alanine} + H_2O + O_2 \rightarrow \text{pyruvate} + NH_3 + H_2O_2 \tag{2}$$

$$2\ H_2O_2 + \text{luminol} \xrightarrow{\text{PO}} \text{4-aminophthalate}^{2-} + 2\ H^+ + N_2 + 2\ H_2O + h\nu \tag{3}$$

A complete and rapid reaction requires an excess of apo-D-AAO over FAD and a limited amount of PO (see below) to protect the apoprotein from inactivation.

The photon yield of the luminol-dependent chemiluminescence (CL) is greatest at rather alkaline pH values (>10). A workable compromise with the activity ranges of D-AAO and PO is a pH of 9.2 in the assay. This pH allows good enzyme activities and a satisfactory light emission. Under the conditions of this assay, the photon output (measured as counts per minute) is proportional to the FAD concentration over a wide range.

Equipment

Commercially available luminometers, such as the Berthold Biolumat LB 9500 T, are well suited for the measurements of small volumes of assay mixtures (0.5 ml). However, any spectrometer using anticoincidence recording of photons in the visible range, such as liquid scintillation spectrometers, may by used. A two-channel recorder, e.g., Pharmacia Rec 482, is connected with the luminometer. The equipment for the preparation of apo-D-AAO is mentioned with the procedure (see below).

Chemiluminescence Mixture

The CL assay mixture contains in 0.5 ml sodium bicarbonate buffer (0.1 M, pH 9.2): 0.1 M sodium chloride, 23 mM D-alanine, 25 μM luminol,

---

[7] A. Hinkkanen and K. Decker, *Anal. Biochem.* **132,** 202 (1983).

and 10 μg per milliliter horseradish peroxidase. It is conveniently made up by mixing 20 ml of a 0.2 M NaHCO$_3$–Na$_2$CO$_3$ buffer (pH 9.2) with 10 ml of a 0.4 M NaCl solution, 5 ml of a 0.184 M D-alanine solution, 0.2 ml of a 5 mM luminol solution, and 0.4 mg of a PO lyophilizate (250 U/mg) dissolved in 1 ml of 0.2 M bicarbonate buffer (pH 9.2). The mixture is filled up to 40 ml with quartz-distilled water. It should be prepared freshly every day.

FAD Reference Solution

Highly purified FAD for standardization can be obtained by extracting D-amino-acid oxidase with cold trichloroacetic acid (TCA).[8] TCA can be removed from the FAD preparation by several ether extractions. Alternatively, FAD can be purified from flavin-contaminated commercial preparations using HPLC[7,9] or simple paper chromatography.[10] The amount of impurities in the FAD reference sample can be checked by determination of the absorbance ratio A_{260}/A_{450} at pH 7.0. The pure substance gives a value of 3.25.[11] The concentration of standard FAD is determined spectrophotometrically using a molar absorbance coefficient ε of 11.3 mM^{-1} cm^{-1} at 450 nm (pH 7).[12] The accuracy of the method is limited by the purity of the reference FAD.

Preparation of Apo-D-Amino-Acid Oxidase

The apoenzyme of D-AAO from hog kidney is most conveniently and reproducibly prepared by dialysis of the holoenzyme against 1 M KBr according to Massey and Curti.[13] Crystalline D-AAO is dissolved in 0.1 M pyrophosphate buffer (pH 8.5) to a final concentration of 0.5 mg/ml. It is dialyzed at 4° for 2 days (three changes) against 0.1 M sodium pyrophosphate buffer (pH 8.5), containing 1 M KBr and 3 mM EDTA. The complete removal of FAD from the apoprotein at the end of the dialysis is ascertained in the following way: CL assays are run with about 1 μg of apoenzyme each in the presence and absence of 20 fmol FAD. The signal-to-noise ratio should be no less than 2 : 1; otherwise the dialysis should be continued. KBr is removed from the protein solution by another dialysis against 0.1 M sodium pyrophosphate buffer (pH 8.5) for 48 hr. Turbidity is

[8] V. Massey and B. E. P. Swoboda, Biochem. Z. 338, 474 (1963).
[9] D. R. Light, C. Walsh, and M. A. Marletta, Anal. Biochem. 109, 87 (1980)
[10] K. Decker and V. D. Dai, Eur. J. Biochem. 3, 132 (1967).
[11] H. Friedmann, in "Methoden der enzymatischen Analyse" (H.-U. Bergmeyer, ed.), 3rd ed., p. 2232. Verlag Chemie, Weinheim, 1974.
[12] O. Warburg and W. Christian, Biochem. Z. 298, 150 (1938).
[13] V. Massey and B. Curti, J. Biol. Chem. 241, 3417 (1966).

removed by centrifugation if necessary. The volume of the apoenzyme solution is adjusted so that it contains 100 μg protein/ml. The apo-AAO should be stored in small portions at $-30°$.

Alternatively, FAD can be removed from the apoprotein by acid treatment of the holoenzyme in saturated ammonium sulfate solution.[12]

Sample Preparation

General. FAD present in free form or bound noncovalently to protein can be extracted from biological material with ice-cold trichloroacetic acid or perchloric acid. FAD is split in strong acid into FMN and 5'-AMP. Contact with strong acids (pH $\leq$ 2) should be as short as possible and at the lowest temperature feasible. FAD solutions may be concentrated by freeze-drying without damage. All operations should be carried out in dim or, preferably, in red light. FAD samples should be stored in the dark at $-30°$.

Bacterial Cells. The cold cell suspension is thoroughly mixed with ice-cold $HClO_4$ (final concentration 5%, v/v). Denatured protein and cell debris are collected by centrifugation and rehomogenized in small amounts of $HClO_4$ ($0°$) several times. The supernatants are combined and immediately neutralized by adding solid $KHCO_3^-$ (foams). The precipitated $KClO_4$ is removed by centrifugation.

Tissues. A useful procedure for the preparation of tissue extracts has been described by Keppler *et al.*[14] Freeze-stop technique is used to obtain the tissue samples from laboratory animals. The frozen tissue is weighed and homogenized with 5 volumes of frozen perchloric acid ($0.6\ M$). After centrifugation, the supernatant is neutralized by addition of solid $KHCO_3$ and $KClO_4$ is removed by centrifugation.

Separation of FAD Pools. Free FAD can be separated from that noncovalently bound to proteins by gel filtration of the centrifuged crude extract from bacterial cells.[3] Cold $HClO_4$ is then used to extract FAD from the protein fraction as described above.

Assay Procedure

The standard procedure for the CL assay uses a 3-ml cuvette that contains 0.5 ml CL mixture and 0.2–2 nM (0.1–1 pmol/assay) FAD. Mixture and sample are preincubated at $37°$ for 3 min. The reaction is started by addition of about 1 μg apo-D-AAO and allowed to reach the maximal photon production (CL_{max}) (Fig. 1). CL_{max} is proportional to the amount of biologically active FAD in the sample.

[14] D. Keppler, J. Rudigier, and K. Decker, *Anal. Biochem.* **38,** 105 (1970).

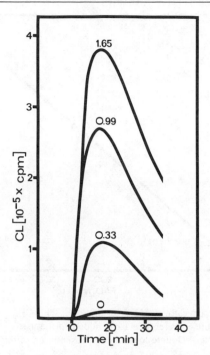

FIG. 1. Kinetics of chemiluminescence. Numbers on the curves indicate picomoles FAD added to the standard assay (see text). Apo-D-AAO was injected at 10 min. CL_{max} is reached about 8 min later.

The amount of apo-D-AAO added to the mixture depends on the content of "active apoenzyme" in the preparation. "Active apoenzyme" (apoprotein that is able to recombine with FAD to form an active enzyme) must be present in the CL mixture in at least a 10-fold excess over FAD. The "activity" value of the apoenzyme preparations is established by determination of CL_{max} obtained in the presence of saturating concentrations of FAD ($\sim$2 nM, dependent on the amount of apoenzyme).

The FAD reference solution is assayed under identical conditions. A graph is prepared by plotting the CL_{max} values obtained versus the amount of FAD (picomoles) added (Fig. 2). The amount of FAD in the unknown samples is determined with the aid of that calibration curve. A satisfactory linearity is achieved between 0.2 and 2 nM FAD. The sensitivity of the assay normally allows dilution of the sample, which greatly decreases the effect of interfering factors. Dilutions of samples should be made shortly before the measurements.

The presence of factors in the biological samples potentially interfer-

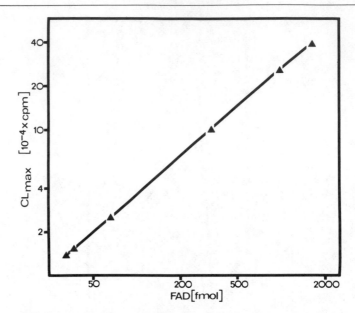

FIG. 2. Calibration curve for FAD determination. The CL_{max} values (see Fig. 1) are plotted against the amount of reference FAD added to the assay. The background CL_{max} (1×10^4 cpm) is subtracted. Double-logarithmic presentation of data.

ing with the photon yield or the recombination capacity of the apoenzyme for FAD may be detected in the following way. Two parallel assays with the biological samples are prepared; to one of them a known amount of reference FAD is added and the recovery determined. If necessary, a recovery factor has to be introduced in the calculation of the actual FAD content of the sample. The background emission (assay blank) is obtained by replacing the FAD in the assay with water. This value is subtracted from the sample values. The relative standard deviation of 10 independent measurements of the sample (0.75 pmol FAD each) was 0.038 (3.8%).

Acid-extracted biological samples may contain traces of H_2O_2 and thus evoke a short burst of CL when added to the CL mixture. In this case, the CL after sample addition is allowed to decay to the background level before the addition of apo-D-AAO. All measurements are carried out in the absence of light.

Inactivation of Apo-D-Amino-Acid Oxidase during the Assay

In the presence of H_2O_2 and PO, a time-dependent inactivation of apo-D-AAO takes place.[7] This is thought to be due to active oxygen species

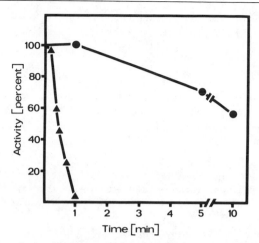

FIG. 3. Kinetics of peroxidase-dependent inactivation of apo-D-amino-acid oxidase by H_2O_2. Each point represents CL_{max} in the presence of 2.3 pmol FAD. Apo-D-AAO was preincubated with 1 μM H_2O_2 in the presence (▲) or absence (●) of peroxidase for the times indicated on the abscissa. Activity is given as percentage of a sample not preincubated.

evolving in the PO-catalyzed reaction (Fig. 3). An apoenzyme which has been preincubated in the presence of hydrogen peroxide alone can still be reactivated by FAD, whereas incubation in the presence of peroxide plus PO completely inactivates the apo-D-AAO. The presence of excess FAD and luminol during the preincubation partially protects the apoenzyme.[7]

This inactivation of apo-D-AAO does not interfere with the lumino-

FAD CONTENT OF BIOLOGICAL SAMPLES

Organism	Content (nmol/g wet wt.)	Reference
Rat liver	34.9	7
Saccharomyces cerevisiae	15.5	4
Pseudomonas fluorescens	51.7	4
Arthrobacter oxidans rf⁻	66.0[a]	7
	44.5[b]	7
Escherichia coli	15.1	4
Clostridium butyricum	72.0	4
Clostridium kluyveri	1490	15

[a] Riboflavin-requiring mutant grown in presence of 15 μM riboflavin.
[b] Same as above, grown with 2 μM riboflavin.

metric FAD assay if the conditions given above, in particular the relation of apo-D-AAO and PO, are strictly followed.

Comments

The luminometric method has been used to determine the FAD contents of rat liver and of some bacterial cells.[3,7] The table lists these values and some obtained by other authors.[15]

Acknowledgment

The work of the authors was supported by grants of the Deutsche Forschungsgemeinschaft, Bonn-Bad Godesberg, Federal Republic of Germany, through SFB 206.

[15] M. Brühmüller and K. Decker, *Anal. Biochem.* **71**, 550 (1976).

[32] Heavy Riboflavin Synthase from *Bacillus subtilis*

By ADELBERT BACHER

Riboflavin synthase catalyzes the formation of one molecule each of riboflavin and 5-amino-6-ribitylamino-2,4(1H,3H)-pyrimidinedione from two molecules of 6,7-dimethyl-8-ribityllumazine.[1] The enzyme has been purified to variable extent from several organisms such as baker's yeast,[2] *Ashbya gossypii*,[3] *Eremothecium ashbyii*,[4] and spinach.[5] The yeast enzyme has been obtained in apparently pure form, and its reaction mechanism has been studied in considerable detail.[1,6] A molecular weight of 70,000–80,000 was suggested on the basis of sedimentation experiments.

Various *Bacillus* and *Clostridium* species studied in this laboratory displayed two riboflavin synthase activities of widely different molecular

---

[1] G. W. E. Plaut, *in* "Comprehensive Biochemistry" (M. Florkin and E. H. Stotz, eds.), Vol. 21, p. 11. Elsevier, Amsterdam, 1971.
[2] G. W. E. Plaut and R. A. Harvey, this series, Vol. 18, Part B, p. 515.
[3] G. W. E. Plaut, *J. Biol. Chem.* **238**, 2225 (1963).
[4] H. Mitsuda, K. Nakajima, T. Nadamoto, and Y. Yamada, this series, Vol. 66, p. 307.
[5] H. Mitsuda, F. Kawai, and Y. Suzuki, this series, Vol. 18, Part B, p. 539.
[6] G. W. E. Plaut and R. L. Beach, *in* "Flavins and Flavoproteins" (T. P. Singer, ed.), p. 737. Elsevier, Amsterdam, 1976.

weight.[7] The light enzyme from *Bacillus subtilis* has a molecular weight of 70,000 and appears similar to the yeast enzyme. On the other hand, the heavy enzyme of this microorganism is a 1,000,000-Da protein with rather unusual structural properties.[7,8]

Assay Method

Principle. The formation of riboflavin from 6,7-dimethyl-8-ribityllumazine is monitored photometrically at 470 nm.

Preparation of 6,7-Dimethyl-8-ribityllumazine. Palladium on charcoal (10%, 150 mg) in 50 ml of water is stirred in a hydrogen atmosphere for 10 min. 5-Nitroso-6-ribitylamino-2,4-(1*H*,3*H*)-pyrimidinedione[2] (1 g, 3.45 mmol) is added, and the solution is hydrogenated at room temperature and atmospheric pressure until the absorption of hydrogen is terminated. Freshly distilled diacetyl (5 ml) is added and the suspension is kept for 1 hr at room temperature under a nitrogen atmosphere. The catalyst is removed by filtration, and the filtrate is concentrated to a yellow syrup under reduced pressure. The syrup is dissolved in 10 ml of water, and the solution is placed on a column of Dowex 50W X8 (H^+ form, 200–400 mesh, 2.5 × 35 cm) which is subsequently developed with deionized water. Fractions of 25 ml are collected. Fractions 12–80 are combined and evaporated to a small volume (about 10 ml) under reduced pressure. Ethanol (10 ml) is added and the solution is kept in the refrigerator. Bright yellow crystals form overnight which are collected and dried over P_2O_5. Yield, 720 mg, 64%.

Reagents

Phosphate buffer, 0.1 *M*, pH 7.4, containing 10 m*M* sodium sulfite and 10 m*M* EDTA

6,7-Dimethyl-8-ribityllumazine, 6 m*M*

Trichloroacetic acid, 15%

Procedure. Assay mixtures contain 1.0 ml of 0.1 *M* phosphate buffer (pH 7.4) containing 10 m*M* EDTA and 10 m*M* sodium sulfite, 0.1 ml of 6 m*M* 6,7-dimethyl-8-ribityllumazine, and protein. They are incubated at 37°. At the start and at the end of incubation, aliquots of 0.5 ml are retrieved, and protein is precipitated by the addition of 0.2 ml of 15% trichloracetic acid. The samples are centrifuged, and absorbance is measured at 470 nm. The amount of riboflavin formed is calculated from the

[7] A. Bacher, R. Baur, U. Eggers, H. Harders, M. K. Otto, and H. Schnepple, *J. Biol. Chem.* **255**, 632 (1980).

[8] A. Bacher, H. C. Ludwig, H. Schnepple, and Y. BenShaul, *J. Mol. Biol.* **186** (in press).

increase of optical density at 470 nm (molar absorbance of riboflavin, 9300). For continuous measurements, the reaction is monitored by a registrating spectrophotometer equipped with a thermostatted cell.

Definition of unit. One unit of enzyme activity catalyzes the formation of 1 nmol riboflavin/hr at 37°.

Enzyme Purification[7,8]

Principle

The enzyme can be conveniently isolated from the genetically derepressed mutant H94 of *Bacillus subtilis* whose enzyme level is about 70-fold elevated as compared to the wild strain. Thus, the enzyme constitutes approximately 2% of the cellular protein. It should be noted that about 80% of the total activity in the crude extract corresponds to light enzyme.

Procedure

Culture Medium. The culture medium contains $(NH_4)_2SO_4$, 2 g; Na_2HPO_4, 7.2 g; KH_2PO_4, 3.5 g; sodium citrate $\cdot$ 5H$_2$O, 1.1 g; MgSO$_4$ $\cdot$ 7H$_2$O, 0.2 g; CaCl$_2$ $\cdot$ 6H$_2$O, 0.45 g; FeCl$_3$ $\cdot$ 6H$_2$O, 2.5 mg; MnSO$_4$ $\cdot$ H$_2$O, 3.4 mg; KCl, 1 g; DL-tryptophan, 80 mg; glucose, 5.0 g; casein hydrolysate (Merck), 2 g; deionized water, 1 liter.

Fermentation. Culture medium (500 liters) in a batch fermenter is inoculated with 5 liters of cell suspension grown in Erlenmeyer flasks. The suspension is grown with stirring and aeration (10 m^3/hr) for 16 hr at 37°. The suspension is chilled and the cells are harvested by centrifugation. The cell paste is stored at $-20°$.

Preparation of Cell Extract. Frozen bacterial cells (50 g) are thawed with gentle stirring in 100 ml of 75 mM phosphate (pH 6.3) containing 10 mM EDTA and 10 mM sodium sulfite. The suspension is ultrasonically treated in a chilled beaker (3 × 3 min). The suspension is centrifuged in a JA 10 rotor at 8000 rpm for 1 hr. The pellet is discarded.

Heat Treatment. The crude cell extract is incubated at 40° for 2 hr. The resulting suspension is cooled to 4° and centrifuged in a JA 10 rotor at 8000 rpm for 1 hr. The pellet is discarded. All subsequent steps are performed at 4°.

Ammonium Sulfate Precipitation. Ammonium sulfate (170 g/liter) is added to the supernatant with stirring, while the pH is kept at 6.5 by the addition of 6 M NH$_4$OH. The suspension is centrifuged, and the pellet is discarded. Ammonium sulfate (122 g/liter) is added slowly to the supernatant with stirring, and the suspension is centrifuged. The pellet is dissolved in 17 ml of 0.1 M phosphate buffer (pH 7.0) containing 10 mM EDTA and 10 mM sodium sulfite. The solution is dialyzed against the same buffer.

Ultracentrifugation. The solution is centrifuged in a Ti 60 rotor (Spinco) at 33,000 rpm and 4° for 16 hr. The supernatant is removed using a peristaltic pump and discarded. The pellet is dissolved in 20 ml of 0.1 M phosphate buffer (pH 7.0) containing 10 mM EDTA and 10 mM sodium sulfite. The solution is centrifuged again under the same conditions. The pellet is dissolved in 15 ml of 0.1 M phosphate buffer (pH 7.0) containing 10 mM EDTA and 10 mM sodium sulfite.

Sephacryl S-300 Chromatography. The solution is placed on a column of Sephacryl S-300 superfine (2.5 × 90 cm) which is developed with 0.1 M phosphate buffer (pH 7.0) containing 10 mM EDTA and 10 mM sodium sulfite (flow rate, 20 ml/hr). Fractions of 10 ml are collected. Heavy riboflavin synthase is eluted in fractions 24–29, and light riboflavin synthase is subsequently eluted in fractions 34–38. Fractions 24–29 are pooled and concentrated by pelleting in a Ti 60 rotor (33,000 rpm, 4°, 16 hr). The pellet is dissolved in 7.0 ml of 0.1 M phosphate (pH 7.0) containing 10 mM EDTA and 10 mM sodium sulfite.

Preparative Polyacrylamide Gel Electrophoresis. A column of polyacrylamide gel (2.5 × 10 cm) is prepared in a preparative polyacrylamide gel electrophoresis apparatus from a solution containing 0.14 M phosphate (pH 7.2), 4% acrylamide, 0.11% bisacrylamide, and 0.1% tetramethylethylenediamine. Polymerization is started by the addition of 0.05% ammonium persulfate. The electrode buffer contains 0.1 M phosphate (pH 7.2), and is recycled continuously. The gel is subjected to preelectrophoresis at a field strength of 5 V/cm for 12 hr. A solution containing 10,000 units of enzyme from the previous step, 20% sucrose and 0.01% bromphenol blue is placed on top of the gel. Electrophoresis is continued at 5 V/cm. The elution chamber is flushed with 0.1 M phosphate (pH 7.0) containing 10 mM EDTA and 10 mM sodium sulfite at a rate of 12 ml/hr. Fractions of 9 ml are collected. Bromphenol blue, which serves as tracking dye, is eluted in fractions 10–12. The enzyme is eluted in fractions 35–44. These fractions are pooled and concentrated by ultracentrifugation (rotor Ti 60, 33,000 rpm, 16 hr). Typical results from this purification procedure are presented in Table I.

Crystallization

Large crystals suitable for X-ray analysis can be grown by the vapor diffusion method at 20° from 1.3 M sodium/potassium phosphate (pH 8.7)

TABLE I
PURIFICATION OF HEAVY RIBOFLAVIN SYNTHASE
FROM
B. subtilis MUTANT H94

Procedure	Total activity (nmol/hr)	Specific activity (nmol/mg · hr)
Cell extract	430,000[a]	210[a]
Ammonium sulfate	330,000[a]	360[a]
Ultracentrifugation	130,000[a]	310[a]
Sephacryl S-300	45,000	600
Electrophoresis[b]	35,000	2000

[a] Mixture of light and heavy riboflavin synthase.
[b] Preparative polyacrylamide gel electrophoresis is performed in batches of 10,000 units.

containing 5-nitroso-6-ribitylamino-2,4(1*H*,3*H*)-pyrimidinedione.[9] The initial solution contains 0.7 *M* sodium/potassium phosphate (pH 8.7), 0.3 m*M* 5-nitroso-6-ribitylamino-2,4(1*H*,3*H*)-pyrimidinedione, and 2 mg of protein per ml. The final protein concentration after equilibration of the diffusion chambers is about 4 mg/ml.

Isolation of Subunits

A solution of heavy riboflavin synthase is centrifuged in a 5–20% sucrose gradient containing 0.1 *M* Tris buffer, pH 8.0 (SW 27 rotor, 4°, 20,000 rpm, 15 hr). The gradients are fractionated and monitored for protein and riboflavin synthase activity. The bottom half of the gradient contains the bulk of protein, but no riboflavin synthase activity. These fractions are pooled, dialyzed against 0.1 *M* phosphate (pH 7), and concentrated by pelleting in a Ti 60 rotor (33,000 rpm, 4°, 16 hr). The pellet is dissolved in 0.1 *M* phosphate (pH 7.0), yielding a solution of electrophoretically pure β subunits.

Fractions near the top of the gradient which contain riboflavin synthase activity are pooled, dialyzed against 0.1 *M* phosphate (pH 7.0), and concentrated by ultrafiltration (Amicon PM10 membrane). The resulting solution of α subunit trimers contains a small amount of β subunits.

Properties

The enzyme sediments in the analytical ultracentrifuge as a single, symmetrical boundary at a velocity of $s_{20,w}^{0} = 26.5$ S. It migrates as a

[9] R. Ladenstein, H. C. Ludwig, and A. Bacher, *J. Biol. Chem.* **258,** 11981 (1983).

TABLE II
PROPERTIES OF HEAVY RIBOFLAVIN SYNTHASE
FROM B. subtilis

$s^\circ_{20,w}$	26.5 S
Molecular weight	1,000,000
Subunit composition	$\alpha_3\beta_{60}$
$A^{1\%,1\ cm}_{280\ nm}$	7.0
Specific activity[a]	2000 nmol/mg · hr
K^a_M	130 μM

[a] Formation of riboflavin from 6,7-dimethyl-8-ribityllumazine.

single band in analytical polyacrylamide gel electrophoresis and in cellulose acetate electrophoresis. Other properties are summarized in Table II.

Stability. The enzyme is stable for periods of several months when stored as solution in 0.1 M phosphate (pH 7.0) containing 10 mM EDTA and 10 mM sodium sulfite. It is not recommended to keep pure enzyme samples in the freezer. At $-20°$, slow decomposition occurs by formation of large β subunit aggregates devoid of α subunits.

Catalytic Activity. The dismutation of 6,7-dimethyl-8-ribityllumazine yielding riboflavin and 5-amino-6-ribitylamino-2,4(1H,3H)-pyrimidinedione is catalyzed by the α subunits of the enzyme.[7] Since the enzyme contains only 7% of α subunits by weight, its V_{max} is modest (2,000 nmol/mg · hr) as compared to light riboflavin synthase from B. subtilis, which consists exclusively of α subunits (V_{max} = 50,000 nmol/mg · hr). Kinetic analysis yields a linear Lineweaver–Burk plot indicating a Michaelis constant of K_m = 130 μM.

The catalytic role of the β subunits is not yet completely understood. A role in the synthesis of 6,7-dimethyl-8-ribityllumazine had been proposed earlier by studies with β subunit deficient mutants.[10] Additional information was obtained in connection with enzyme studies on the flavinogenic yeast *Candida guilliermondii*. Cell extracts of this organism catalyze the formation of 6,7-dimethyl-8-ribityllumazine from 5-amino-6-ribitylamino-2.4(1H,3H)-pyrimidinedione or its 5'-phosphate.[11–13] The reaction requires the addition of four carbon atoms which are derived from a pentose phosphate via an intermediate (Compound X) of hitherto

[10] A. Bacher and B. Mailänder, *J. Bacteriol.* **134,** 476 (1978).
[11] E. M. Logvinenko, G. M. Shavlovsky, A. E. Zakal'sky, and I. V. Zakhodylo, *Biokhimiya* **47,** 931 (1982).
[12] E. M. Logvinenko, G. M. Shavlovsky, and N. Y. Tsarenko, *Biokhimiya* **49,** 45 (1984).
[13] P. Nielsen, G. Neuberger, H. G. Floss, and A. Bacher, *Biochem. Biophys. Res. Commun.* **118,** 814 (1984).

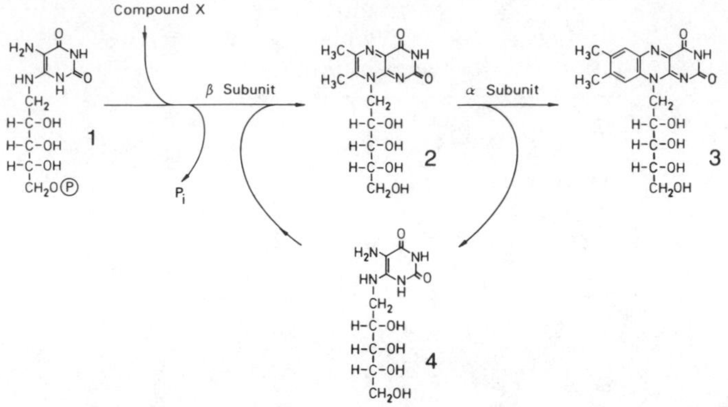

FIG. 1. Reactions catalyzed by heavy riboflavin synthase. **1**, 5-Amino-6-ribitylamino-2,4(1*H*,3*H*)-pyrimidinedione; **2**, 6,7-dimethyl-8-ribityllumazine; **3**, riboflavin; **4**, 5-amino-6-ribitylamino-2,4(1*H*,3*H*)-pyrimidinedione 5'-phosphate.

unknown structure.[13,14] Heavy riboflavin synthase from *B. subtilis* as well as isolated β subunits catalyze the formation of 6,7-dimethyl-8-ribityllumazine from Compound X and one of the pyrimidine derivatives shown in Fig. 1.[14] The details of this reaction are not yet known.

Structure. The enzyme has a molecular weight of 1,000,000. It has an unusual subunit stoichiometry consisting of 60 β subunits (M_r 16,000) and three α subunits (M_r 23,500).[7] The complete amino acid sequence of the β subunit has been determined.[15] The enzyme is an approximately spherical molecule with a radius of 78 Å as shown by electron microscopy of single molecules and of crystals.[8,16] Immunochemical evidence indicates that the α subunits are located in the interior of the molecule. This suggests that the 60 β subunits form a capsid-like structure with icosahedral 532 symmetry, which contains the three α subunits in the central cavity.

Well-ordered crystals have been obtained which yield X-ray reflections extending to a resolution of 3.5 Å.[9] The crystals have the space group P6₃22, and the enzyme molecules are packed in hexagonal layers. The noncrystallographic symmetry has been shown to be 532, in agreement with the proposed icosahedral structure of the enzyme molecule.[16]

[14] A. Bacher, P. Nielsen, G. Neuberger, and H. G. Floss, *in* "Flavins and Flavoproteins" (R. C. Bray, P. C. Engel, and S. G. Mayhew, eds.), p. 799. de Gruyter, Berlin, 1985.

[15] H. C. Ludwig, F. Lottspeich, A. Henschen, R. Ladenstein, and A. Bacher, *in* "Flavins and Flavoproteins" (R. C. Bray, P. C. Engel, and S. G. Mayhew, eds.), p. 379. de Gruyter, Berlin, 1985.

[16] R. Ladenstein, B. Meyer, R. Huber, H. Labinschinski, K. Bartels, H. D. Bartunik, L. Bachmann, H. C. Ludwig, and A. Bacher, *J. Mol. Biol.* **186** (in press).

Dissociation and Reaggregation. Heavy riboflavin synthase dissociates in 0.1 *M* Tris buffer at pH values of ≥ 7, yielding trimers of α subunit which appear identical with light riboflavin synthase. Under these conditions, the β subunits reaggregate into large molecules with the shape of hollow spheres and with a diameter of about 290 Å. Various molecular species appear to coexist in a state of equilibrium. The average molecular weight of these aggregates is $3-6 \times 10^6$.[8,17]

The dissociation of the enzyme under the conditions described above can be suppressed by the substrate analog 5-nitroso-6-ribitylamino-2,4(1*H*,3*H*)-pyrimidinedione, which enhances the stability of the enzyme considerably. The compound can also induce a reaggregation of isolated β subunits leading to the formation of spherical molecules with the stoichiometry β_{60} and a molecular weight of about 960,000.[8]

Ligand Binding. Structural analogs of 5-amino-6-ribitylamino-2,4(1*H*,3*H*)-pyrimidinedione and 6,7-dimethyl-8-ribityllumazine are bound by both α and β subunits. Beta subunits bind only ribityl-substituted compounds, whereas α subunits can also bind xylityl derivatives. 8-Ribityllumazines with oxo substituents in position 6 and/or 7 are tightly bound by both types of subunits ($K_D < 1 \mu M$). Riboflavin binds tightly to the α subunits, but only weakly to the β subunits.[17]

[17] A. Bacher and H. C. Ludwig, *Eur. J. Biochem.* **127**, 539 (1982).

[33] Separation of Flavins and Flavin Analogs by High-Performance Liquid Chromatography

By ROBERT P. HAUSINGER, JOHN F. HONEK, and CHRISTOPHER WALSH

Structural analogs of flavin coenzymes differing in redox potential and reactivity have been used to probe the mechanism for a number of flavoenzymes.[1-3] Studies of this type require flavin derivatives of high purity and at the proper coenzyme level (e.g., FMN, FAD level). This chapter describes an improved enzymatic procedure to convert riboflavin analogs

[1] D. R. Light and C. Walsh, *J. Biol. Chem.* **255**, 4264 (1980).
[2] C. Walsh, J. Fisher, R. Spencer, D. W. Graham, W. T. Ashton, J. E. Brown, R. D. Brown, and E. F. Rogers, *Biochemistry,* **17**, 1942 (1978).
[3] V. Massey and P. Hemmerich, *Biochem. Soc. Trans.* **8**, 246 (1980).

METHODS IN ENZYMOLOGY, VOL. 122

to the FAD level using *Brevibacterium ammoniagenes* extracts.[4] HPLC applications are described for the purification of flavin analogs and for the study of naturally occurring 5-deazaflavins in methanogenic bacteria.[5]

Flavin Analogs

Over the past decade a number of reports have appeared using ring-modified flavins to reconstitute specific apoflavoenzymes to probe mechanisms. For example much attention has focused on 5-deaza- and 1-deazaflavins at the FMN and FAD levels[6]; the 5-deaza analogs are restricted to two-electron redox chemistry in the ground state,[7] while the 1-deaza analogs have been useful in comparing modes of O_2 activation in flavoprotein oxidases and monooxygenases.[8]

Riboflavin: X = Y = N
5-Deazariboflavin: X = CH, Y = N
1-Deazariboflavin: X = N, Y = CH

Massey and co-workers, among others, have conducted systematic studies of a variety of other flavin analogs at the FMN and FAD levels with several apoflavoenzymes. These studies have included 8-thio-,[9] 2-thio-,[10] and 4-thioflavins[11] among others.

8-Thioriboflavin   2-Thioriboflavin: X = S, Y = O
                             4-Thioriboflavin: X = O, Y = S

[4] R. Spencer, J. Fisher, and C. Walsh, *Biochemistry* **15**, 1043 (1976).
[5] L. D. Eirich, G. D. Vogels, and R. S. Wolfe, *Biochemistry* **17**, 4583 (1978).
[6] L. B. Hersh and C. Walsh, this series, Vol. **66**, p. 277.
[7] J. Fisher, R. Spencer, and C. Walsh, *Biochemistry* **15**, 1054 (1976).
[8] R. Spencer, J. Fisher, and C. Walsh, *Biochemistry* **16**, 3594 (1977).
[9] V. Massey, S. Ghisla, and E. G. Moore, *J. Biol. Chem.* **254**, 9640 (1979).
[10] A. Claiborne, V. Massey, P. F. Fitzpatrick, and L. M. Schopfer, *J. Biol. Chem.* **257**, 174 (1982).
[11] V. Massey, A. Claiborne, M. Biemann, and S. Ghisla, *J. Biol. Chem.* **259**, 9667 (1984).

In all of these studies the synthetic analogs at the riboflavin level have been converted regiospecifically to the 5'-FMN species and then to the FAD level in a single pot enzymatic incubation with a flavokinase/FAD synthetase mixture from extracts of *B. ammoniagenes*. Typical incubations produce 50–100 nmol coenzyme analogs,[4] which are then most conveniently purified by HPLC methods. Analogs at the FMN level are readily prepared from the FAD analogs by treatment with phosphodiesterase from *Naja naja*. Alternatively, less specific chemical phosphorylation of the riboflavin analogs can be performed which requires HPLC purification of the 5'-FMN species.[12] A summary of thin-layer, paper, liquid, and affinity chromatography methods for purifying flavins has recently been published.[13] Here is presented a detailed procedure for generating and purifying the FAD analogs, which includes improvements over the original methodology.[4,14]

An inoculum of *B. ammoniagenes* (ATCC 6872) was prepared by suspending the lyophilized cells in 10 ml nutrient broth (8g/liter, Difco) and gently shaking overnight at 30°. The bacteria were propagated by using dextrose stabs, and for large-scale growth the following media was utilized: glucose, 20 g/liter; Bacto-peptone (Difco), 10 g/liter; yeast extract (Difco), 3.0 g/liter; urea, 5.0 g/liter; KH_2PO_4, 2.0 g/liter; K_2HPO_4, 2.0 g/liter; $MgSO_4 \cdot 7H_2O$, 0.50 g/liter; $CaCl_2 \cdot 2H_2O$, 0.10 g/liter; $FeSO_4 \cdot 7H_2O$, 0.010 g/liter. The final pH was adjusted to 7.2 with KOH. The cells were grown for 24 hr at 30° with vigorous aeration and stirring at 500 rpm using a 14-liter Chemapac fermenter. FAD excretion resulted in a yellow growth media. The bacteria were harvested at late log phase by centrifugation at 11,000 g for 20 min. The green cell pellet was resuspended in 0.1 M potassium phosphate (pH 7.5) buffer and repelleted to yield 300 g cells wet weight. Enzyme activity was stable over several weeks for cells stored at −15°.

The bacterial cells (50 g) were difficult to disrupt, requiring two passes through a French press at 19,000 psi in a 1:1 suspension with 0.1 M potassium phosphate buffer. The broken cells were centrifuged at 31,000 g for 30 min, yielding a deep yellow supernatant solution which was brought to 50% ammonium sulfate. After 15 min, the mixture was centrifuged for 15 min at 27,000 g to separate insoluble contaminants from the soluble FAD synthetase activity. The supernatant solution was adjusted to 80% ammonium sulfate and centrifuged at 27,000 g for 15 min, yielding a green pellet (2.6 g). This enzyme complex was suspended in 0.1 M

[12] P. Nielson, P. Rauschenbach, and A. Bacher, *Anal. Biochem.* **130**, 359 (1983).
[13] M. A. Marletta and D. R. Light, *in* "Modern Chromatographic Analysis of the Vitamins" (A. P. De Leenheer, E. E. S. Lambert, and G. M. De Ruyter, eds.), p. 413. Dekker, New York, 1984.
[14] R. Spencer, Ph.D. Thesis, Massachusetts Institute of Technology, Cambridge (1978).

TABLE I
FLAVIN ANALOGS AVAILABLE AT THE FAD
COENZYME LEVEL

FAD analog	Redox potential (mV)	References
4-Thio	−55	11
2-Thio	−126	1,10
7,8-Dichloro	−126	15
7-Chloro-8-demethyl	−128	1
9-Aza	−135	1
8-Chloro-7-demethyl	−144	1
8-Chloro	−152	16
7-Bromo	−154	17
8-Methylthio	−204	1
FAD	−208	—
6-Methyl	−219	1
3-Deaza	−240	1
6-Hydroxy	−265, −305	18,19
1-Deaza	−280	8
8-Mercapto	−290	9
5-Deaza	−310	4,7
8-Hydroxy	−340	20
8-Methylamino	?	1
8-Methylacetate	?	21
8-Methylacetamide	?	21

potassium phosphate buffer (pH 7.5, 15 ml) and frozen in aliquots at −70°. Excellent FAD synthetase activities were obtained with this resuspended pellet; further purification, which has been described,[4] is not required.

To obtain FAD level analogs, the riboflavin derivatives (2.6 μmol) were dissolved in potassium phosphate buffer (30 mM, pH 7.5, 30 ml, warming solution as necessary) containing 120 μmol ATP, 270 μmol MgCl$_2$, and 1 ml of the enzyme complex. The incubation mixture was filtered (0.2 μm) and incubated with shaking at 37°. The progress of the reactions was monitored at 12, 24, and 48 hr (using normal and reverse-phase thin-layer chromatography) with most riboflavin analogs being converted by 48 hr. It is important to note that filtration of the reaction mixture was essential for FAD analog synthesis. Bacterial contamination apparently prevents the conversion of riboflavin analogs.

Table I summarizes the FAD analogs which have been obtained using this and earlier procedures.[15-21] In contrast to these analogs, the 7,8-didemethyl-8-hydroxy-5-deazariboflavin (F$_o$) and 5-methyl-F$_o$ analogs,

[15] V. Massey and T. Nishino, "Flavins and Flavoproteins (K. Yagi and T. Yamano, eds.), p. 1. University Park Press, Baltimore, Maryland, 1980.
[16] E. G. Moore, E. Cardemil, and V. Massey, *J. Biol. Chem.* **253**, 6413 (1978).
[17] C. Thorpe and V. Massey, *Biochemistry* **22**, 2972 (1983).

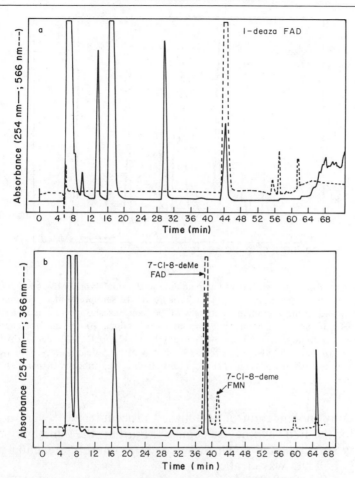

FIG. 1. Purification of FAD-level flavin analogs by HPLC. The incubation mixtures containing the FAD analogs were prepared for HPLC as described in the text. Chromatography of these samples (3 mg) was performed at 2.0 ml/min on a C_{18} reverse-phase semipreparative column (Alltech, 10 × 250 mm) in buffer containing 5 mM ammonium acetate (pH 7.2) and 5% methanol. After 15 min isocratic elution to remove nucleotide contaminants, a 15-min linear gradient from 5 to 20% methanol was carried out. Isocratic elution continued for 10 min followed by a 15-min linear gradient from 20 to 100% methanol to remove unreacted substrate and other products. The profiles shown are for incubation mixtures generating (a) 1-deaza-FAD and (b) 7-chloro-8-demethyl-FAD.

[18] S. G. Mayhew, C. D. Whitefield, S. Ghisla, and M. S. Schuman-Jorns, *Eur. J. Biochem.* **44,** 579 (1974).

[19] R. Hille, J. A. Fee, and V. Massey, *J. Biol. Chem.* **256,** 8933 (1981).

[20] S. Ghisla, V. Massey, and S. G. Mayhew, *in* "Flavins and Flavoproteins" (T. P. Singer, ed.), p. 334. Elsevier, Amsterdam, 1976.

[21] G. Zanetti, V. Massey, and B. Curti, *Eur. J. Biochem.* **132,** 201 (1983).

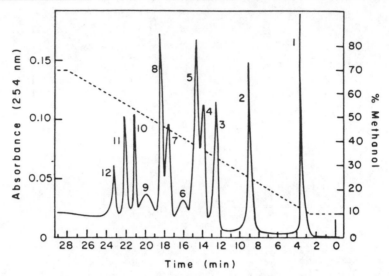

FIG. 2. Separation of a variety of flavin analogs at the riboflavin, FMN, and FAD levels. Flavin analogs were separated using a LiChrosorb RP-18 semipreparative column (Altex, 1 × 25 cm) and a linear gradient elution with methanol as shown, in 5 mM ammonium acetate (pH 6.0) buffer at 4 ml/min. The sample contained 10–15 nmol of each of the following flavin analogs: (1) ATP, (2) 8-hydroxy-FAD, (3) FAD, (4) 1-deaza-FAD, (5) 5-deaza-FAD, (6) FMN, (7) 1-deaza-FMN, (8) riboflavin, (9) 5-deaza-FMN, (10) 1-deazariboflavin, (11) 6-methylriboflavin, (12) 5-deazariboflavin. Reprinted with permission from Light et al.[25]

prepared by modification of established procedures,[22–24] were not substrates for the FAD synthetase complex. Further, the 6-methylriboflavin was very slow reacting under these conditions and the reaction with 8-hydroxyriboflavin was erratic.

The products of the FAD synthetase reactions were purified by HPLC after Amicon PM10 filtration (to remove protein), lyophilization, resuspension, and prefiltration through C$_{18}$ Sep-Paks (Waters). The structure of the particular FAD analog determined the optimal chromatographic conditions; however, a generally applicable procedure was developed as shown in Fig. 1. Dual-wavelength monitoring was used to detect flavins and other products of the reactions.

The HPLC system summarized here allows rapid and efficient separation and isolation of all the flavin coenzyme analogs at the riboflavin,

[22] W. E. Ashton, R. D. Brown, F. Jacobson, and C. Walsh, J. Am. Chem. Soc. 101, 4419 (1979).
[23] W. Ashton, R. D. Brown, and R. L. Tolman, J. Heterocycl. Chem. 15, 489 (1978).
[24] W. T. Ashton and R. D. Brown, J. Heterocycl. Chem. 17, 1709 (1980).

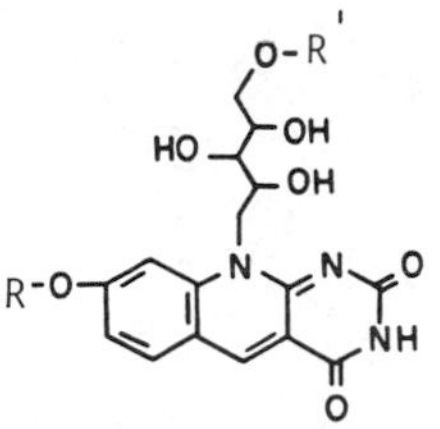

	$\underline{R}$	$\underline{R}'$
F_O	H–	–H
$F+$	H–	$-PO_3^{2-}$
F_{420}	H–	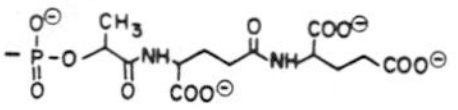 (structure)
F_{390}-A	AMP–	same as F_{420}
F_{390}-G	GMP–	same as F_{420}

FIG. 3. Structure of 5-deazaflavins.

FMN, and FAD levels after enzymatic incubation with flavokinase/FAD synthetase. It is ideally adapted for flavin analogs labeled with radioactive (e.g., ^3H, ^{14}C) or stable isotopes (^{13}C, ^{15}N). The power of this technique is demonstrated by the resolution of several flavin derivatives at various coenzyme levels[25] shown in Fig. 2.

5-Deazaflavins

In addition to the typical FMN and FAD flavin coenzymes, methanogenic bacteria possess high concentrations (~ 1 μmol/g dry weight) of 5-deazaflavin species.[26,27] Natural 5-deazaflavins arc also present in *Streptomycetes*[28,29] and *Mycobacterium*.[30] Under anaerobic conditions the major methanogen 5-deazaflavin is coenzyme F_{420}, a low-potential, two-electron redox carrier.[5,31] Hydrolysis products of F_{420}, called $F+$ and F_O, have also been characterized.[26] F_O, the riboflavin-level chromophore, has been chemically synthesized[22] and is excreted at high concentration into the growth media of active cells.[32] Two other 5-deazaflavins, termed F_{390}-A and F_{390}-G, have recently been described as occurring in cells exposed to oxidative conditions[33] (Fig. 3).

The *Methanobacterium thermoautotrophicum* 5-deazaflavins are all readily separated by reverse-phase HPLC (Fig. 4). In addition, F_{420} spe-

[25] D. R. Light, C. Walsh, and M. A. Marletta, *Anal. Biochem.* **109,** 87 (1980).
[26] L. D. Eirich, G. D. Vogels, and R. S. Wolfe, *J. Bacteriol.* **140,** 20 (1979).
[27] P. van Beelen, W. J. Geerts, A. Pol, and G. D. Vogels, *Anal. Biochem.* **131,** 285 (1983).
[28] A. P. M. Eker, A. Pol, P. van der Meyden, and G. D. Vogels, *FEMS Microbiol. Lett.* **8,** 161 (1980).
[29] J. D. R. McCormick and G. O. Morton, *J. Am. Chem. Soc.* **104,** 4014 (1982).
[30] T. Naraoka, K. Momoi, K. Fukasawa, and M. Goto, *Biochim. Biophys. Acta* **797,** 377 (1984).
[31] F. Jacobson and C. Walsh, *Biochemistry* **23,** 979 (1984).
[32] R. Kern, P. J. Keller, G. Schmidt, and A. Bacher, *Arch. Microbiol.* **136,** 191 (1983).
[33] R. P. Hausinger, W. H. Orme-Johnson, and C. Walsh, *Biochemistry* **24,** 1629 (1985).

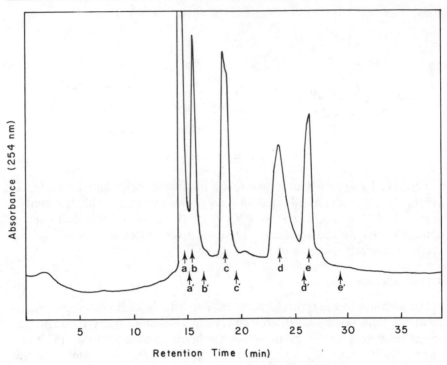

FIG. 4. Reverse-phase HPLC of methanogen 5-deazaflavins. A 30-min linear gradient from 0 to 50% methanol in 50 mM ammonium formate (pH 7.0) buffer was used for chromatography at 2.0 ml/min on a 10-μm C_{18} column (Alltech, 10 × 250 mm). The absorbance profile at 254 nm is shown for samples of (a) F_{390}-G, (b) F_{390}-A, (c) F_{420}, (d) F_o, and (e) F+. In addition, the elution positions are shown for samples reduced to the 1,5-dihydro level by NaBH$_4$ (a'–e', respectively).

cies containing four and five glutamyl residues isolated from *Methanosarcina barkeri* are resolved from the diglutamyl-F_{420} species.[27] Preparative-scale isolation of these species is readily accomplished using HPLC conditions similar to those in Fig. 4.[32,33]

We have recently utilized HPLC separations to demonstrate that exposure of methanogenic cells to oxidative conditions leads to the adenylylation and guanylylation of F_{420} at the 8-hydroxy position, producing F_{390}-A and F_{390}-G.[33] These species, of unknown function, lack the characteristic fluorescence of F_{420} and have absorbance maxima at 390 rather than 420 nm as in the coenzyme. The F_{390} species were digested with snake venom phosphodiesterase and the products were isolated by using HPLC. The product retention times (Table II) provided initial evidence that F_{390}-A was composed of F_{420} and AMP, and that F_{390}-G was

TABLE II
IDENTIFICATION OF F_{390} DIGESTION PRODUCTS
BY HPLC

Sample	Retention times (min)	
	Reverse-phase[a]	Anion-exchange[b]
F_{390}-A	23.5	—
F_{390}-G	21.3	—
PDE-F_{390}-A1[c]	28	—
PDE-F_{390}-A2	18	7.5
PDE-F_{390}-G1	29	—
PDE-F_{390}-G2	11.5	13.0
F_{420}	28	—
AMP	19	7.5
GMP	12	13.2

[a] A 30-min linear gradient from 0 to 30% methanol was used for sample chromatography at 2.0 ml/min on a 10-μm column (Alltech, 10 × 250 mm) equilibrated in 50 mM ammonium formate (pH 7.0).
[b] Isocratic chromatography was performed using a Whatman Partisil 10 SAX 10-μm column (4.6 × 250 mm) at 0.67 ml/min in 50 mM potassium phosphate buffer (pH 3.5).
[c] PDE, Snake venom phosphodiesterase-digested sample.

made up of F_{420} and GMP. Further experiments using the products isolated from HPLC were used to confirm these findings.[33]

The oxidized and air-stable dihydro-F_oH_2 species have been shown to be readily separated by HPLC.[25] As shown in Fig. 4, the other methanogen 5-deazaflavins similarly exhibit different retention times for their two redox forms. Since the 5 position in the 5-deazaflavins is exchange stable, in analogy to the C-4 of NAD(P)H, one can monitor hydrogen transfer into this locus in redox transformations[4] (Fig. 5).

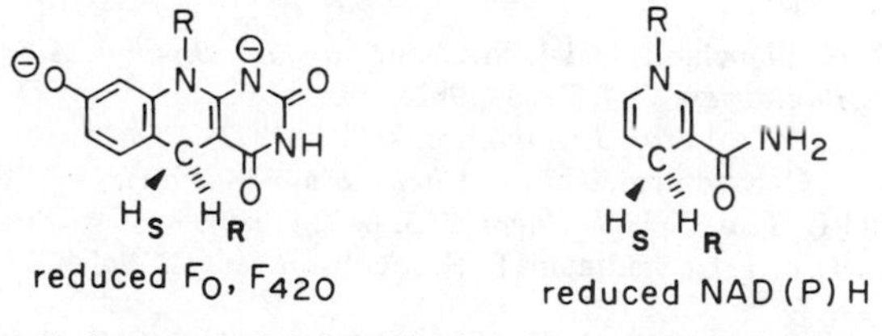

FIG. 5. Structure of 5-deazaflavins compared with NAD(P)H.

TABLE III
CHROMATOGRAPHIC PROPERTIES OF F_o ANALOGS

| Compound | HPLC[a] | | Silica gel[b] | RPS[c] |
	A^d	B^e		
7,8-Didemethyl-8-hydroxy-5-deaza-riboflavin (F_o)	9.0	5.7	—	0.33
5-CH_3-F_o	9.9	7.2	0.11	0.10
10-N-Dealkyl-5-deaza-F_o	44.4	15.8	0.40	—
Riboflavin	27.6	13.2	—	0.17
8-Hydroxy-8-demethyl-riboflavin	6.6	5.4	—	0.30

[a] HPLC was performed utilizing a semipreparative Alltech C_{18} column (10 × 250 mm).
[b] Merck Kieselgel 60F_{254} analytical TLC plates were used; solvent system: $CHCl_3$ (90%)–MeOH (10%).
[c] Analtech RPS-F (C_{18}) analytical reverse-phase plates were used; solvent system: water (70%)–MeOH (30%).
[d] Solvent system: 25% methanol–75% 5 mM $NH_4^+OAc^-$.
[e] Solvent system: 40% methanol–60% 5 mM $NH_4^+OAc^-$.

A number of methanogenic bacterial enzymes catalyze two-electron redox transformations of F_o and F_{420}. F_{420}-reducing hydrogenases,[34-36] formate dehydrogenase,[37,38] and NADPH oxidoreductase[39] have been purified essentially to homogeneity. For the methanogen oxidoreductase[40] and the hydrogenase,[41] studies on 5-deazaflavin specificity have been conducted. Several synthetic analogs of the methanogen 5-deazaflavin have been tested for substrate reactivity with the *Methanococcus vannielii* NADPH : F_{420} oxidoreductase.[41] Table III summarizes the chromatographic properties of several synthetic F_o analogs.[22,23] Of particular interest is the effect on chromatographic properties brought about by the struc-

[34] F. S. Jacobson, L. Daniels, J. A. Fox, C. T. Walsh, and W.H. Orme-Johnson, *J. Biol. Chem.* **257**, 3385 (1982).
[35] S.-L. C. Jin, D. K. Blanchard, and J.-S. Chen, *Biochim. Biophys. Acta* **748**, 8 (1983).
[36] S. Yamazaki, *J. Biol. Chem.* **257**, 7926 (1982).
[37] N. L. Schauer and J. G. Ferry, *J. Bacteriol.* **150**, 1 (1982).
[38] J. B. Jones and T. C. Stadtman, *J. Biol. Chem.* **256**, 656 (1981).
[39] S. Yamazaki and L. Tsai, *J. Biol. Chem.* **255**, 6462 (1980).
[40] S. Yamazaki, L. Tsai, T. C. Stadtman, F. S. Jacobson, and C. Walsh, *J. Biol. Chem.* **255**, 9025 (1980).
[41] S. Yamazaki, L. Tsai, and T. C. Stadtman, *Biochemistry* **21**, 934 (1982).

tural alterations; e.g., the introduction of an 8-hydroxy group increases polarity, whereas removal of the ribityl side chain or alkylation at C-5 increases the hydrophobicity. These trends can be used in developing HPLC purification procedures for other F_0 analogs.

In conclusion, HPLC is a tremendous tool which can be used to purify and separate flavin analogs at various coenzyme levels. These analogs have been and continue to be useful probes of enzyme mechanisms. It is hoped that the methods described here will stimulate further work in this field. In addition, the development of 5-deazaflavin analogs may yield insights into the mechanisms of enzymes which use the natural 5-deazaflavin coenzymes.

Acknowledgments

We wish to thank Dr. Tadhg Begley, Dr. Wallace Ashton (Merck & Co.), and Ms. Helen Getto for providing several flavin analogs. This work was supported by Grant GM31574 from the National Institutes of Health, by National Institutes of Health Postdoctoral Fellowship GM08527 (to R.P.H.), and by FCAC (Quebec) Postdoctoral Fellowship (to J.F.H.).

[34] Preparation, Properties, and Separation by High-Performance Liquid Chromatography of Riboflavin Phosphates

By Peter Nielsen, Peter Rauschenbach, and Adelbert Bacher

Riboflavin 5'-phosphate is synthetically prepared in large amounts for use in pharmaceutical preparations. This compound and its structural analogs also have an important role in investigations of flavoenzyme mechanisms. In spite of considerable efforts, however, the preparation of pure 5'-FMN[1] remains a problem. Chemical phosphorylation of riboflavin invariably yields a complex mixture of various riboflavin phosphates.[2] Although this has been recognized for some time, a more detailed investigation was hampered by the lack of satisfactory analytical methods.

---

[1] Abbreviations used: 2'-FMN, riboflavin 2'-phosphate; 3'-FMN, riboflavin 3'-phosphate; 4'-FMN, riboflavin 4'-phosphate; 5'-FMN, riboflavin 5'-phosphate; 2',4'-RDP, riboflavin 2',4'-diphosphate; 2',5'-RDP, riboflavin 2',5'-diphosphate; 3',4'-RDP, riboflavin 3',4'-diphosphate; 3',5'-RDP, riboflavin 3',5'-diphosphate; 4',5'-RDP, riboflavin 4',5'-diphosphate; FAD, flavin adenine dinucleotide.

[2] G. Scola-Nagelschneider and P. Hemmerich, *Eur. J. Biochem.* **66**, 567 (1976).

Efforts to purify 5'-FMN by thin-layer chromatography of crude synthetic product has had only limited success.[2] Pure material could be obtained by chromatography on immobilized apoflavodoxin from *Megasphaera elsdenii*.[3] However, this method is laborious and not generally available. Other methods for the preparation of 5'-FMN include the denaturation of flavoproteins and the phosphorylation of riboflavin by the enzyme flavokinase.[4,5] These procedures are laborious, and the enzymatic phosphorylation of riboflavin analogs is limited by the rather narrow substrate specificity of flavokinase.

Reverse-phase HPLC can serve as a simple and widely available method for sensitive and accurate determination of isomeric riboflavin phosphates and structural analogs.[6] With this tool in hand, simple and reproducible methods for the preparation of various isomers in pure form could be developed.

Analytical HPLC Separation of Flavin Phosphates

Principle. A variety of isomeric riboflavin monophosphates and diphosphates can be separated by reverse-phase HPLC using isocratic ammonium formate eluants.

Reagents

Eluant 1: 17% methanol, 0.1 M ammonium formate, 0.1 M formic acid
Eluant 2: 20% methanol, 0.1 M ammonium formate
Eluant 3: 13% methanol, 50 mM ammonium formate, 50 mM magnesium formate, pH 4.0
Eluant 4: 27% methanol, 5 mM tetrabutylammonium formate, pH 3.5

Procedure. Analytical HPLC is performed with columns of Nucleosil 100-10 C$_{18}$ from Macherey and Nagel, Düren, West Germany (4 × 250 mm; 4000 theoretical plates as determined with phenylacetic acid as solute and 30% methanol adjusted to pH 3.0 with formic acid as solvent at a flow rate of 1.0 ml/min). Optimum separation is achieved at a flow rate of 1.0–2.0 ml/min. Injection volumes should not exceed 300 μl. The quantity of total flavin phosphates should not exceed 1 μmol. The effluent is monitored by photometry or fluorometry. The minimum amount detectable by

[3] S. G. Mayhew and M. J. J. Strating, *Eur. J. Biochem.* **59**, 539 (1975).
[4] S. G. Mayhew, *in* "Flavins and Flavoproteins" (H. Kamin, ed.), p. 185. University Park Press, Baltimore, Maryland, 1971.
[5] A. H. Merill and D. B. McCormick, this series, Vol. 66, Part E [40], p. 287.
[6] P. Nielsen, P. Rauschenbach, and A. Bacher, *Anal. Biochem.* **130**, 359 (1983).

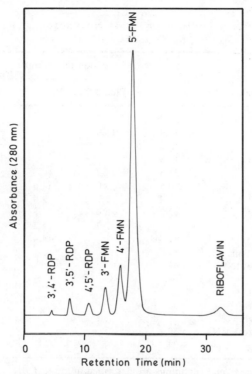

FIG. 1. HPLC of commercial FMN (Sigma). Column, Nucleosil 100-10 C_{18}, 250 × 4 mm; inject volume, 3 μl (40 μg); eluant, 100 mM ammonium formate and 100 mM formic acid in 17% methanol (eluant 1).

photometry at 436 nm is <100 pmol (corresponding to a signal five times the photometer noise of 2% at 5×10^{-3} absorbance units full scale). Less than 30 pmol can be detected photometrically at 254 nm. With fluorometric detection, sensitivity is better than 1 pmol (excitation, 470 nm; emission, 530 nm).

Comments. The chromatogram of a commercial FMN sample is shown in Fig. 1. All commercial FMN samples analyzed contain only about 75% 5'-FMN.[6]

Typical retention times are summarized in Table I. Eluant 1 separates all isomeric monophosphates of riboflavin. However, 2'-FMN comigrates with riboflavin in this eluant. The use of eluant 2 allows the baseline separation of riboflavin and 2'-FMN. Eluant 3 is superior to other eluants for the analytical separation of riboflavin diphosphates. The reverse-phase HPLC method is also useful for the separation of the phosphoric acid derivatives of a variety of riboflavin analogs and of FAD.

TABLE I
RETENTION TIMES OF RIBOFLAVIN PHOSPHATES IN
REVERSE-PHASE HPLC[a]

Compound	Retention time (min)			
	Eluant 1	Eluant 2	Eluant 3	Eluant 4
Riboflavin	32.2	21.2		5.0
5'-FMN	17.9	11.0		12.5
4'-FMN	15.7	9.5		11.3
3'-FMN	13.3	6.8		10.2
2'-FMN	31.8	18.2		
4',5'-RDP	10.6	8.0	12.6	31.6
3',5'-RDP	7.4	2.3	9.0	25.2
2',5'-RDP	13.2	6.0	17.3	
3',4'-RDP	4.5	2.3	5.2	17.8
2',4'-RDP	11.4		14.4	
5-Deazariboflavin	84.7			
5-Deaza-FMN	48.3			
5-Deaza-3',5'-RDP	24.3			
Hexafluoro-riboflavin	56.7			
Hexafluoro-FMN	30.1			
FAD	24.7			

[a] Experimental conditions are described in the text.

The various phosphates of riboflavin can also be separated by ion-pair chromatography on reverse-phase HPLC columns (eluant 4).[6] A similar approach was reported by Entsch and Sim.[7] However, we found ion-pair chromatography to be less reproducible than the reverse-phase technique.[6] It should also be noted that buffer substances appropriate for ion-pair chromatography are not easily removed from solutes in preparative work.

Preparative HPLC Separation

Principle. By the use of a preparative HPLC column it is possible to isolate milligram amounts of 5'-FMN in high purity (>95%) from crude preparations of FMN. Recovery rates are high. This is significant in work with rare analogs of riboflavin and with isotope-labeled samples.

[7] B. Entsch and R. G. Sim, *Anal. Biochem.* **133,** 401 (1983).

Procedure. Preparative separations are performed with a column of LiChrosorb RP-18 from Merck AG, Darmstadt, West Germany (10 μm, 250 × 16 mm; 2000 theoretical plates as determined with phenylacetic acid as solute and 30% methanol adjusted to pH 3.0 with formic acid as eluant at a flow rate of 25 ml/min). Eluants contain 0.1 M ammonium formate and 0.1 *M* formic acid in aqueous methanol (12.5% methanol for riboflavin monophosphates; 10% methanol for riboflavin diphosphates; 20% methanol for 5-deazaflavin phosphates). Flow rates of 10–40 ml/min give optimum separation. Injection volumes should not exceed 6 ml. The quantity of total flavin phosphates injected should not exceed 20 μmol. The effluent is monitored photometrically. Fractions are collected and concentrated by evaporation under reduced pressure. Ammonium formate is removed by repeated lyophilization.

Conventional Ion-Exchange Chromatography

Principle. Partial separation of isomeric compounds can be achieved by conventional ion-exchange chromatography. This technique is particularly useful in combination with analytical HPLC to monitor the exact composition of fractions. Ion exchange permits the preparation of large amounts of enriched material suitable for subsequent purification by preparative HPLC. Some compounds can even be prepared in pure or almost pure form by conventional chromatography.

Method A. FMN (sodium salt from Sigma, F 6720, 256 mg, 0.52 mmol) is dissolved in 55 ml of 30% 2-propanol. The solution is applied to a column of DEAE-Sephacel (Pharmacia Inc., acetate form, 2.5 × 21.5 cm). The column is washed with 75 ml of 30% 2-propanol. The flow rate is 26 ml/hr. Flavin phosphates are eluted with a discontinuous gradient of triethylammonium acetate (pH 7.0) in 30% 2-propanol using an Ultrograd gradient maker (LKB). Fractions are analyzed by analytical HPLC (Fig. 2). Appropriate fractions are combined and evaporated to dryness under reduced pressure. The buffer salt is removed by lyophilization.

This procedure separates riboflavin monophosphates efficiently from riboflavin and riboflavin diphosphates. The riboflavin monophosphates are eluted in a single band. However, 4'-FMN is considerably enriched in the leading edge, and 3'-FMN is enriched in the trailing edge.

Method B.[8] Commercial FMN (260 mg, 0.52 mmol) is dissolved in 45 ml of water. The solution is adjusted to pH 7.0 by the addition of 0.1 *M* sodium hydroxide and applied to a column of DEAE-cellulose DE-52

C. G. van Schagen and F. Müller, *Eur. J. Biochem.* **120**, 33 (1981).

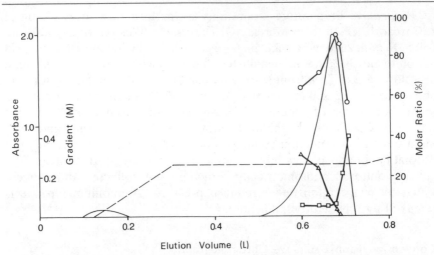

FIG. 2. Chromatography of commercial FMN on DEAE-cellulose by Method A. Column of DEAE-Sephacel, acetate form, 2.5 × 21.5 cm; gradient of triethyammonium acetate (pH 7) in 30% 2-propanol. Absorbance at 515 nm (——); gradient of triethylammonium acetate (– –); 3′-FMN (□); 4′-FMN (△); 5′-FMN (○).

(Whatman, 25 × 20.3 cm, chloride form). The flow rate is 48 ml/hr. The column is washed with 700 ml of water. Flavin phosphates are eluted with 25 mM ammonium carbonate, pH 7.8 (Fig. 3). The trailing edge of the monophosphate band contains almost pure 5′-FMN. Fractions are combined and evaporated to dryness under reduced pressure. The remaining residue is lyophilized.

Preparation of Riboflavin Monophosphates

5′-FMN. Enriched fractions from DEAE-cellulose chromatography (Method B) are combined, yielding 23 μmol of 95% pure 5′-FMN. Preparative HPLC separation of this material affords 5′-FMN with >99% purity.

4′-FMN. Enriched fractions from DEAE-cellulose chromatography (Method A) are combined and concentrated, yielding a mixture of 63% 5′-FMN and 34% 4′-FMN (0.33 mmol). Preparative HPLC affords 4′-FMN with >99% purity.

3′-FMN. The preparation of this compound is based on the observation that 5′-FMN and 4′-FMN are rapidly hydrolyzed by alkaline phosphatase, whereas 3′-FMN reacts slowly.[6] Commercial FMN (4.0 g, 8.0 mmol) is suspended in 200 ml of 0.1 M Tris hydrochloride (pH 8.5) containing 5 mM MgCl$_2$. Alkaline phosphatase (Boehringer, Grade II, 350 U) is added and the reaction mixture is incubated at 37° for 26 hr. The

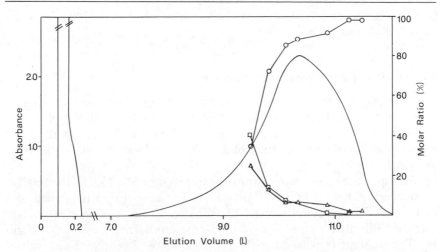

FIG. 3. Chromatography of commercial FMN on DEAE-cellulose by Method B. Column of DEAE-cellulose DE-52, chloride form, 2.5 × 23.5 cm; elution with 25 mM ammonium carbonate (pH 7.8). Absorbance at 445 nm (——); 3'-FMN (△); 4'-FMN (□); 5'-FMN (○).

precipitate consisting of riboflavin is removed by centrifugation. The supernatant is concentrated to a volume of 100 ml by evaporation under reduced pressure. Alkaline phosphatase (70 U) is added and incubation at 37° is continued for 17 hr. The suspension is centrifuged. The supernatant contains 43% 3'-FMN (0.42 mmol), 6% 4'-FMN, and 49% riboflavin. The solution is placed on a column of DEAE-cellulose DE-52 (Whatman; 4.5 × 31 cm; acetate form equilibrated with 30% 2-propanol). The flow rate is 70 ml/hr. The column is washed with 30% 2-propanol. Riboflavin monophosphates are eluted with 0.11 M triethylammonium acetate (pH 6.5) in 30% 2-propanol. Fractions from the trailing edge of the riboflavin monophosphate band contain pure 3'-FMN. They are combined, concentrated under reduced pressure, and lyophilized. Yield, 0.47 mmol; purity, >99%.

2'-FMN. Only trace amounts of 2'-FMN are present in commercial FMN. However, the compound can be obtained by acid-catalyzed isomerization of 3'-FMN.[9,10] A solution of crude 3'-FMN (0.29 mmol, prepared by phosphatase treatment of commercial FMN as described above) in 11 ml of 0.1 M hydrochloric acid is incubated at 50° for 51 hr. The reaction is quenched by the addition of 2 ml of 2 M ammonium formate. The suspension is centrifuged. The supernatant is applied to a preparative

[9] P. Nielsen, Ph.D. Thesis, Technical University of Munich (1983).
[10] P. Nielsen, J. Harksen, and A. Bacher, *Eur. J. Biochem.* (in press).

HPLC column (eluant 2) and the 2'-FMN fraction is collected. Yield: 57 μmol (19%); purity, 98%.

Preparation of Riboflavin Diphosphates

Principle. Enriched material is obtained by ion-exchange chromatography. Subsequent HPLC separation of appropriate fractions yields pure 3',5'-RDP and 4',5'-RDP. Other riboflavin diphosphates are synthesized by acid-catalyzed isomerization of 3',5'-RDP and subsequent HPLC separation.[9]

Ion-exchange Chromatography. Commercial FMN (5.0 g, 10 mmol) is suspended in 400 ml of 30% 2-propanol. The suspension is applied to a column of DEAE-cellulose DE-52 (Whatman, 4.5 × 27.5 cm, acetate form, equilibrated with 30% aqueous 2-propanol). The flow rate is 125 ml/hr. The column is washed with 2 liters of 30% 2-propanol and subsequently with 6 liters of 0.11 M triethylammonium acetate (pH 6.0) containing 30% 2-propanol. This procedure elutes riboflavin and riboflavin monophosphates. The column is subsequently developed with 0.15 M triethylammonium acetate (pH 6.0). Three orange yellow bands are eluted separately. Fractions from 3.5 to 6.3 liters (fraction I) and from 7.8 to 10.1 liters (fraction II) are collected and concentrated. Fraction I contains 83% 4',5'-RDP (0.14 mmol); fraction II contains 92% 3',5'-RDP (0.20 mmol).

3',5'-RDP and 4',5'-RDP. The fractions described above are subjected to preparative HPLC separation, yielding pure 3',4'-RDP and 4',5'-RDP.

Other Riboflavin Diphosphates. A solution of crude 3',5'-RDP (0.22 mmol prepared by ion-exchange chromatography as described above) in 13 ml of 0.17 M hydrochloric acid is incubated at 50° for 48 hr. The solution is applied directly to a preparative HPLC column (eluant 1). Fractions are collected, yielding 5.7 μmol 3',4'RDP (98% purity), 50 μmol 3',5'-RDP (98% purity), 33 μmol 4',5'-RDP (96% purity), 41 μmol 2',5'-RDP (98% purity), and 4.1 μmol 2',4'-RDP (84% purity).

Phosphorylation of Riboflavin Analogs

Principle. Riboflavin analogs are phosphorylated by the chlorophosphoric acid method as described by Scola-Nagelschneider and Hemmerich.[2] The product mixture is immediately purified by preparative HPLC. Unreacted starting material can be recovered.

Chlorophosphoric Acid.[2] Freshly distilled $POCl_3$ (7.3 ml, 80 mmol) is cooled in ice water and stirred vigorously in a 25 ml flask. Water (2.9 ml, 0.16 mol) is added over a period of 2–3 hr using a peristaltic pump. The solution is allowed to warm up to room temperature for 15 min. During this time a stream of dry nitrogen is slowly bubbled through the liquid in order to remove hydrochloric acid.

5-Deazariboflavin 5'-Phosphate. 5-Deazariboflavin (7.0 mg; 18.6 μmol) is treated with 0.5 ml (6.7 mmol) of freshly prepared chlorophosphoric acid for 19 hr at room temperature. The reaction is terminated by the addition of 5 ml of water. NH_4OH is added to a final pH of 3.0. The solution is applied directly to a preparative HPLC column. Fractions are collected and evaporated to dryness under reduced pressure. The residue is lyophilized. Yield: 5.0 μmol (26.7%); purity, 95%.

7α,7α,7α,8α,8α,8α-Hexafluororiboflavin 5'-Phosphate. Hexafluororiboflavin (7.0 mg; 14.5 μmol) is treated with chlorophosphoric acid and the product is purified as described above.[11] Yield, 2.1 μmol (14.1%); purity, 95%.

Structure and Properties

The structures of the compounds under study were unequivocally established by periodate titration[12] (Fig. 4, Table II) and by a kinetic study of their acid-catalyzed isomerization.[9,10] Chemical properties are summarized in Table II. Thin-layer chromatography data are summarized in Table III. Isomeric monophosphates could not be adequately separated by thin-layer chromatography. However, the method is sometimes useful to distinguish monophosphates from diphosphates and from unphosphorylated flavins.

3',5'-RDP binds tightly to apoflavodoxin from *M. elsdenii*. The dissociation constant of 9.7 nM is about two magnitudes larger as compared to the natural cofactor, 5'-FMN.[6] Apoflavodoxin reconstituted with 3',5'-RDP catalyzes the transfer of reduction equivalents from H_2 via hydrogenase from *Clostridium kluyveri* to metronidazol.[9] Some other compounds under study also bind to apoflavodoxin, but the complexes show no enzyme activity. Some of the compounds have limited catalytic activities with luciferase from *Photobacterium fisheri* (Table II).

Stability

Riboflavin phosphates as solids or in aqueous solution at pH 7.0 are stable for at least several months when stored at −20° and protected from light. Rapid isomerization occurs in acid solution (pH <2) at elevated temperature. The thermodynamic equilibrium is characterized by the presence of about 65% 5'-FMN, 11% 4'-FMN, 8% 3'-FMN, and 15% 2'-

[11] P. Nielsen, A. Bacher, D. Darling, and M. Cushman, *Z. Naturforsch., C: Biosci.* **38C,** 701 (1983).

[12] J. S. Dixon and D. Lipkin, *Anal. Chem.* **26,** 1092 (1954).

TABLE II

PROPERTIES OF FLAVINS AND FLAVIN PHOSPHATES

Compound	Purity[a] (%)	Periodate consumption (mol/mol flavin)	Phosphate content (mol/mol flavin)	Enzymatic hydrolysis[b]	Apoflavodoxin binding		Luciferase activity[d]
					RF[c] (%)	K_D (nM)	
Riboflavin		2.80					100
5'-FMN	>99[e]	2.12	1.04	46	0.9	0.11	3.36
4'-FMN	>99[e]	1.02	0.91	6.7			0.16
3'-FMN	99	1.07	1.03	0.1			0.22
2'-FMN	98	2.14	0.95	0.5			0.31
4',5'-RDP	99	0.89	1.98	100			0.55
3',5'-RDP	99	0	1.87	75	4.6	9.7	
2',5'-RDP	99		2.13	104			0.16
3',4'-RDP	95	0	2.03	2.6			
2',4'-RDP	84	0	2.13	14			
5-Deaza-FMN	95				6.3	0.16	0.10
5-Deaza-3',5'-RDP	98				17.9	2.6	0.10
Hexafluoro-FMN	95				4.8	3.2	0

[a] Determined by analytical HPLC.
[b] Relative velocity of enzymatic hydrolysis with alkaline phosphatase from bovine intestine.
[c] RF, Residual fluorescence after titration with a 2-fold excess of apoflavodoxin from *M. elsdenii.*
[d] Relative catalytic activity with luciferase from *Photobacterium fisheri;* 5'-FMN = 100%.
[e] No impurity detectable.

TABLE III
THIN-LAYER CHROMATOGRAPHY OF RIBOFLAVIN PHOSPHATES[a]

Compound	R_f							
	I	II	III	IV	V	VI	VII	VIII
Riboflavin	0.42	0.54	0.41	0.81	0.74	0.93	0.57	0.53
5'-FMN	0.27	0.11	0.62	0.55	0.14	0.82	0.34	0.28
4'-FMN	0.30	0.12	0.64	0.56	0.21	0.82	0.39	0.29
3'-FMN	0.28	0.10	0.62	0.55	0.21	0.84	0.36	0.29
2'-FMN	0.31	0.11	0.68	0.55	0.12	0.82	0.36	0.31
4',5'-RDP	0.16	0.04	0.67	0.52	0.05	0.46	0.27	0.09
3',5'-RDP	0.15	0.04	0.61	0.52	0.05	0.39	0.23	0.05
2',5'-RDP	0.16	0.04	0.62	0.52	0.04	0.32	0.25	0.14
3',4'-RDP	0.15	0.04	0.65	0.52	0.04	0.37	0.16	0.10
5-Deaza-FMN	0.36	0.27		0.54		0.85	0.43	0.36
5-Deaza-3',5'-RDP		0.15		0.42		0.18		0.10
Hexafluoro-5'-FMN	0.51	0.14		0.68			0.39	0.45

[a] Precoated cellulose or silica gel plates (Merck AG, Darmstadt, West Germany) were developed with the solvent systems indicated. I, *n*-Butanol/ethanol/water (50:15:35, v/v)/cellulose; II, *n*-butanol/ethanol/water (50:15:35, v/v)/silica gel; III, *tert*-butanol/water (60:40, v/v)/cellulose; IV, *tert*-butanol/water (60:40, v/v)/silica gel; V, collidine/water (75:25, v/v)/cellulose; VI, collidine/water (75:25, v/v)/silica gel; VII, *n*-butanol/acetic acid/water (50:20:30, v/v)/cellulose; and VIII, *n*-butanol/acetic acid/water (50:20:30, v/v)/silica gel.

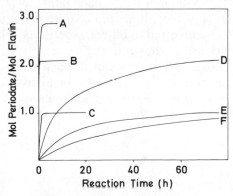

FIG. 4. Periodate consumption by riboflavin phosphates. Aqueous solutions containing 35–50 μM flavin and 160 μM sodium periodate were incubated in a photometric cell at 23°. The rate of periodate consumption was recorded at 230 nm.[12] A, riboflavin; B, 2'-FMN; C, 3'-FMN, D, 5'-FMN; E, 4'-FMN; F, 4',5'-RDP.

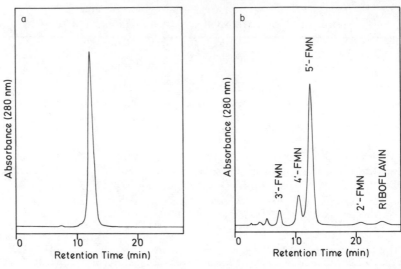

Fig. 5. HPLC of (a) pure 5'-FMN and (b) a 5'-FMN sample heated at 121° for 15 min in 0.1 M HCl. Column, Nucleosil, 100-10 C_{18}; eluant, 100 mM ammonium formate in 20% methanol, pH 6.3 (eluant 2).

FMN. The rate constants for the various isomerization reactions have been determined under a variety of experimental conditions.[9,10] According to the official method of the American Association of Analytical Chemists, flavins are extracted from biological samples by treatment with 0.1 M HCl at 121° for 15–30 min.[13] It is known that this treatment leads to hydrolysis of FAD. However, it should also be noted that a substantial fraction of 5'-FMN is converted to other isomeric phosphates under these experimental conditions, as shown in Fig. 5.

At pH values above 2, hydrolysis of the phosphoric acid ester bond becomes the dominant reaction. The rate of hydrolysis is maximal at about pH 4.[9] At pH values above 7, the isoalloxazine ring is rapidly destroyed at elevated temperature.[14] No phosphate migration was observed.[9]

[13] W. Horwitz, ed., "Official Methods of Analysis," 12th Ed. Assoc. Off. Anal. Chem., Washington, D.C., 1975.
[14] D. E. Guttman and T. E. Platek, *J. Pharm. Sci.* **56,** 1423 (1967).

[35] Competitive Binding Assays for Riboflavin and Riboflavin-Binding Protein

By Harold B. White, III

Principle. Riboflavin-binding protein from chicken egg tightly and specifically binds riboflavin in preference to FMN or FAD. In a solution containing limiting amounts of binding protein, radiolabeled riboflavin will readily equilibrate with the endogenous bound and unbound riboflavin pools. In contrast to protein-bound riboflavin, free riboflavin is not retained by anion-exchange resins such as DEAE-cellulose; thus, the two riboflavin pools can be separated and analyzed. The equilibrium distribution of radiolabeled riboflavin in the free and bound pools is determined by the total amounts of riboflavin and its binding protein, and by the final specific radioactivity of the riboflavin. By monitoring the amount of bound, radiolabeled riboflavin as a function of unlabeled riboflavin or the amount of riboflavin-binding protein, respectively, the vitamin or the protein can be assayed.[1]

Isotope Dilution Assay for Riboflavin

Reagents

Tris–HCl, 25 mM, pH 7.5. Used as a diluent for all the solutions below.

D-[2-^{14}C]Riboflavin, 10 μg/ml (161 μCi/mg). Solutions are stored frozen in the dark.

D-Riboflavin, 50 μg/ml. Prepared from recrystallized riboflavin and used as a standard. Stored in a brown bottle at 4°. Since riboflavin preparations often contain impurities, it is advisable to check the concentration spectrophotometrically at several wavelengths. The molar extinction coefficients at 260, 375, and 450 nm are 27,700, 10,600, and 12,200, respectively.[2]

Aporiboflavin-binding protein, 0.1 mg/ml. Riboflavin-binding protein is easily purified from egg white.[3] Based on a molecular weight of

---

[1] S. E. Lotter, M. S. Miller, R. C. Bruch, and H. B. White, III, *Anal. Biochem.* **125,** 110 (1982).

[2] K. Yagi, *Methods Biochem. Anal.* **10,** 319 (1962).

[3] M. S. Miller and H. B. White, III, this volume [36].

29,000–30,000[4] 1.0 μg of aporiboflavin-binding protein will bind 12.5–12.9 ng of riboflavin.

NaCl, 1.0 M and 4.0 M

Preparation of DEAE-Cellulose Minicolumns. "Pipet tips" (Sarstedt, No. 91.787) are suspended vertically in test tube racks and a small plug of glass wool is seated in the tip of each. One milliliter of a thick slurry of DE-52 (Whatman) equilibrated with 25 mM Tris–HCl (pH 7.5) is added to each minicolumn. The columns are ready to use when the meniscus enters the resin bed. It is convenient to construct Plexiglas column racks which nest above racks that hold 24 scintillation vials (Fisher, 03-337-13) so that eluates can be collected directly into scintillation vials. The minicolumns may be reused many times, provided they are washed with 2.0 ml of 4 M NaCl, reequilibrated with 4.0 ml of 25 mM Tris–HCl (pH 7.5), and kept moist.

Incubation Conditions. Varying amounts of (0–1.5 μg) of unlabeled riboflavin are added to a series of glass test tubes (10 × 75 mm) and will be used to generate a standard curve. Similarly, varying amounts of sample solutions are added to another series of glass test tubes. To each tube is added 100 μl of aporiboflavin-binding protein solution and 20 μl of ^{14}C-labeled riboflavin solution. Tris–HCl buffer is then added to a total volume of 1.0 ml. The reaction mixtures are thoroughly mixed and incubated at least 5 min at room temperature.

Blanks and Reagent Controls. Two additional reaction mixtures should be prepared in duplicate. One contains excess aporiboflavin-binding protein to verify that the assays were conducted with limiting protein. This control also monitors the total bindable radioactivity used per assay. If the bound counts in the control decrease over a period of weeks, it indicates that the riboflavin solution is decomposing. The other reaction mixture contains a 100-fold excess of unlabeled riboflavin (≥400 μl) to be used as a background blank.

Separation of Free and Bound Riboflavin. Riboflavin-binding protein is negatively charged and binds tightly to DEAE-cellulose while riboflavin is uncharged and will not bind. The content of each reaction mixture is poured directly into a corresponding DE-52 minicolumn and the eluate is discarded. Each assay tube is rinsed with two successive 1.0-ml buffer washes and the contents applied to the appropriate columns. Any remaining free riboflavin is then washed off the columns with three 2-ml buffer washes. To prevent mixing, successive washes are applied to columns only after the preceding wash has completely drained and the meniscus has entered the resin.

[4] Y. Hamazume, T. Mega, and T. Ikenaka, *J. Biochem. (Tokyo)* **95**, 1633 (1984).

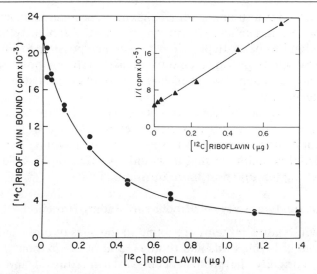

FIG. 1. Standard curve for the determination of riboflavin by isotope dilution. Increasing amounts of unlabeled riboflavin compete with labeled riboflavin for a limited number of binding sites on riboflavin-binding protein. Free and bound riboflavin are separated on small DEAE-cellulose columns. Riboflavin bound to riboflavin-binding protein is batch-eluted into scintillation vials. The amount of protein-bound riboflavin is determined and plotted as a function of added unlabeled riboflavin. The duplicate values have been averaged and the reciprocal values have been plotted to yield the linear standard curve shown in the inset. (Taken from Lotter et al.[1] with permission of the publisher.)

The rack of minicolumns is then positioned above a rack of scintillation vials. Protein-bound riboflavin is eluted directly into the vials with four successive 0.5-ml washes of 1.0 M NaCl. Ten milliliters of ACS liquid counting solution (Amersham) is added to each vial and the samples are counted in a liquid scintillation counter to determine the amount of protein-bound radioactivity. Riboflavin binds sufficiently tightly to riboflavin-binding protein (K_D 1 nM) that dissociation and loss of bound riboflavin before elution of the protein are insignificant under the assay conditions.[1]

Analysis of Data. After subtracting background, a standard curve is constructed as is shown in Fig. 1. The linear plot of the standard curve,[5] (counts per minute)$^{-1}$ vs amount of unlabeled riboflavin, is particularly useful in determining the amount of riboflavin in samples.

Applications and Limitations of the Assay. When homogeneous aporiboflavin-binding protein is used as described here, there is no need for a

[5] R. P. Ekins, *in* "Radiochemical Methods in Analysis" (D. J. Coomber, ed.), p. 349. Plenum, New York, 1975.

standard curve because unknown riboflavin values can be calculated directly; however, in practice it is convenient to use a standard curve which serves as a double check. In principle it should be possible to conduct this assay using diluted egg white directly as a source of riboflavin-binding protein as has been done by Fazekas et al.[6] In that case a standard curve would be necessary.

This assay has advantages over fluorometric assays[7] in that turbid solutions such as milk can be assayed directly.[1] Although this assay has not been applied to a variety of tissue extracts, it seems, based on the results with urine and milk, that it should be widely applicable for solutions containing 0.5 μg/ml or more of riboflavin.

Radioligand Binding Assay for Riboflavin-Binding Protein

Riboflavin-binding proteins are found in the eggs of birds and reptiles, in the plasma of laying birds and reptiles, and in the plasma of pregnant mammals. Normally the protein occurs as a mixture of apo and holo forms. The amount of bound riboflavin varies from near 100% saturation in chicken egg yolk to near 0% in the egg whites of some birds. The indeterminate amount of endogenous riboflavin will dilute added radiolabeled riboflavin such that the binding of the radioligand is not a simple linear function of the concentrations of the binding protein. By taking into account this isotope dilution, the assay described below can determine both the total amount of riboflavin-binding protein and its fractional saturation.

Reagents. All of the reagents used in the preceding assay for riboflavin are used in this assay except the specific activity of the D-[2-^{14}C]riboflavin is 16.1 μCi/mg.

Sample Preparation. Egg white, egg yolk, and plasma are diluted in 25 mM Tris–HCl (pH 7.5). The yolk and plasma are diluted 1:4 and egg white is diluted 1:1. The samples are homogenized by stirring. Particulate material is removed by a short (10 min) centrifugation at 40,000 g.

Incubation Conditions. The experimental protocol is a variation of that described for the assay of riboflavin. Increasing amounts, representing 7–10 different concentrations of a diluted sample, are added to a series of glass tubes (10 × 75 mm). To each tube is added 20 μl of [^{14}C]riboflavin solution and Tris–HCl buffer is added to a final volume of 1.0 ml. The reaction solutions are mixed, incubated, and the bound riboflavin isolated exactly as described for the assay of riboflavin.

Blanks and Controls. Since riboflavin is sparingly soluble, it is not practical in most cases to add a 100-fold excess of unlabeled riboflavin to

[6] A. G. Fazekas, C. E. Menendes, and R. S. Rivlin, *Biochem. Med.* **9**, 167 (1974).
[7] J. A. Tillotson and M. M. Bashor, *Anal. Biochem.* **107**, 214 (1980).

obtain a background blank. Blanks obtained by omitting protein tend to be slightly higher than the extrapolated value at zero protein or assays of egg whites deficient in riboflavin-binding protein.

The total amount of bindable [^{14}C]riboflavin in an assay (dpm_T) is determined by adding an excess of aporiboflavin-binding protein solution to a standard assay mixture. This value is critical for the analysis of the data and thus duplicates are averaged.

Although the amount of riboflavin-binding protein can be calculated from the specific radioactivity of the added riboflavin, it is useful to generate a standard curve using aporiboflavin-binding protein.

Theory. The derivation and analysis that follow assume (1) the total amount of riboflavin in each assay tube is in excess of the available binding sites, (2) the free and bound riboflavin are at isotopic equilibrium, and (3) the binding is tight ($K_D < 10$ nM).

The specific radioactivity of riboflavin in each assay tube can be expressed as dpm_T/Rf_T where Rf_T is the total amount of riboflavin. The numerator is expressed in units of radioactivity and the denominator is expressed in units of moles. Equation (1) below simply expresses the fact that the specific radioactivity of protein-bound riboflavin (B) is the same as the total (T) riboflavin pool at equilibrium.

$$\text{dpm}_B/\text{Rf}_B - \text{dpm}_T/\text{Rf}_T \qquad (1)$$

Consider that the sum of labeled and unlabeled riboflavin bound to riboflavin-binding protein is equivalent stoichiometrically to the total binding protein added to the assay, $\text{Rf}_B = \text{RBP}_T$, and that the total amount of riboflavin is composed of added radiolabeled riboflavin (*Rf) and unlabeled endogenous riboflavin initially bound to protein; $\text{Rf}_T = $ *Rf + Rf. Substitution of these equalities yields Eq. (2):

$$\text{dpm}_B/\text{RBP}_T = \text{dpm}_T/(\text{*Rf} + \text{Rf}) \qquad (2)$$

Equation (2) can be rearranged to the linear Eq. (3):

$$\text{dpm}_T/\text{dpm}_B = \text{*Rf}/\text{RBP}_T + \text{Rf}/\text{RBP}_T \qquad (3)$$

For any unknown sample the fractional saturation Rf/RBP_T will be constant. The unknown in Eq. (3) is RBP_T. The measured variable in the assay is volume (V). Since $\text{RBP}_T = V[\text{RBP}]$, the substitution gives Eq. (4):

$$\frac{\text{dpm}_T}{\text{dpm}_B} = \frac{\text{*Rf}}{[\text{RBP}]}\frac{1}{V} + \frac{\text{Rf}}{\text{RBP}_T} \qquad (4)$$

This equation has the form $y = mx + b$ where $\text{Rf}_T/[\text{RBP}]$ is the slope m, and Rf/RBP_T, the fractional saturation, is the y intercept b.

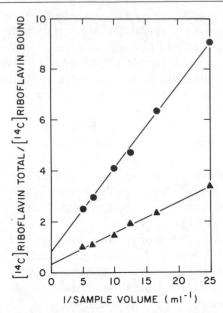

FIG. 2. Assay for riboflavin-binding protein in the yolk (●) and white (▲) of eggs from a White Leghorn chicken. Both samples were diluted 1 : 1 (v/v) with distilled water, homogenized by stirring, centrifuged to remove solids, and assayed as described in the text. The slope of each line is inversely proportional to the concentration of riboflavin-binding protein in each sample. The value of the y intercept equals the proportion of riboflavin-binding protein in the sample that contains bound riboflavin. Given the specific radioactivity of the [14C]riboflavin or using pure riboflavin-binding protein as a standard, the amount of riboflavin-binding protein in the samples can be calculated. For the egg in this analysis the yolk contained 0.44 mg/ml of binding protein that was 84% saturated with riboflavin while the albumen contained 1.35 mg/ml of the protein which was 34% saturated. (Taken from Lotter et al.[1] with permission of the publisher.)

Analysis of Data. After correcting for quenching and background, the ratio of bound radioactivity to total bindable radioactivity is calculated for each reaction mixture and plotted vs V^{-1}, as is shown in Fig. 2 for chicken egg yolk and egg white samples. Typically a straight line will be generated from which a slope can be easily determined. Points with dpm_T/dpm_B values between 1 to 4 are most reliable. (Note that values of 1.0 occur when riboflavin becomes the limiting reagent. These points should not be included in the analysis.) Curvature can result from errors in determining dpm_T or the background blank values. Typically the correlation coefficients for the data points are above 0.98. Thus the slope and the corresponding protein concentration are well determined. The extrapolated y intercept is very sensitive to slight changes in the slope and thus the fractional saturation is less precisely determined.

[36] Isolation of Avian Riboflavin-Binding Protein

By MARK S. MILLER and HAROLD B. WHITE, III

Chicken riboflavin-binding protein (RBP) is a very acidic, heat-stable phosphoglycoprotein.[1] It binds one molecule of riboflavin with a dissociation constant of 1.3 nM.[2] It is produced in the liver[3] and oviduct[4] of the laying hen and is deposited respectively in the yolk and white of eggs. The yellowish cast of a typical egg white is due to riboflavin bound to this protein. Although a single estrogen-responsive gene controls the synthesis of the protein in liver and oviduct, the protein isolated from yolk has a more complicated carbohydrate composition[5] and appears to be missing seven carboxy-terminal amino acids present in the egg white protein.[6] Serum RBP, the secretory form of the liver protein, differs slightly in carbohydrate composition from the yolk protein.[7]

The isolation of RBP from egg white, yolk, and hen plasma will be covered in this chapter. Since avian RBP is highly anionic, purification by conventional ion-exchange chromatography is quite effective. Riboflavin-binding proteins have been isolated from other sources, including mammalian tissue.[8-10] These proteins have properties quite different from their avian counterparts, and can best be purified by the affinity methods covered elsewhere in this series.[11]

Egg White Riboflavin-Binding Protein

The first isolation of RBP protein from egg white was reported in 1958 by Rhodes *et al.*[12] The following year, the same group published a more complete report on the "flavoprotein–apoprotein system of egg white."[1] The purification relied mainly on the affinity of RBP for DEAE-cellulose

[1] M. B. Rhodes, N. Bennett, and R. E. Feeney, *J. Biol. Chem.* **234**, 2054 (1959).
[2] J. Becvar and G. Palmer, *J. Biol. Chem.* **257**, 5607 (1982).
[3] J. A. Froehlich, A. H. Merrill, Jr., C. O. Clagett, and D. B. McCormick, *Comp. Biochem. Physiol. B* **66B**, 397 (1980).
[4] S. Mandeles and E. D. Ducay, *J. Biol. Chem.* **237**, 3196 (1962).
[5] M. S. Miller, E. G. Buss, and C. O. Clagett, *Biochim. Biophys. Acta* **677**, 225 (1981).
[6] Y. Hamazume, T. Mega, and T. Ikenaka, *J. Biochem. (Tokyo)* **95**, 1633 (1984).
[7] M. S. Miller, R. C. Druch, and H. B. White, III, *Biochim. Biophys. Acta* **715**, 126 (1982).
[8] A. H. Merrill, Jr., J. A. Froehlich, and D. B. McCormick, *J. Biol. Chem.* **254**, 9362 (1979).
[9] K. Muniyappa and P. R. Adiga, *Biochem. J.* **187**, 537 (1980).
[10] A. H. Merrill, Jr., J. A. Froehlich, and D. B. McCormick, *Biochem. Med.* **25**, 198 (1981).
[11] A. H. Merrill and D. B. McCormick, this series, Vol. 66, p. 338.
[12] M. B. Rhodes, P. R. Azari, and R. E. Feeney, *J. Biol. Chem.* **230**, 399 (1958).

under conditions in which most other proteins do not bind. Since that initial report, variations of the isolation procedure have been published by Farrell et al.,[13] Murthy et al.,[14,15] and Becvar and Palmer.[2] Each of these procedures depends on the affinity of the protein for DEAE-cellulose.

Egg white RBP can be purified to apparent homogeneity in three steps: batch adsorption to DEAE-cellulose, salt precipitation, and chromatography on CM- or DEAE-cellulose. Although all steps should be carried out at 5° if possible, the exceptional stability of RBP allows for considerable handling at room temperature. Throughout the purification procedures, the strong absorbance of protein-bound riboflavin at 455 nm may be used as a visual indicator for the holoprotein.[1,16] The molar absorption coefficient of holo-RBP at 455 nm is 11,000 M^{-1} cm^{-1}.[17] Native egg white RBP is normally only about 30–35% saturated with riboflavin.[18] In the procedure used by Becvar and Palmer,[2] RBP was saturated by addition of excess riboflavin prior to purification.

Preparation of the Egg Whites. Prior to adsorption onto DEAE-cellulose, the egg white must be carefully prepared. Egg white consists of thick and thin albumen portions. The viscosity of the thick albumen can be conveniently reduced by sonication.[13] Forcing the egg white through several layers of cheesecloth may also be used to effect dissolution of the thick albumen. The pH and ionic strength of the egg white must be adjusted to allow for maximum binding of RBP to the DEAE-cellulose, with minimum binding of the other proteins. This can be accomplished by dialysis against 0.1 M sodium acetate buffer (pH 4.3),[13] or by dilution with an equal volume of low ionic strength buffer.[2] The pH of the final solution should be near 4.3 since very few other egg white proteins bind to DEAE at this pH. It is important that the ionic strength of the solution be 0.1 or less. If there is any doubt, the ionic strength should be checked by conductance. The insoluble precipitate formed during this stage is removed by centrifugation[13] or by filtration through several layers of cheesecloth and Miracloth (Calbiochem).[2]

Batch Adsorption to DEAE-Cellulose. DEAE-cellulose, which had been equilibrated with the starting buffer (0.1 M sodium acetate buffer, pH 4.3), is added directly to the pale yellow egg white solution. Farrell *et*

[13] H. M. Farrell, Jr., M. F. Mallette, E. G. Buss, and C. O. Clagett, *Biochim. Biophys. Acta* **194,** 433 (1969).

[14] U. S. Murthy, S. K. Podder, and P. R. Adiga, *Biochim. Biophys. Acta* **434,** 69 (1976).

[15] U. S. Murthy and P. R. Adiga, *Indian J. Biochem. Biophys.* **14,** 118 (1977).

[16] M. Nishikimi and K. Yagi, *J. Biochem. (Tokyo)* **66,** 427 (1969).

[17] A. Kozik, *Biochim. Biophys. Acta* **704,** 542 (1982).

[18] R. E. Feeney and R. G. Allison, eds. "Evolutionary Biochemistry of Proteins," p. 66. Wiley (Interscience), New York, 1975.

al.[13] found that the best purification was achieved at this stage when the amount of DEAE-cellulose was limited. They recommended that 1 g of DE-32 (Whatman) be used per 14 g of protein. As a guideline, approximately 10 g of Whatman DE-32 or 30 g of preswollen Whatman DE-52 should be used for 1 liter of egg white. The mixture is equilibrated with stirring for 30 min and then vacuum filtered on a Büchner funnel. The DEAE-cellulose, which is bright yellow with adsorbed RBP, is washed with several liters of starting buffer until the effluent is clear, slurried in starting buffer, and packed in a 2.5-cm diameter column. (Note: The column dimensions here and in subsequent steps are appropriate for preparations starting with 1–4 liters of egg white.) The column is washed with starting buffer until the A_{280} of the column effluent returns to baseline. RBP is eluted from the column with 1 M NaCl in starting buffer. Yellow fractions with A_{280} greater than 1.0 are collected and pooled.

Salt Precipitation. The pooled eluate from DEAE-cellulose is next subjected to $(NH_4)_2SO_4$ precipitation. Farrell *et al.*[13] recommended this step and we found that it is useful for removing ovomucoid as well as for concentrating the RBP. Farrell *et al.*[13] collected the 60–80% saturated $(NH_4)_2SO_4$ precipitate but we have found less loss of material with a 55–85% cut. The solution is brought to 55% saturation by slow addition of solid $(NH_4)_2SO_4$ and stirred for 1 hr. Following centrifugation for 20 min at 16,000 g, the white precipitate is discarded. The supernatant is taken to 85% saturation with solid $(NH_4)_2SO_4$, equilibrated for 1 hr, and centrifuged as above. The almost colorless supernatant is discarded. The yellow precipitate is redissolved in a minimum volume of water and dialyzed against distilled water overnight.

Apoprotein Purification by CM-Cellulose Chromatography. The final step in the procedures of Farrell *et al.*[13] and Rhodes *et al.*[1] is chromatography on CM-cellulose. It should be noted that prior to this step in the isolation very little of the riboflavin bound to RBP is lost. The binding of riboflavin is pH dependent (pK 3.85).[13,19] When the CM-cellulose step is used, the pH of the solution is adjusted to below 3.2 and riboflavin completely dissociates from RBP. If it is desirable to purify the native holoprotein, proceed to the next section.

The dialyzed RBP solution is adjusted to pH 3.14 by addition of 10% HCl and applied to a 2.5 × 30-cm column of CM-cellulose (CM-23 or CM-52, Whatman) which had been equilibrated with the starting buffer (25 mM sodium acetate buffer, pH 3.14). The column is washed with starting buffer until the yellow band (riboflavin and residual holo-RBP) is com-

[19] G. Blankenhorn, *in* "Flavins and Flavoproteins" (K. Yagi and T. Yamano, eds.), p. 405. University Park Press, Baltimore, Maryland, 1980.

pletely eluted from the column. Colorless apo-RBP is eluted with 25 mM sodium acetate buffer (pH 5.8). Fractions with A_{280} greater than 1.0 are pooled. Farrell et al.[13] found that recycling the pH 3.14 eluate one or two times through the regenerated CM-cellulose column was required for efficient recovery of the protein. The pooled fractions containing pure apo-RBP are exhaustively dialyzed against distilled water and lyophilized.

Holoprotein Purification by DEAE-Cellulose Chromatography. The dialyzed solution from the salt precipitation step is applied to 2.5 × 60-cm column of DEAE-cellulose (DE-52, Whatman) which had been equilibrated with starting buffer (0.1 M sodium acetate buffer, pH 4.3). The column is developed with a linear salt gradient formed from 500 ml each of starting buffer alone and starting buffer containing 1 M NaCl. The column is monitored at both 280 and 455 nm. Fractions with A_{455} greater than 0.1 are pooled, dialyzed exhaustively against distilled water, and lyophilized.

Final Notes on Egg White RBP Purification. Although the steps outlined above (batch DEAE, salt precipitation, and CM- or DEAE-cellulose chromatography) usually result in pure preparations of RBP, occasionally an additional step is required. If the protein does not appear homogeneous by the assay methods detailed elsewhere in this volume[20] or by polyacrylamide gel electrophoresis, a gel filtration step may be employed. The sample, concentrated by lyophilization, is applied to a BioGel column and eluted with a low ionic strength buffer. We have had success using a 2.5 × 80-cm column of BioGel P-100 and 25 mM sodium acetate buffer (pH 4.3).

Yolk Riboflavin-Binding Protein

The isolation of RBP protein from egg yolk was first published by Ostrowski et al.[21] and by Zak and Ostrowski.[22] Improved methods were subsequently reported by Miller,[23] Miller et al.,[5] and Murthy et al.[24] Since the properties of yolk RBP are similar to those of its egg white counterpart, much of the isolation procedure is identical. The major difference is in the initial preparation of the yolk prior to DEAE batch treatment.

Preparation of Egg Yolk. Yolks are separated from egg white and carefully washed in distilled water to remove any adhering egg white. The yolk membranes are ruptured by forcing the yolk through several layers of cheesecloth. In one study in which it was imperative that there was no chance of contamination with egg white RBP, yolk was obtained directly

---

[20] H. B. White, III, this volume [35].
[21] W. Ostrowski, B. Skarzynski, and Z. Zak, *Biochim. Biophys. Acta* **59**, 515 (1962).
[22] Z. Zak and W. Ostrowski, *Acta Biochim. Pol.* **10**, 427 (1963).
[23] M. S. Miller, Ph.D. Thesis, Pennsylvania State University, University Park (1976).
[24] U. S. Murthy, K. Sreekrishna, and P. R. Adiga, *Anal. Biochem.* **92**, 345 (1979).

from ovarian follicles.[25] An equivalent volume of 0.1 M sodium acetate buffer (pH 4.3) is added to the yolk and stirred until the mixture is homogeneous. The mixture is centrifuged at 12,000 g for 1 hr to remove the phosvitin/lipovitellin-containing yolk granules. The supernatant is dialyzed against running water and then against 0.1 M sodium acetate buffer (pH 4.3). Insoluble material is removed by centrifugation at 12,000 g for 30 min.

Batch Adsorption to DEAE-Cellulose. This step is the same as outlined for egg white RBP. Use about 8 g of moist, equilibrated DEAE-cellulose (DE-52, Whatman) per dozen egg yolks. Since egg yolk readily clogs most filter paper, we recommend the use of a fast-flowing hardened filter paper, such as Whatman No. 541, when washing the cellulose on a Büchner funnel.

Subsequent Purification Steps. Following batch elution from DEAE-cellulose, purification is completed by $(NH_4)_2SO_4$ precipitation (55–85% cut) and gradient elution from DEAE-cellulose as described for egg white holo-RBP. Once again, if an additional purification step is required, gel filtration on BioGel P-100 as described previously is recommended.

Serum Riboflavin-Binding Protein

RBP was first identified in the serum of laying hens by Blum.[26,27] It was shown by Winter *et al.*[28] to be the product of the same gene as the egg white and yolk RBP, and it was shown by Farrell *et al.*[29] to be serologically indistinguishable from the egg white and yolk proteins.

Serum RBP was first purified by Murthy and Adiga.[30] By using DEAE-cellulose and Sephadex G-100 chromatography, they isolated 5 mg of RBP from 44 ml of pooled serum from estrogen-stimulated male chicks. Small amounts of serum RBP (1–2 mg) were also purified by Merrill and McCormick using flavin affinity chromatography.[31] A larger scale purification method was reported by Miller *et al.* in which 270 mg of the protein was isolated from 20 liters of plasma.[7] This method consists of the four steps outlined below.

[25] M. S. Miller, M. T. Mas, and H. B. White, III, *Biochemistry* **23**, 569 (1984).
[26] J. C. Blum, *Proc. Int. Congr. Nutr., 7th, 1966* Vol. 5, p. 550 (1967).
[27] J. C. Blum, "Le métabolisme de la riboflavine chez la poule pondeuse." F. Hoffman-La Roche et Cie., Paris, 1967.
[28] W. P. Winter, E. G. Buss, C. O. Clagett, and R. V. Boucher, *Comp. Biochem. Physiol.* **22**, 897 (1967).
[29] H. M. Farrell, E. G. Buss, and C. O. Clagett, *Int. J. Biochem.* **1**, 168 (1970).
[30] U. S. Murthy and P. R. Adiga, *Biochem. J.* **170**, 331 (1978).
[31] A. H. Merrill and D. B. McCormick, *Anal. Biochem.* **89**, 87 (1978).

Preparation of Plasma and Salt Precipitation. Fresh blood is collected from laying hens and immediately treated with 0.1 volume of 0.5 M sodium citrate buffer (pH 5.5) to prevent coagulation. Plasma is collected by centrifugation of the citrated whole blood.

Due to the large volume and high ionic strength of the starting material, $(NH_4)_2SO_4$ is used as the initial purification step. We found that a 45–85% $(NH_4)_2SO_4$ cut resulted in the best yield. The 45–85% cut is prepared as described for egg white riboflavin-binding protein, except that the final centrifugation step is for 1 hr at 16,000 g. This lengthy centrifugation is necessary to pack the precipitate which floats on top of the 85% supernatant. The precipitate is carefully skimmed from the clear solution, resuspended in a minimum volume of water, and dialyzed against running water overnight. The pH of the retentate is adjusted to 4.3 with acetic acid and insoluble material is removed by centrifugation.

Batch Adsorption to DEAE-Cellulose. This step is carried out as described for egg white RBP. Use about 10 g of moist, equilibrated DEAE-cellulose (DE-52, Whatman) per liter of plasma. The pooled RBP fractions, eluted from the column with 1 M NaCl, are dialyzed against distilled water prior to the next step.

DEAE-Cellulose Chromatography. The partially purified RBP solution is applied to a column of DEAE-cellulose (2.5 × 60 cm for a 20-liter preparation) and chromatographed with a linear salt gradient as described for the egg white holoprotein. The pooled RBP fractions are dialyzed and lyophilized.

Gel Filtration. The concentrated RBP peak from the previous step is applied to a 2.5 × 80-cm column of BioGel P-100 and chromatographed with 25 mM sodium acetate buffer (pH 4.3). Although this step is optional for the purification of the egg white and yolk proteins, it is needed to complete the purification of serum RBP.

Properties of Riboflavin-Binding Protein

The purified preparations of RBP should be examined for homogeneity and assayed for riboflavin-binding capacity. Pure egg white RBP migrates as a single band during electrophoresis on SDS–polyacrylamide gels with an apparent molecular weight of 34,000–36,000.[6,7,15,32] Yolk RBP and serum RBP are indistinguishable from egg white RBP by this technique.[7] Molecular weights of egg white and yolk RBP obtained from sedimenta-

[32] T. F. Kumosinski, H. Pessen, and H. M. Farrell, Jr., *Arch. Biochem. Biophys.* **214**, 714 (1982).

tion equilibrium data range from 32,000 to 37,000.[1,13,32–34] The molecular weight of egg white RBP, calculated from amino acid sequence data and including phosphate and carbohydrate components, is 29,200.[6] The yolk and serum forms of RBP have molecular weights of 29,000–30,000 based on their compositions.[6,7]

Pure RBP often reveals charge heterogeneity when examined by polyacrylamide gel electrophoresis in nondenaturing buffers[2] or by isoelectric focusing.[2,3] Rhodes et al.[1] reported the pI of egg white RBP as 3.9–4.1. They found two forms of the apoprotein, identical in binding capacity but differing in phosphate content, which were separated by CM-cellulose chromatography. Froehlich et al.[3] found three forms of the liver and yolk proteins with isoelectric points of 3.85 (liver), 4.00 (yolk), and 3.91 (liver and yolk). They suggested that the difference might be due to heterogeneity in phosphate or sialic acid contents. Egg white and yolk RBP both contain seven to eight residues of phosphate[1,35] all of which occur as phosphoserine within a 13-amino acid segment of a single tryptic peptide.[6,25] Electrophoresis of this peptide reveals considerable heterogeneity, with two major forms containing seven and eight residues of phosphate as well as several minor forms.[7]

Egg white and yolk RBP contain two asparagine-linked oligosaccharides.[36] Although both proteins contain mannose, N-acetylglucosamine, galactose, and sialic acid, the low content of the latter two sugars in egg white RBP suggests that this species contains ovomucoid-type structures while the yolk protein might have the complex-type oligosaccharide structures.[5,6] The carbohydrate composition of serum RBP is similar to that of yolk RBP except that it contains fucose and additional residues of several other components.[7]

Riboflavin-binding capacity may be assayed by fluorescence quenching[13,28] or by competitive binding[37] as described elsewhere in this volume.[20] A rapid spectrophotometric method was suggested by Kozik.[17] Assuming a molar absorption coefficient for the apoprotein of 49,000 M^{-1} cm^{-1} at 280 nm,[38] the amount of RBP in a mixture of holo- and apo-RBP can be determined from the A_{280} after subtracting out the contribution of protein-bound riboflavin at this wavelength (A_{280} corrected $= A_{280} -$

[33] W. Ostrowski and A. Krawczyk, Acta Chem. Scand. 17, S241 (1963).
[34] W. Ostrowski, Z. Zak, and A. Krawczyk, Acta Biochim. Pol. 15, 241 (1968).
[35] M. S. Miller, M. Benore Parsons, and H. B. White, III, J. Biol. Chem. 257, 6818 (1982).
[36] T. Mega and T. Ikenaka, Anal. Biochem. 119, 17 (1982).
[37] S. E. Lotter, M. S. Miller, R. C. Bruch, and H. B. White, III, Anal. Biochem. 125, 110 (1982).
[38] M. Nishikimi and Y. Kyogoku, J. Biochem. (Tokyo) 73, 1233 (1973).

$1.55A_{455}$). After saturating the protein with riboflavin and removing the excess by chromatography on Sephadex G-25, the amount of protein-bound riboflavin is determined from the A_{455}, using a molar absorption coefficient 11,000 M^{-1} cm^{-1}. Since the stoichiometry of binding is 1 : 1, the maximum binding capacity of RBP is 12.5–12.9 μg of riboflavin per milligram of protein based on a molecular weight of 29,000–30,000.

[37] Fluorometric Titration of Urinary Riboflavin with an Apoflavoprotein

By JERRY ANN TILLOTSON and MARK M. BASHOR

Principle. Riboflavin, the flavin moiety of the flavodoxin in egg white, has a distinct fluorescence which is quenched when it combines with an equal molar quantity of the egg white apoprotein.[1,2] This quenching of the flavin fluorescence makes possible direct titration of urinary riboflavin with the purified apoprotein.[3] The concentration of urinary riboflavin is calculated from the difference in fluorescence before and after the apoprotein titration.

Methods

Flavoprotein Preparation. The flavoprotein is isolated from one extra large chicken egg at room temperature as follows. The egg white is homogenized to a uniform viscosity with a glass–Teflon tissue homogenizer. Sodium chloride is added and dissolved to a final concentration of 20% (w/v).[4] An equal volume of aqueous 5% phenol is added with constant stirring.[4] The copious precipitate is removed by filtration (Whatman No. 1 filter paper) and the filtrate is centrifuged at 1500 g for 10 min. The clear yellow supernatant is dialyzed 24–48 hr against 5–10 changes of 100 volumes of deionized water. The material is then dialyzed for 24 hr against 20 volumes of 0.05 M Tris (pH 7.5). If necessary, the dialyzed material is clarified by centrifugation. The supernatant is applied to a 1.8 × 10-cm column of DEAE-Sephadex A-50 previously equilibrated with 0.05 M Tris

---

[1] M. B. Rhodes, N. Bennett, and R. E. Feeney, *J. Biol. Chem.* **234**, 2054 (1959).

[2] M. Nishikimi and Y. Kyogoku, *J. Biochem.* (*Tokyo*) **73**, 1233 (1973).

[3] J. A. Tillotson and M. M. Bashor, *Anal. Biochem.* **107**, 214 (1980).

[4] W. P. Winter, E. G. Buss, C. O. Clagett, and R. V. Boucher, *Comp. Biochem. Physiol.* **22**, 897 (1967).

(pH 7.5). A bright yellow band (flavoprotein) forms at the top of the column as the sample is applied. The column is washed with 0.05 M Tris–0.1 M NaCl (pH 7.5) until the absorbance of the effluent at 280 nm returns to baseline. The flavoprotein is then eluted with 0.05 M Tris–0.4 M NaCl (pH 7.5). The fractions containing the flavoprotein (bright yellow) are combined and stored frozen at $-10°$ until the apoprotein is prepared.

Apoprotein Preparation. The flavoprotein is dialyzed for 24–48 hr against 6–10 changes of 40 volumes of 6 mM HCl.[1] The apoprotein is then dialyzed for 24 hr against 40 volumes of 0.05 M Tris (pH 7.5) and stored frozen in 2-ml aliquots containing approximately 2 mg protein. The frozen apoprotein is stable for at least 6 months. The apoprotein concentration is estimated using a molar extinction coefficient of 5×10^4 M^{-1} cm^{-1} at 280 nm.[2,5]

Riboflavin Standard Curve. The riboflavin stock solution contains 18.9 mg riboflavin dissolved in 500 ml deionized water containing 1 ml concentrated hydrochloric acid. The stock solution is stable for 2 months at 4°. The stock solution is diluted 1 : 200 with water to prepare the working standard. Aliquots of the working standard solution are diluted to 10 ml with 0.05 M Tris (pH 7.5) for the standard curve solutions (0.9–40 ng/ml). The working standard and the standard curve solutions are prepared daily.

Urine Samples. Initially the urine sample is diluted 1 : 10 with 0.05 M Tris (pH 7.5). Additional adjustments in the dilutions may be necessary if the fluorescence is too high or low or if >150 μl of the aproprotein is required for the titration.

Assay. A sample of the apoprotein (~ 1 mg/ml) is thawed for each assay and diluted 1 : 4 with 0.05 M Tris (pH 7.5) and kept in ice. The diluted apoprotein is added to the sample in 10-μl aliquots with a Hamilton repeating dispenser.

Three milliliters of a standard or diluted sample is added to a fluorometric cell and the fluorescence is measured. The excitation wavelength is 450 nm and the emission wavelength is 520 nm. The apoprotein is added in 10-μl aliquots, the sample is mixed, and the fluorescence is measured. The apoprotein is added until no change is measured in the fluorescence of the sample after three successive additions of the apoprotein (Fig. 1). The difference in fluorescence before and after the apoprotein titration is proportional to the amount of riboflavin in the sample. The standard curve is prepared by plotting the concentration of riboflavin (ng/ml) in the standard curve solution against the difference in fluorescence units for each standard. The concentration of the urinary riboflavin is calculated from

[5] U. S. Murthy, S. K. Podder, and P. R. Adiga, *Biochim. Biophys. Acta* **434,** 69 (1976).

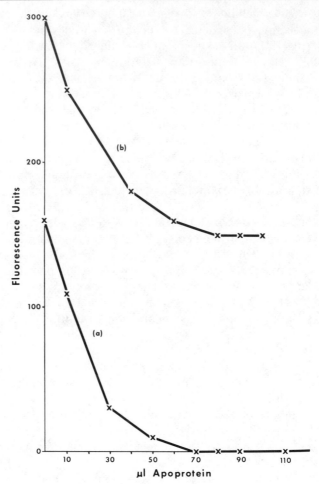

FIG. 1. The fluorescence of a riboflavin standard without (a) and (b) with added urine during apoflavoprotein titration.

the standard curve. The difference in fluorescent units measured for each sample must be in the range of the standard curve for quantitation.

Comments

This assay is a rapid, sensitive, and inexpensive measurement of urinary riboflavin. Several dilutions of a sample including duplicates can be prepared and titrated within minutes because the binding of the riboflavin to the apoprotein is instantaneous. Riboflavin losses are minimized because pretreatment of the sample, except for dilution, is not necessary. The riboflavin concentration of 350–400 samples can be measured with the apoprotein isolated and purified from a single egg.

The measurement of riboflavin by the apoprotein titration is dependent on the difference of the fluorescence of the sample before and after the titration. Figure 1 shows that the nonspecific urine fluorescence does not interfere with the apoprotein titration because the difference in fluorescence measured before and after the apoprotein titration of a riboflavin standard with and without added urine is the same. While the measurement of the concentration of apoprotein is unnecessary, the concentration of the purified protein must be sufficient ($\sim$1 mg/ml) so that the dilution of the sample is minimal. Correction for a change in the sample volume is not necessary if 150 μl or less of the apoprotein is added. The largest variations were measured in urine samples containing small amounts of riboflavin. These samples are readily identified and triplicate fluorescent measurements can be made. In our laboratory the recovery of riboflavin added to urine before dilution of the sample was 90–105%[3] regardless of the background fluorescence.

[38] Sensitive Radiosubstrate/High-Performance Liquid Chromatography Assays for Flavocoenzyme-Forming Enzymes

By SANG-SUN LEE and DONALD B. McCORMICK

Flavokinase (riboflavin kinase) has been determined by spectrophotometrically measuring FMN formed after extractions with benzyl alcohol to remove most of the substrate riboflavin[1] and by counting radioactivity incorporated into FMN from [^{14}C]riboflavin after separation by paper chromatography.[2] FAD synthetase (FMN adenylyltransferase) has been determined by spectrophotometrically measuring product from D-amino-acid oxidase reconstituted by FAD formed[3] and by fluorometrically quantitating FAD formed after phenol extraction and paper chromatography.[4]

[1] D. B. McCormick, *J. Biol. Chem.* **237**, 959 (1962).
[2] A. H. Merrill, R. Addison, and D. B. McCormick, *Proc. Soc. Exp. Biol. Med.* **158**, 572 (1978).
[3] C. DeLuca, this series, Vol. 6, p. 342.
[4] D. B. McCormick, *Biochem. Biophys. Res. Commun.* **14**, 493 (1964).

METHODS IN ENZYMOLOGY, VOL. 122

In the present method, high-performance liquid chromatography (HPLC) was used to separate radioactive flavocoenzymes from a denatured incubation mixture. The radioactivity of the separated products was used to quantitate the enzyme activities.

Assay Method of Flavokinase (ATP: Riboflavin 5'-Phosphotransferase, EC 2.7.1.26)

$$[\text{}^{14}\text{C}]\text{Riboflavin} + \text{ATP} \rightarrow [\text{}^{14}\text{C}]\text{FMN} + \text{ADP}$$

Principle. Using [^{14}C]riboflavin and ATP as substrates, [^{14}C]FMN formed by flavokinase present in any tissue extract is separated by HPLC and quantitated by counting of radioactivity.

Reagents

[2-^{14}C]Riboflavin, 0.5 mM, diluted to approximately 10 mCi/mmol with unlabeled riboflavin. Riboflavin is purified before use by chromatography on a DEAE-Sephadex A-50 column equilibrated with potassium phosphate buffer (0.02 M, pH 7).

ATP, 10 mM

ZnCl$_2$, 10 mM

FMN, 1 mM

Potassium phosphate buffer, 0.75 M, pH 8

Ammonium acetate buffer, 5 mM, pH 6

Procedure. An incubation mixture contains 20 μl of riboflavin, 10 μl each of ATP, ZnCl$_2$ and phosphate buffer, and solution containing flavokinase plus water to a total volume of 100 μl. Final concentrations are 0.1 mM riboflavin, 1 mM ATP, 1 mM Zn^{2+}, 75 mM phosphate, and sufficient enzyme solution (usually 40–80 μg of protein from high-speed supernatants). After 0 and 60 min of incubation in the dark with shaking at 37°, 10 μl of FMN (10 nmol) is added as a carrier, and the reaction is terminated by boiling for 1 min. The assay mixture is centrifuged for 2 min (Microfuge B, Beckman Instruments) to remove denatured protein. Riboflavin, FMN, and FAD in the supernatant are separated by a modification of Light *et al*.[5] Separation is achieved with a μ-Bondapak C$_{18}$ column (0.39 × 30 cm, 10 μm, Waters Associates, Inc.) using a linear gradient of 5 mM ammonium acetate buffer to 50% methanol and 5 mM ammonium acetate buffer developed over 30 min at a flow rate of 1.5 ml/min. The effluent stream is monitored at 450 nm. Fractions are collected every 30 sec and radioactivity is determined in a volume of scintillation fluid (ACS,

[5] D. R. Light, C. Walsh, and M. A. Marletta, *Anal. Biochem.* **109**, 87 (1980).

Amersham Corp.) which is five times the fraction volume by a scintillation counter (LS-3133 T, Beckman Instruments). The radioactivity of the HPLC fractions associated with FMN elution time and with values three times above the background level is summed and quantitated.

Definition of Unit and Specific Activity. One unit is that amount of enzyme which catalyzes the synthesis of 1 nmol of FMN in 1 hr at 37° under these conditions. Specific activity is defined as units per milligram of protein determined by the method of Lowry et al.[6]

Assay Method of FAD Synthetase (ATP : FMN Adenylyltransferase, EC 2.7.7.2)

$$FMN + [^3H]ATP \rightarrow [^3H]FAD + \text{pyrophosphate}$$

Principle. Using $[^3H]ATP$ and FMN as substrates, $[^3H]FAD$ formed is measured after incubation and separation by HPLC.

Reagents

[2,8-^{3}H]ATP, 5 mM, diluted to approximately 10 mCi/mmol with unlabeled ATP

FMN, 1 mM

MgSO$_4$, 10 mM

FAD, 1 mM

Potassium phosphate buffer, 0.75 M, pH 7.5

Procedure. An incubation mixture contains 10 μl each of ATP, FMN, MgSO$_4$ and phosphate buffer, and solution containing FAD synthetase plus water to a total volume of 100 μl. Final concentrations are 0.5 mM ATP, 0.1 mM FMN, 1 mM Mg^{2+}, 75 mM phosphate, and enzyme solution (usually 40–80 μg of protein from high-speed supernatants). After 0 and 60 min of incubation in the dark with shaking at 37°, 10 μl of FAD (10 nmol) is added as a carrier. The rest of the procedure is the same for the flavokinase assay except the radioactivity of the HPLC fractions associated with an FAD elution time is evaluated.

Definition of Unit and Specific Activity. One unit is that amount of enzyme which catalyzes the synthesis of 1 nmol of FAD in 1 hr at 37° under these conditions. Specific activity is defined as units per milligram of protein.

[6] O. H. Lowry, N. J. Rosebrough, A. L. Farr, and R. J. Randall, *J. Biol. Chem.* **193**, 265 (1951).

[39] ^{1}H NMR Spectral Analysis of the Ribityl Side Chain of Riboflavin and Its Ring-Substituted Analogs[1]

By Gary Williamson and Dale E. Edmondson

Although flavin coenzymes have been extensively studied by multinuclear NMR spectral techniques (^{1}H, ^{13}C, ^{15}N),[2-6] most of the work has centered on the nuclei in the heterocyclic isoalloxazine ring. The resonances due to the side-chain protons on the D-ribityl ring usually appear as an unresolved mixture of multiplets which have not been rigorously analyzed. A detailed description of the ^{1}H NMR spectra of the ribityl side chain would be useful for structural studies of flavin coenzymes which appear to have alterations in the ribityl side chain. Such alterations have been observed for the 2′,5′-anhydroflavins,[7] 8-formylriboflavins,[8] the various phosphorylated isomers of FMN[9] and naturally occurring flavins such as schizoflavins,[10] and the flavin coenzyme of methanol oxidase.[11]

With the availability of high-field NMR instruments, the enhanced resolution obtained permits a more detailed analysis of the various coupled resonances. In addition, the ready availability of computer programs which permit simulation of up to a seven-spin system allows direct comparison of the observed spectrum with simulated spectra.

We describe here the results of such an analysis using several riboflavin analogs in D$_2$O solutions at various pH values as demanded for solubilization of the analog of interest. Confirmation of spectral assignments are also described by selectively monodeuterating the 5′-CH$_2$ group of the riboflavin side chain.

[1] The unpublished work reported here was supported by the National Institutes of Health GM-29433. Funds for the purchase of the NMR instrument used in this work were provided by the National Science Foundation and by Emory University.

[2] F. J. Bullock and O. Jardetzky, J. Org. Chem. 30, 2056 (1965).
[3] H. J. Grande, C. G. van Schagen, T. Jarbandhan, and F. Muller, Helv. Chim. Acta 60, 348 (1977).
[4] K. Kawano, N. Ohishi, A. T. Suzuki, Y. Kyogoku, and K. Yagi, Biochemistry 17, 3854 (1978).
[5] C. G. van Shagen and F. Müller, Eur. J. Biochem. 120, 33 (1981).
[6] C. T. Moonen, J. Vervoort, and F. Müller, Biochemistry 23, 4859 (1984).
[7] D. E. Edmondson, Biochemistry 16, 4308 (1977).
[8] D. E. Edmondson, Biochemistry 13, 2817 (1974).
[9] G. Scola-Nagelschneider and P. Hemmerich, Eur. J. Biochem. 66, 567 (1976).
[10] S. Tachibana and T. Murakami, this series, Vol. 66, p. 333.
[11] R. H. Abeles, personal communication.

Riboflavin Analogs

Riboflavin was purchased from the Sigma Chemical Co., St. Louis, MO and used without further purification. 8α-N-Imidazolylriboflavin was synthesized by the condensation of 8α-bromotetraacetylriboflavin[12] with imidazole, and was purified after acid hydrolysis of the side-chain acetyl groups as described by Williamson and Edmondson.[13] N-3-Carboxymethylriboflavin was generously provided by Dr. D. B. McCormick of this department. All flavin analogs were homogeneous as judged by thin-layer chromatography or by HPLC analysis.

$5'$-R,S-Monodeuteroriboflavin was prepared by the enzymatic oxidation of riboflavin to its $5'$-aldehyde followed by reduction by $NaBD_4$. A partially purified preparation of the vitamin B_2-aldehyde-forming enzyme was obtained from the mycelia of *Schizophyllum commune* ATCC 20165, using a procedure modified slightly from that described by Tachibana and Oka[14] in that the dried mycelia were broken using a "bead beater" (Biospec Products, Bartlesville, OK).

The 50–90% ammonium sulfate fraction was used and exhibited a specific activity of 0.01 units/mg. The enzyme preparation is quite specific for riboflavin, as $N(3)$-carboxymethylriboflavin and 8α-N-imidazolylriboflavin are not oxidized to their respective $5'$-aldehydes on incubation with the enzyme in the presence of 2,6-dichlorophenolindophenol.

Riboflavin (160 mg) and 2,6-dichlorophenolindophenol (Sigma) (260 mg) are dissolved in 2 liters of 50 mM sodium acetate buffer (pH 5.5). The crude vitamin B_2-aldehyde-forming enzyme preparation (0.4 units) is added and the mixture stirred in the dark for 18 hr at 30°. The flavin is then extracted into 500 ml of benzyl alcohol, and the benzyl alcohol fraction washed once with water. After addition of 2 volumes of diethyl ether, the flavin is extracted into distilled water and the aqueous phase washed once with ether to remove any traces of benzyl alcohol. The aqueous flavin solution is reduced in volume by rotary evaporation and taken to dryness by lyophilization. The riboflavin $5'$-aldehyde was judged to be greater than 90% pure on TLC (silica gel plates, 1-butanol : pyridine : H_2O, 6 : 4 : 3 v/v/v). In this system, riboflavin $5'$-aldehyde exhibited an R_f of 0.65 and riboflavin an R_f of 0.70. An aldehydic proton was observed at 18.83 pm in the ¹H NMR spectrum (4.5 M DCl in D_2O) and the product gave a positive test with phenylhydrazine.

The product (36 mg) is converted to R,S-monodeuteroriboflavin by

[12] W. H. Walker, T. P. Singer, S. Ghisla, and P. Hemmerich, *Eur. J. Biochem.* **26**, 279 (1972).
[13] G. Williamson and D. E. Edmondson, *Biochemistry*, in press.
[14] S. Tachibana and M. Oka, *J. Biol. Chem.* **256**, 6682 (1981).

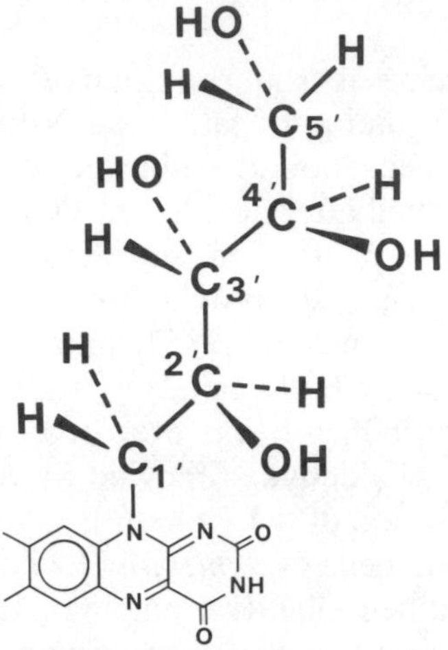

FIG. 1. Structure of riboflavin with the ribityl side chain emphasized.

reduction with 80 mg NaBD$_4$ in 125 ml of 0.1 M potassium phosphate, pH 8.0 at 25°, for 30 min. The riboflavin product, after benzyl alcohol extraction and reextraction into water as described above, showed an identical mobility to authentic riboflavin on TLC.

NMR Spectral Data

All ^{1}H NMR spectra were acquired using a Nicolet 360 MHz spectrometer in 5-mm NMR tubes at ambient temperature. A sweep width of 4000 Hz was used with 16,000 data points. The number of acquisitions varied between 4 and 128, depending on the sample concentration. Flavin samples were dissolved in "100%" D$_2$O (Sigma) solutions containing 4.5 M DCl (for riboflavin at 50 mg/ml) or 0.1 M potassium phosphate, pH 6.8 [for N(3)-carboxymethylriboflavin at 10 mg/ml]. 8α-N-Imidazolylriboflavin solutions were made by dissolving 10 mg of the flavin hydrochloride salt in 1 ml of D$_2$O (final pH, 1.5). All chemical shifts were measured relative to an internal standard of 0.05% 3-(trimethylsilyl)tetradeuterosodium propionate (Wilmad Glass Co.).

^{1}H NMR spectral simulations were performed using the NMCSIM program on the computer of the Nicolet spectrometer.

^{1}H NMR Spectral Data and Spectral Assignments

The structure in Fig. 1 shows the detailed stereochemistry and numbering of the D-ribityl side chain of riboflavin, which will be useful to the

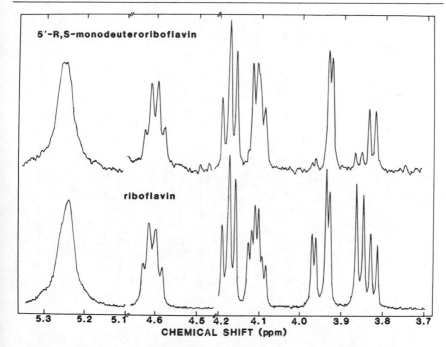

FIG. 2. ¹H NMR spectra at 360 MHz of the ribityl side-chain protons of 5'-R,S-mono-deuteroriboflavin and of riboflavin. See text for experimental details.

reader in following the description and assignment of the proton NMR spectra discussed below.

A comparison of the proton spectra of riboflavin with that of 5'-R,S-monodeuteroriboflavin is shown in Fig. 2. It is evident that monodeuteration in the 5' position has resulted in extensive alterations in the four resonances between 3.8 and 3.9 ppm, in the two doublets between 3.9 and 4.0 ppm, and in loss of splitting in the sextet centered about 4.1 ppm. No influence of the deuteration is observed in the triplet at 4.18 ppm, the quartet at 4.6 ppm, and the unresolved doublet at 5.24 ppm. The resonances between 3.8 and 4.0 ppm are assigned to the 5'-methylene protons which are magnetically inequivalent and which demonstrate geminal coupling in addition to coupling to the single 4' proton. Both resonances are still observed in the deuterated riboflavin spectrum since reduction with NaBD$_4$ is nonstereospecific and thus both the 5'a and 5'b protons are present in an amount 50% that of the parent proteo compound. Deuteration, however, eliminates the geminal coupling, thus resulting in a simplified spectrum.

The loss in splitting of the sextet at 4.1 ppm upon 5'-deuteration leads

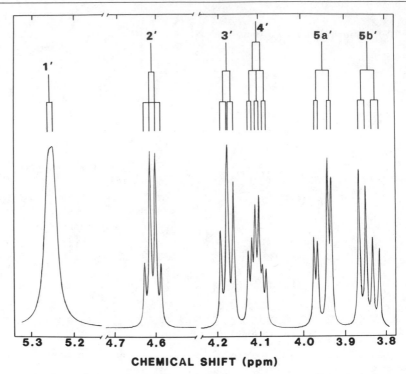

FIG. 3. Computer-simulated 360 MHz ^{1}H NMR spectrum of the ribityl side-chain protons of the oxidized riboflavin cation.

to its assignment as the 4′ proton. Selective decoupling experiments are also consistent with this assignment. The complex splitting pattern observed is due to coupling of the 4′ proton to both the two nonequivalent 5′ protons and also to the single 3′ proton.

The 3′ proton resonance at 4.18 ppm is readily identified since the triplet structure is the result of coupling to the single 4′ and single 2′ protons and by the fact that 5′-deuteration has no influence on its splitting pattern. Of the two remaining resonances, the quartet at 4.6 ppm integrates to one-half the intensity of that at 5.24 ppm, which results in their respective assignments to be due to the single, 2′ proton and the two 1′ protons. The observed quartet for the 2′ proton is the result of coupling to the single 3′ proton and to the magnetically equivalent 1′ protons. No geminal coupling of the two 1′ protons is observed in the case of riboflavin but, as discussed below, is observed with N(3)-carboxymethylriboflavin.

As a further proof of the validity of the above assignments, simulations of the riboflavin side-chain spectrum were performed. The only

TABLE I

PROTON–PROTON COUPLING CONSTANTS FOR THE RIBITYL SIDE CHAIN OF SELECTED
RIBOFLAVIN ANALOGS

Protons involved	Coupling constant (Hz)		
	Riboflavin[a]	N(3)-Carboxy-methylriboflavin[a]	8α-(N)-Imidazo-lylriboflavin[a]
1' (Hₐ) – 1' (H_b)	0	12.5	0
1' (Hₐ) – 2' (H)	4.8	9.6	5.4
1' (H_b) – 2' (H)	4.8	2.6	5.4
2' (H) – 3' (H)	5.1	4.8	4.7
3' (H) – 4' (H)	6.2	6.9	6.7
4' (H) – 5' (Hₐ)	3.1	2.9	3.0
4' (H) – 5' (H_b)	6.2	6.4	6.7
5' (Hₐ) – 5' (H_b)	12.2	12.0	11.9

[a] See text for solvent conditions used.

spectrum simulated which resulted in a satisfactory agreement with the experimental spectrum is shown in Fig. 3. The various coupling constant values used in the simulation and observed experimentally are given in Table I. Chemical shift values for the ribityl side-chain protons are given in Table II.

TABLE II

CHEMICAL SHIFT VALUES FOR THE RIBITYL PROTONS OF SELECTED
RIBOFLAVIN ANALOGS

Proton position	Chemical shift (ppm)[a]		
	Riboflavin[b]	N(3)-Carboxy-methyl-riboflavin	8α-(N)-Imidazo-lylriboflavin[b]
1' (Hₐ)	5.24	4.99	4.88
1' (H_b)	5.24	4.77	4.88
2' (H)	4.61	4.37	4.29
3' (H)	4.18	3.93	3.87
4' (H)	4.11	3.97	3.97
5' (Hₐ)	3.95	3.88	3.84
5' (H_b)	3.84	3.74	3.72

[a] Values for resonances in the absence of any coupling.
[b] See text for solvent conditions used.

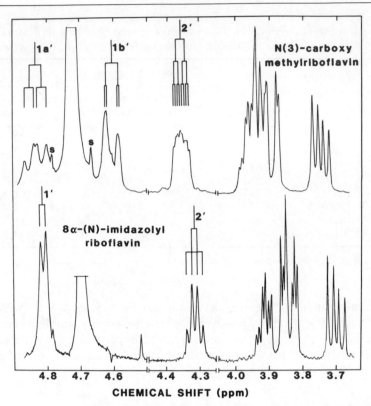

FIG. 4. ¹H NMR spectra (360 MHz) of the ribityl side-chain protons of N(3)-carboxy-methylriboflavin and 8α-(N)-imidazolylriboflavin. See text for experimental details.

More complex splitting patterns are observed for the ribityl side-chain spectra of two water-soluble riboflavin analogs: N(3)-carboxymethylriboflavin and 8α-(N)-imidazolylriboflavin (Fig. 4). Of interest is the finding that the 1'-methylene protons of the N(3)-carboxymethyl analog exhibit magnetic inequivalence which results in geminal coupling (Fig. 4, top). This leads to a more complex splitting pattern for the 1' protons and for the 2' proton. Alterations in the chemical shift values of the other ribityl protons (presumably due to changes in the pH and ionic character of the solvent) lead to overlap of the 3' and 4' protons with the 5'a proton. Simulations of the N(3)-carboxymethylriboflavin spectrum showed a satisfactory fit with the experimental results with only minor alterations in the respective coupling constants (Table I) but more significant alterations in respective chemical shifts.

No magnetic inequivalence of the 1'-methylene protons is apparent for 8α-(N)-imidazolylriboflavin (Fig. 4, bottom). As in the case of N(3)-carboxymethylriboflavin, alterations in the chemical shifts of the ribityl protons resulted in overlap of the resonances due to the protons at the 3', 4', and 5'a positions. Satisfactory spectral simulations were achieved with only minor alterations in the respective coupling constants (Table I) and with more substantial chemical shift alterations (Table II).

Concluding Remarks

The results presented here will hopefully serve as a useful guide for the interpretation of ¹H NMR spectral properties of flavin analogs that have undergone modifications in the ribityl side chain. It could be argued that such spectral analysis might profitably be applied to the phosphorylated coenzyme forms of riboflavin: FMN and FAD. Attempts have been made in that direction in this laboratory and in other studies.[15] The additional splitting due to coupling to the $I = 1/2$ phosphorus nucleus results in even more complex spectra. Satisfactory resolution of the 3', 4', and 5' protons would require the availability of NMR spectrometers of higher field and therefore resolution (500 MHz). Since such high-resolution instruments are not routinely available and instruments in the 300 MHz region are, we felt a more detailed analysis of the riboflavin form to be of more general concern and usefulness.

Of interest is the finding that the observed coupling constants of the protons in the riboflavin side chain are relatively insensitive to ring substitution or to pH of the aqueous solution. The only major difference involved the 1'-methylene protons which appear to be magnetically equivalent under acidic conditions but not under neutral conditions. Geminal coupling of the 1'-methylene protons has been observed in the proton spectra of FMN and FAD at neutral pH.[15] It could thus be argued to be a result of isoalloxazine ring "stacking" in aqueous solutions, as is known to occur for flavin analogs which are soluble at neutral pH values. Under conditions of ring protonation at acid pH values, such ring stacking would not occur due to charge repulsion and, as observed, no 1'-methylene geminal coupling is observed.

Of further interest is the observed geminal coupling of the 5'-methylene protons which occurs for all of the flavin analogs reported here. This magnetic inequivalence is a result of restricted rotation of the 5'-methylene group on the NMR time scale and could result from either intra- or

[15] M. Kainosho and Y. Kyogoku, *Biochemistry* 11, 741 (1972).

intermolecular interactions. No alterations in 5'-geminal coupling is observed for 8α-(N)-imidazolylriboflavin over a concentration range of 1.05–35.6 mM, which suggests the interactions restricting the mobility of the 5'-methylene group to be intramolecular.

Finally, we would like to point out the utility of spectral simulations in the interpretations of complex splitting patterns. Such programs are commercially available (such as for the Nicolet instrument) and are invaluable as an aid in spectral interpretation.

Section VIII

Pteridines, Analogs, and Pterin Coenzymes (Folate)

[40] Purification of Rat Liver Folate-Binding Protein : Cytosol I[1]

By ROBERT J. COOK and CONRAD WAGNER

Folate-binding protein : cytosol I (FBP-CI) was originally identified in rat liver cytosol following an intraperitoneal dose of [³H]folic acid (pteroylglutamic acid, PteGlu).[2] It was subsequently purified to apparent homogeneity and shown to contain $H_4PteGlu_5$.[3]

Assay Method

During the initial stages of purification FBP-CI is detected by virtue of the naturally bound ligand, [³H]$H_4PteGlu_5$. Subsequent stages in the purification procedure release the bound ligand and the protein is then detected by its ability to bind the released $H_4PteGlu_5$.

Principle. Binding of [³H]folates is measured by a centrifugal procedure which rapidly separates macromolecules from ligand using BioGel P-2 (Bio-Rad) columns. Protein may also be desalted prior to measuring binding of [³H]folates by this method. The procedure for this assay has been previously published.[3,4]

Reagents. BioGel P-2 (200–400 mcsh) is preswollen in 10 mM potassium phosphate (pH 7.0) containing 10 mM 2-mercaptoethanol and 1 mM sodium azide. The gel is gravity-packed (2 ml) in small plastic columns (4.5-ml screening tubes, Whale Scientific, Denver, CO). Other BioGel P gels up to P-10 have been successfully used for this binding assay and desalting procedure.

³H-Labeled Ligand. The natural ligand, $H_4PteGlu_5$, labeled with ³H may be obtained either by extracting the bound folate from FBP-CI after the Sephacryl S-200 step of the purification or as the dissociated ligand at the DEAE-cellulose step of the purification. The [³H]folate is synthesized endogenously by dosing rats intraperitoneally with [G-³H]PteGlu (25 μCi) at 48 and 24 hr prior to sacrifice. A portion of the FBP-CI isolated from the Sephacryl S-200 column is concentrated to at least 10,000 cpm/ml, using an Amicon concentrator, is made 1 M with respect to 2-mercapto-

[1] This work was supported by NIH Grant No. AM15289 and the Medical Research Service of the Veterans Administration.
[2] M. M. Zamierowski and C. Wagner, *J. Biol. Chem.* **252**, 933 (1977).
[3] R. J. Cook and C. Wagner, *Biochemistry* **21**, 4427 (1982).
[4] A. J. Wittwer and C. Wagner, *Proc. Natl. Acad. Sci. U.S.A.* **77**, 4484 (1980).

ethanol, and is then heated in a boiling water bath for 10 min. Denatured protein is removed by centrifugation and the supernatant containing [³H]folate is stored in 1-ml aliquots at $-20°$ under N_2. The specific activity of the [³H]folate may be determined by treating with conjugase, isolating the monoglutamate derivative by HPLC, and the quantitatively assaying with *Lactobacillus casei*.[5,6]

Procedure. Protein samples and [³H]folate (500–1000 cpm) in a total volume of 100–200 μl are incubated at 4° for 10 min. Protein is then separated from unbound [³H]folate by centrifugation through a BioGel P-2 column at 620 g (R_{max}) for 3 min.[4] The eluate is collected in tubes, mixed with 5 ml ACS (Amersham), and counted in a liquid scintillation spectrometer. The amount of radioactivity in the eluate indicates [³H]folate binding. The columns are regenerated by washing twice with 1.5 ml of 10 mM potassium phosphate (pH 7.0) containing 1.0 mM sodium azide and centrifuging at 35 g (R_{max}) for 3 min.

Enzyme Purification

Preparation of Rat Liver Cytosol. The procedure described is for three or four male Sprague–Dawley rats (150–300 g), which yield 30–40 g wet weight of liver. The rats are dosed intraperitoneally with 25 μCi of [G-³H]PteGlu (0.5 ml) 48 and 24 hr prior to sacrifice. The animals are killed by decapitation and the livers rapidly excised and rinsed in cold (0–4°) potassium phosphate-buffered sucrose (10 mM potassium phosphate, pH 7.0, 0.25 M sucrose, 10 mM 2-mercaptoethanol, 1 mM sodium azide, and 0.1 mM phenylmethylsulfonyl fluoride, PMSF). All subsequent steps are performed at 0–4°. The liver is weighed, minced with scissors, and homogenized in 4 volumes of buffered sucrose in a Potter-Elvehjem homogenizer with a medium-fitting Teflon pestle (three strokes, 1300 rpm). The homogenate is centrifuged at 5000 g (R_{av}) for 10 min and the pooled supernatant is then centrifuged at 220,000 g (R_{av}) for 90 min (50,000 rpm in a Beckman 50.2 Ti rotor). The resulting supernatant may be used directly or stored at $-20°$.

Sephacryl S-200 Chromatography. The high-speed supernatant is concentrated to 40–50 ml in an Amicon concentrator equipped with a PM30 membrane. The concentrate is applied to a column (5 × 90 cm) of Sephacryl S-200 equilibrated with 10 mM potassium phosphate (pH 7.0) containing 10 mM 2-mercaptoethanol, 1 mM sodium azide, and 0.1 mM PMSF. The column is eluted with equilibration buffer in an upward direc-

[5] S. D. Wilson and D. W. Horne, this volume [43].
[6] S. D. Wilson and D. W. Horne, *Anal. Biochem.* **142,** 529 (1984).

tion and 10-ml fractions are collected. FBP-CI is identified by counting 100-μl aliquots in 5 ml ACS. FBP-CI elutes at approximately 650 ml, equivalent to a M_r of 210,000. Tubes containing FBP-CI are pooled and concentrated to 20–25 ml using an Amicon concentrator fitted with a PM30 membrane.

DEAE-Cellulose Chromatography. The FBP-CI concentrate is applied to a column (2 × 20 cm) of DEAE-cellulose equilibrated with 0.1 *M* potassium phosphate (pH 6.1) containing 0.1 *M* 2-mercaptoethanol and 1 m*M* sodium azide. The column is washed with 50 ml of equilibration buffer and then eluted with a linear gradient of 0–0.5 *M* KCl in equilibration buffer (400 ml total volume). Fractions (4 ml) are collected at a flow rate of 0.5 ml/min and aliquots (100 μl) are counted in 5 ml ACS to identify the dissociated $H_4PteGlu_5$ peak, which elutes at approximately 0.3–0.4 *M* KCl. The $H_4PteGlu_5$ is used to assay fractions for folate-binding activity as described under Procedure. The folate binding peak elutes early during the KCl gradient (0.05–0.1 *M*). The tubes containing folate-binding activity are pooled, concentrated, and dialyzed against 10 m*M* potassium phosphate (pH 7.0) containing 10 m*M* 2-mercaptoethanol and 1 m*M* sodium azide until the conductivity is equivalent to that of the dialysis buffer.

The column in this step is eluted with buffer containing 0.1 *M* 2-mercaptoethanol to protect the dissociated [^{3}H]$H_4PteGlu_5$. The tubes containing the highest counts are stored at −20° under nitrogen in 1-ml aliquots and used to assay folate binding.

5-HCO-H_4PteGlu Affinity Chromatography. The affinity gel is prepared by covalently linking 5-HCO-H_4PteGlu (Lederle) to AH-Sepharose 4B (Pharmacia) as described previously.[4,7] The concentrated (5 ml), dialyzed folate-binding peak from DEAE-cellulose is applied to a column (1 × 3 cm) of affinity gel equilibrated in 10 m*M* potassium phosphate (pH 7.0) containing 10 m*M* 2-mercaptoethanol and 1 m*M* sodium azide. The column is washed with equilibration buffer until absorption at 280 nm has decreased to a minimum and is then eluted with a linear gradient of 0–1 *M* KCl in equilibration buffer (total volume 100 ml). The column is washed with 1 *M* KCl until the absorption of 280 nm is at a minimum. The folate-binding activity is eluted from the column with a linear gradient of 0–10 m*M* PteGlu in equilibration buffer (pH 8.0) containing 1 *M* KCl (total volume 50 ml). Fractions (5 ml) are collected at a flow rate of 0.4 ml/min and assayed for folate-binding activity. To assay for [^{3}H]folate binding activity the fraction eluted with PteGlu must be desalted. Aliquots (200 μl) are desalted twice through BioGel P-2 columns

[7] S. Waxman and C. Schreiber, *Biochemistry*, **25**, 5422 (1975).

PURIFICATION OF FOLATE-BINDING PROTEIN : CYTOSOL I[a]

Step	Volume (ml)	Total protein (mg)	[³H]Folate-binding activity (cpm/mg)	Purification (-fold)
Cytosol	138	1070	500	1
Sephacryl S-200	86	186	4,250	8
DEAE-cellulose	80	40	28,000	56
Affinity gel 1	25	353	39,900	80
Affinity gel 2	30	153	90,900	182

[a] From Cook and Wagner.[3] Reprinted with permission from the American Chemical Society (copyright 1982).

by the centrifugal method described for the binding assay.[4] The amount of binding of [³H]folate is then determined.

At this stage the protein usually gives a single band on SDS–polyacrylamide gel electrophoresis. If there are any contaminants the affinity gel step should be repeated. A summary of the purification is given in the table.

The PteGlu bound to FBP-CI can be removed by dialysis against 10 mM potassium phosphate (pH 7.0) containing 1 mM sodium azide. Alternatively, FBP-CI can be desalted on a BioGel P-6 column of sufficient size to separate the protein from unbound PteGlu and KCl, and then applied to a column (1 × 5 cm) of DEAE-cellulose equilibrated with 10 mM potassium phosphate (pH 7.0) containing 1 mM sodium azide and eluted with a linear gradient of 0–0.2 M KCl (total volume, 100 ml).

Properties

FBP-CI has a native M_R of 210,000 as determined by gel filtration. Pure protein gives a single band on SDS–polyacrylamide gel electrophoresis with an M_r of 100,000, suggesting FBP-CI is a dimer. The endogenous folate ligand is tightly bound $H_4PteGlu_5$. Binding studies showed that the fully reduced folate monoglutamates were the most effective at competing with the natural ligand, $H_4PteGlu_5$, for binding, while PteGlu was 20-fold less effective. The oxidized polyglutamates $PteGlu_3$ and $PteGlu_5$ did not inhibit the binding of the natural ligand.

FBP-CI has no known enzyme activity. Previous studies[2] showed FBP-CI was not serine transhydroxymethylase (glycine hydroxymethyltransferase, EC 2.1.2.1), glutamate formiminotransferase (EC 2.1.2.5), or tetrahydropteroylglutamate methyltransferase (5-methyltetrahydrofolate–homocysteine methyltransferase, EC 2.1.1.13). The trifunctional formyl–methenyl–methylene synthetase (EC 6.3.4.3; EC 3.5.4.9; EC 1.5.1.5) requires H_4PteGlu, has an M_r of 226,000 in sheep liver,[8] 150,000 in pig liver,[9] and is known to be dimeric. However, this activity elutes at an equivalent M_r of 225,000 on Sephacryl S-200 and is separable from FBP-CI on DEAE-cellulose.[3] The enzyme 5,10-methylenetetrahydrofolate reductase (EC 1.7.99.5) was also eliminated, as it elutes at an M_r of 160,000–170,000 on Sephacryl S-200.

It is probable that FBP-CI has some enzymatic function involving the bound H_4PteGlu$_5$ ligand.

[8] J. K. Paukert, L. D'ari-Strauss, and J. Rabinowitz, *J. Biol. Chem.* **251**, 5104 (1976).
[9] L. V. L. Tan, E. J. Drury, and R. E. MacKenzie, *J. Biol. Chem.* **252**, 1117 (1977).

[41] Dimethylglycine Dehydrogenase and Sarcosine Dehydrogenase: Mitochondrial Folate-Binding Proteins from Rat Liver[1]

By ROBERT J. COOK and CONRAD WAGNER

It was shown that after an intraperitoneal[2] or oral dose[3] of [^{3}H]pteroylglutamate (PteGlu) that rat liver mitochondria contained protein(s), of molecular weight 90,000, which bound folate polyglutamates. Further purification revealed the folate was bound to two closely related enzymes, dimethylglycine dehydrogenase (EC 1.5.99.2) and sarcosine dehydrogenase (EC 1.5.99.1).[4-6] These two enzymes perform sequential oxidative demethylation reactions in the choline degradation pathway, which occurs exclusively in liver mitochondria.[7] Both enzymes contain covalently bound FAD,[8] and tightly, but not covalently, bound H_4PteGlu$_5$.[4-6] Earlier

[1] This work was supported by NIH Grant No. AM15289 and the Medical Research Service of the Veterans Administration.
[2] M. M. Zamierowski and C. Wagner, *J. Biol. Chem.* **252**, 933 (1977).
[3] R. J. Cook and J. A. Blair, *Biochem. J.* **178**, 651 (1979).
[4] A. J. Wittwer and C. Wagner, *Proc. Natl. Acad. Sci. U.S.A.* **77**, 4484 (1980).
[5] A. J. Wittwer and C. Wagner, *J. Biol. Chem.* **256**, 4102 (1981).
[6] A. J. Wittwer and C. Wagner, *J. Biol. Chem.* **256**, 4109 (1981).
[7] C. G. MacKenzie, *in* "Amino Acid Metabolism" (W. McElroy and H. B. Glass, eds.), p. 684. Johns Hopkins Univ. Press, Baltimore, Maryland, 1955.
[8] R. J. Cook, K. S. Misono, and C. Wagner, *J. Biol. Chem.* **259**, 12475 (1984).

studies were carried out with partially purified enzyme from rat liver[9,10] and monkey liver.[10] The enzymes have now been purified to homogeneity from rat liver[5] and pig liver.[11]

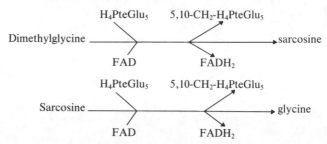

It is proposed that a methylene group is generated in both of these reactions as a result of the oxidation of the N-methyl group. *In vivo,* the enzymes presumably use the bound $H_4PteGlu_5$ to accept the methylene groups with the formation of $5,10\text{-}CH_2\text{-}H_4PteGlu_5$. During purification, however, bound $H_4PteGlu_5$ is lost and the enzyme reaction proceeds with the formation of formaldehyde. These two enzymes are initially separated by DEAE-cellulose chromatography which removes the bound folate, and their subsequent purification is achieved by exploiting their folate-binding capabilities.

Assay Method

Principle. The activity of these two enzymes may be estimated by linking the reduction of FAD to phenazine methosulfate (PMS) and then following the reduction of 2,6-dichlorophenolindophenol (DCPIP) at 600 nm.[10] Enzyme assays of crude tissue extracts or mitochondrial sonicates require the addition of KCN (1 mM) to inhibit nonspecific reduction of the dye. Substrate blanks should also be assayed under these conditions. The assay is run at room temperature.

Reagents

Potassium phosphate buffer, 1.0 M, pH 7.0
Dimethylglycine hydrochloride, 1.0 M, pH 7.0
Sarcosine hydrochloride, 1.0 M, pH 7.0
PMS, 22 mM (prepared fresh)
DCPIP, 3 mM
KCN, 10 mM, pH 7.0

---

[9] C. G. MacKenzie and D. D. Hoskins, this series, Vol. 5 [100b].
[10] W. R. Frizell and C. G. MacKenzie, this series, Vol. 17, Part A, [141].
[11] D. J. Steenkamp and M. Husain, *Biochem. J.* **203,** 707 (1982).

TABLE I

PURIFICATION OF DIMETHYLGLYCINE DEHYDROGENASE AND SARCOSINE DEHYDROGENASE

Enzyme and purification step	Total protein (mg)	Total units (nmol/min)	Specific activity[b] (nmol/min·mg)	Purification (-fold)	Yield (%)
Dimethylglycine dehydrogenase					
Mitochondrial extract	4727	9,129	1.93	1	100
Sephadex G-150	1470	7,800	5.31	2.75	85.4
DEAE-cellulose 1	200	8,048	40.2	20.9	88.2
Affinity gel[a]	—	—	—	—	—
BioGel P-6	37.6	4,824	128.3	66.5	52.8
DEAE-cellulose 2	12.3	3,300	268.3	139.0	36.1
Sarcosine dehydrogenase					
Mitochondrial extract	4727	13,862	2.93	1	100
Sephadex G-150	1470	10,862	7.39	2.52	78.4
DEAE-cellulose 1	56.7	7,690	135.6	46.3	55.5[b]
Affinity gel[a]	—	—	—	—	—
BioGel P-6	18.6	5,757	309.5	105.6	41.5
DEAE-cellulose 2	10.6	2,957	279.0	95.2	21.3

[a] Not determined.

[b] The large decrease in sarcosine dehydrogenase at this stage is due to the separation from dimethylglycine dehydrogenase which will convert sarcosine to glycine.

Procedure. Into a 1-cm light path cuvette (1.0- to 1.5-ml capacity) is pipetted 100 μl each of potassium phosphate buffer, substrate, and PMS, 33 μl DCPIP, and enzyme to a final volume of 1 ml. The reduction of DCPIP is followed at 600 nm. A blank rate of DCPIP reduction by enzyme without substrate is also determined and subtracted from the rate in the presence of substrate. Enzyme activity is expressed as nmol/min and specific activity as nmol/min·mg.

Enzyme Purification

The procedure described is for 200–300 g wet wt. of rat liver and is a modified version of Wittwer and Wagner.[5] A purification scheme is given in Table I.

Isolation of Mitochondria. Rats are killed by decapitation and the livers are rapidly excised, coarsely chopped, rinsed in cold (0–4°) buffer (0.25 M sucrose, 10 mM potassium phosphate, pH 7.0, 1 mM sodium azide, and 0.1 mM phenylmethylsulfonyl fluoride, PMSF) and weighed. All subsequent steps are carried out at 0–4°. The liver is finely chopped and homogenized in a Potter-Elvehjem homogenizer using a medium-fitting Teflon pestle (three strokes, 1300 rpm) in 3 volumes of buffer. Mitochondria are isolated by the method of Bustamante[12] as modified by Wittwer and Wagner,[5] but without 2-mercaptoethanol as this interferes with the PMS-DCPIP assay. The mitochondrial pellet is resuspended in an equal volume of 10 mM potassium phosphate (pH 7.0) containing 1 mM sodium azide. Mitochondria can be used immediately or stored at $-20°$ for several months with no apparent loss of enzyme activity.

Extraction of Mitochondria. The mitochondrial suspension in 40-ml aliquots is freeze-thawed three times in an ethanol (95%)/dry-ice bath. The freeze-thawed suspension is then disrupted with an MSE sonicator at $0°$ in an ice bath at maximum power (10 sec, 15 times with 30 sec between bursts). The sonicate is centrifuged at 220,000 g (R_{av}) for 90 min (50,000 rpm in a Beckman 50.2 Ti rotor). The supernatant fractions are pooled for the next step.

Sephadex G-150 Chromatography. The mitochondrial supernatant is concentrated to 40–60 ml in an Amicon concentrator using a PM30 membrane. The concentrate is applied to a column (5 × 90 cm) of Sephadex G-150 equilibrated with 10 mM potassium phosphate (pH 7.0) containing 1 mM sodium azide. The column is eluted with the same buffer at 1 ml/min in an upward direction, and fractions (20 ml) are collected and assayed for enzyme activity. The enzyme activity should be determined as the difference between the rate plus and minus substrate in the presence of 1 mM KCN. The two enzyme activities elute in overlapping peaks with the peak of sarcosine dehydrogenate activity two tubes (40 ml) ahead of the dimethylglycine dehydrogenase peak, reflecting the difference in the molecular weights of these two enzymes. The tubes containing enzyme activity are pooled and concentrated to 25–50 ml in an Amicon concentrator using a PM30 membrane.

DEAE-Cellulose Chromatography 1. The concentrate containing enzyme activity is applied to a column (2 × 20 cm) of DEAE-cellulose equilibrated with 10 mM potassium phosphate (pH 7.0) containing 1 mM sodium azide. The column is washed with the same buffer until the absorbance at 280 nm has decreased to a minimum (less than 0.25) and then eluted with a linear gradient of 0–0.5 M KCl in equilibration buffer (500 ml each side). Fractions (10 ml) are collected and assayed for enzyme activity. Dimethylglycine dehydrogenase activity elutes first at approximately 0.05 M KCl, followed by sarcosine dehydrogenase at approximately 0.20 M KCl. At this stage the enzymes are separated and can be purified to homogeneity. The enzymes have also lost their tightly bound folate.

---

[12] E. Bustamante, J. W. Soper, and P. L. Pederson, *Anal. Biochem.* **80,** 401 (1977).

TABLE II
PROPERTIES OF DIMETHYLGLYCINE DEHYDROGENASE AND
SARCOSINE DEHYDROGENASE

Property	Dimethylglycine dehydrogenase	Sarcosine dehydrogenase
M_r		
Gel filtration	86,000	99,000
SDS–PAGE	84,000–90,000	92,000–105,000
Absorption maxima	278, 348, 460 nm	278, 342, 455 nm
Ratio 280 nm : 460 nm	17.4	—
Ratio 280 nm : 455 nm	—	15.5
K_m		
Dimethylglycine[a]	0.05 mM	—
Sarcosine[a]	20.0 mM	0.5 mM

[a] D. H. Porter, unpublished results.

There is no apparent loss of activity of either enzyme when stored at 4° at this stage for 7–14 days in the presence of sodium azide (1 mM).

5-HCO-H₄PteGlu Affinity Chromatography. The affinity gel is prepared by covalently linking 5-HCO-H₄PteGlu to AH-Sepharose 4B (Pharmacia) as described previously.[4,13] Separate columns are used for dimethylglycine dehydrogenase and sarcosine dehydrogenase. The pooled enzyme activity from the first DEAE-cellulose column is concentrated to approximately 20 ml and then dialyzed in an Amicon dialysis apparatus, equipped with a PM30 filter, against 10 mM potassium phosphate (pH 7.0) containing 1 mM sodium azide until the conductivity is equal to that of the dialysis buffer. The dialyzed concentrate is applied to a column (1.5 × 15 cm) of affinity gel and then washed with dialysis buffer at 0.25 ml/min until the absorption at 280 nm has decreased to a minimum. The column is then eluted with a linear gradient of 0–1.0 M KCl in dialysis buffer (0.6 ml/min for 2 hr) and then washed with 1 M KCl until the absorption at 280 nm has reached a minimum. The column is now eluted with a linear gradient of 0–0.25 M PteGlu (pH 8.0) in dialysis buffer containing 1.0 M KCl (0.25 ml/min for 4 hr) and then further eluted with 0.25 M PteGlu. Fractions (10 ml) are collected and assayed for enzyme activity.

BioGel P-6 Chromatography. The fractions from the affinity column are now desalted on a BioGel P-6 column (2.6 × 45 cm) equilibrated and eluted with 10 mM potassium phosphate (pH 7.0) containing 1 mM so-

[13] S. Waxman and C. Schreiber, *Biochemistry* 25, 5422 (1975).

dium azide. Fractions (5 ml) are collected and assayed for enzyme activity. At this stage the enzyme contains bound PteGlu which can be removed by a second DEAE-cellulose column.

DEAE-Cellulose Chromatography 2. The fractions containing enzyme activity from the BioGel column are pooled and applied to a column (1.5 × 15 cm) of DEAE-cellulose equilibrated with 10 mM potassium phosphate (pH 7.0) containing 1 mM sodium azide. The column is washed with 25 ml of equilibration buffer and then eluted with the appropriate linear gradient of KCl in the same buffer (0.6 ml/min for 8 hr). Fractions (5 ml) are collected and monitored at 280 nm and assayed for enzyme activity. A gradient of 0–0.25 M KCl is used for dimethylglycine dehydrogenase and 0–0.4 M KCl is used for sarcosine dehydrogenase. Fractions containing enzyme activity are checked for purity by SDS–polyacrylamide gel electrophoresis.

Properties of the Enzymes

These two enzymes show typical flavoprotein UV-visible absorption spectra.[6] The flavin is covalently linked via the 8α position of the isoalloxazine ring to an imidazole N(3) of a histidine residue.[8] Both enzymes are single polypeptides, and comparison of the amino acid compositions indicates a high degree of structural homology.[14] An outline of the properties of both enzymes is given in Table II.

[14] R. J. Cook and C. Wagner, *in* "Chemistry and Biology of Pteridines" (J. A. Blair, ed.), p. 657. de Gruyter, Berlin, 1983.

[42] Membrane-Associated Folate Transport Proteins

By G. B. HENDERSON and F. M. HUENNEKENS

Folate compounds serve as coenzymes for various enzymes that catalyze reactions of one-carbon metabolism. Eukaryotic cells and certain bacteria, however, are unable to synthesize folates *de novo* and thus have developed transport systems to capture these compounds from the environment and to mediate their uptake into cells. Transport systems for folate compounds also account for the primary route by which a number of antifolates (e.g., methotrexate) enter cells. One mechanism of resistance to these drugs involves changes in the transport system that lead to decreased uptake.

Membrane-associated proteins (referred to as "transporters" or "carriers") are the agents responsible for the binding and translocation processes, but, in addition, energy-coupling mechanisms are necessary for the intracellular accumulation of the folate dianion against the membrane potential. These highly hydrophobic proteins are distinct from hydrophilic folate-binding proteins found in milk, serum, and other fluids, which appear to be utilized for the storage and/or extracellular transfer of folate compounds. Folate transport proteins have been identified in the plasma membranes of various cells, but their purification to homogeneity has been achieved only in a few instances.

Principle. Membrane-associated folate transport proteins can be solubilized with the detergent Triton X-100 and purified by procedures adapted to hydrophobic proteins. A unique step, particularly useful in the purification of folate transporters, is adsorption and elution from microgranular silica. Noncovalent and covalent labeling techniques have been employed to monitor folate transport proteins during purification.

Reagents

Vitamin-free casein hydrolysate (ICN Pharmaceuticals)
Folic acid (Sigma)
[^{3}H]Folic acid (Amersham), purified by TLC on cellulose sheets with 20 mM K HEPES–100 mM KCl (pH 7.0) as the solvent
[^{3}H]Methotrexate (Amersham), purified by TLC as described for [^{3}H]folate
HEPES (Sigma)
Triton X-100 (J. T. Baker)
DEAE-cellulose (Whatman)
Microgranular silica, QuSo G-32 (Philadelphia Quartz)
1-Ethyl-3-(3-dimethylaminopropyl)carbodiimide (Sigma)
Dimethyl sulfoxide (J. T. Baker)
Sephacryl S-300 (Sigma)
RPMI 1640 medium (Flow)
Fetal bovine serum (Flow)
Penicillin/streptomycin (Irving Scientific)
N-Hydroxysuccinimide (Sigma)

Folate Transport Protein from *Lactobacillus casei*

Background

Transport of folate compounds in bacteria has been examined in the greatest detail in *Lactobacillus casei*.[1-8] These cells contain a transport

[1] B. A. Cooper, *Biochim. Biophys. Acta* **208,** 99 (1970).

system that mediates the concentrative uptake of various folate compounds. Folate itself is the substrate for which the system has the highest affinity[8] (K_D = 0.2 nM at pH 6). The transport protein responsible for substrate binding and internalization can be induced to a level of 0.35 nmol/10^{10} cells (21,000 copies/cell), which corresponds to 0.5% of the total membrane protein.[6,9] Each transporter has a turnover number[6] of 0.90/min at 30°. Cations are required for folate binding, suggesting that movement of the anion across the membrane proceeds electroneutrally via a cation-symport mechanism.[7] In addition to the transport protein, another functional component is known to be involved in the folate uptake system.[10]

Procedures

Growth of Cells. *Lactobacillus casei* subsp. *rhamnosus* (ATCC 7469) is propagated in the medium described by Flynn *et al.*,[11] supplemented with 10 g/liter of vitamin-free casein hydrolysate and 2 nM folate. Cells are grown from a 1% inoculum without shaking for 16 hr at 30°. Inoculated or full-grown cultures can be stored (in 100-ml portions) for 6 months at −20° without loss in viability. Cells are harvested by centrifugation for 2 min at 12,000 g and then used immediately or stored at −20° as a pellet. Yield is 7–10 g of cells (wet weight)/liter of culture medium.

Folate Binding by Intact Cells. Cells (30 ml) from a 100-ml culture are centrifuged at 12,000 g (2 min, 4°), washed with 20 ml of 50 mM K HEPES–5 mM MgCl$_2$ (pH 7.5), resuspended in the same buffer to a density of 8 × 10^8/ml (150 Klett units; A_{650} = 1.0), and incubated for 1 hr at 23° to deplete energy reserves. The folate-binding capacity of cells (1.0-ml aliquots) is determined by adding [^{3}H]folate (specific activity, 100–500 cpm/pmol) to a concentration of 1.0 μM and incubating the mixture for 5 min at 4°. The cells are recovered by centrifugation at 12,000 g (5 min, 4°),

[2] G. B. Henderson and F. M. Huennekens, *Arch. Biochem. Biophys.* **164,** 722 (1974).

[3] B. Shane and E. L. R. Stokstad, *J. Biol. Chem.* **250,** 2243 (1975).

[4] G. B. Henderson, E. M. Zevely, and F. M. Huennekens, *J. Biol. Chem.* **252,** 3760 (1977).

[5] G. B. Henderson, E. M. Zevely, R. J. Kadner, and F. M. Huennekens, *J. Supramol. Struct.* **6,** 239 (1977).

[6] G. B. Henderson, E. M. Zevely, and F. M. Huennekens, *J. Bacteriol.* **139,** 552 (1979).

[7] G. B. Henderson and S. Potuznik, *J. Bacteriol.* **150,** 1098 (1982).

[8] G. B. Henderson, *in* "Nutritional, Pharmacologic, and Physiologic Aspects of Folates" (R. L. Blakley and M. V. Whitehead, eds.). Wiley, New York, 1986 (in press).

[9] G. B. Henderson, E. M. Zevely, and F. M. Huennekens, *Biochem. Biophys. Res. Commun.* **68,** 712 (1976).

[10] G. B. Henderson, E. M. Zevely, and F. M. Huennekens, *J. Bacteriol.* **137,** 1308 (1979).

[11] L. M. Flynn, V. B. Williams, B. D. O'Dell, and A. G. Hogan, *Anal. Chem.* **23,** 180 (1951).

washed with 4 ml of cold buffer, resuspended in 100 μl of water, transferred to counting vials with two 4-ml aliquots for scintillation fluid, and analyzed for radioactivity. The result is corrected for nonspecific binding in a control prepared identically, except that excess unlabeled folate (100 μM) is added prior to the [^{3}H]folate. Assuming a 1 : 1 stoichiometry, the binding capacity (i.e., the amount of transport protein) is expressed as nanomoles per 10^{10} cells. The dry weight of 10^{10} cells is ~2.2 mg.

A similar procedure is employed to determine the affinity of the transport protein for folate (expressed as the dissociation constant, K_D, for the protein–folate complex), except that the assay volume is increased to 50 ml.[7] After 1 hr at 4°, cells are collected on 0.45-μm cellulose acetate filters, placed in scintillation fluid, and analyzed for radioactivity. K_D values are calculated from double-reciprocal plots of bound versus free folate concentrations.

Labeling the Transport Protein with [^{3}H]Folate. Bound [^{3}H]folate is used to monitor the transport protein during isolation. Noncovalent labeling[4] is achieved by suspending washed *L. casei* cells (100 g) in 400 ml of 20 mM K phosphate (pH 6.5), adding 5 μmol [^{3}H]folate (specific activity, 5 cpm/pmol), and incubating the mixture for 10 min at 4°. Triton X-100 is then added, and the solubilized protein is purified as described below. Folate remains bound to the transport protein throughout the isolation procedure and does not exchange (at pH 7.5) with unlabeled folate. Alternatively, the binding protein can be labeled covalently with the α,γ-anhydride of [^{3}H]folate.[12] The latter is prepared by the following procedure. [^{3}H]Folate (25 μCi), folic acid (5 μmol), and HCl (5 μmol) are combined, and the mixture is dried *in vacuo* over P_2O_5. The residue is dissolved in 20 ml of anhydrous dimethyl sulfoxide containing 100 μmol 1-ethyl-3-(3-dimethylaminopropyl)carbodiimide (EDC), and the solution is incubated for 60 min at 23°. The reagent is then added to cells (100 g) suspended in 2.5 liters of 5 mM K HEPES–150 mM KCl (pH 7.5), and the mixture is stirred for 16 hr at 4°. The cells are recovered by centrifugation at 12,000 g (2 min, 4°), resuspended in 400 ml of 20 mM K phosphate (pH 6.5), and exposed to unlabeled folate (100 mg) for 30 min at 23° to displace noncovalently bound folate. Under these conditions, 70–80% of the folate transport protein is covalently labeled.

Solubilization of the Transport Protein. Cells (400 ml) containing labeled binding protein are treated with 100 ml of 20% Triton X-100 and the mixture is recycled in a prechilled Maton-Gaulin homogenizer for 5 min at 5000 psi[4,5]; the temperature is maintained below 15° by the addition of ice. After centrifugation at 24,000 g (10 min, 4°), the supernatant fraction is

[12] G. B. Henderson and S. Potuznik, *Arch. Biochem. Biophys.* **216,** 27 (1982).

retained and placed on ice. The pellet is resuspended in 500 ml of 20 mM K phosphate–5% Triton X-100 (pH 6.5), and processed again by the above procedure. The yield of solubilized transport protein is ~80%, based upon radioactivity present in a dialyzed aliquot (1.0 ml) of the combined supernatants.

Chromatography on DEAE-Cellulose. The combined supernatants from the previous step are adjusted to pH 7.5 by the addition of 2 N KOH applied to a 2-liter sintered-glass funnel containing 800 g (wet weight) of DEAE-cellulose that has been equilibrated (at 4°) with 20 mM K phosphate (pH 7.5). The equilibration buffer (~800 ml) is used for elution, and the effluent is collected in 100-ml portions. Fractions are discarded until Triton X-100 (detected by absorbance at 280 nm) begins to emerge from the column. Subsequent fractions are pooled, and the solution is clarified by centrifugation at 40,000 g (10 min, 4°). Bound [³H]folate is determined as described above. Purification, 4-fold; recovery, 90%.

Fractionation with Microgranular Silica. Substantial purification and removal of excess Triton X-100 are achieved by absorption and elution of the transport protein from microgranular silica.[4,5] Silica (5 g) is added to the solution from the above step, and the mixture is stirred at 4° for 16 hr. The adsorbent is recovered by centrifugation at 16,000 g (5 min, 4°), washed twice with 250 ml of 50 mM K phosphate (pH 7.5), and then washed with 250 ml of 0.05 M NaHCO₃–0.05% Triton X-100 (pH 9). The protein is eluted from the silica by addition of 250 ml of 0.2 M Na₂CO₃– 0.2% Triton X-100, followed by rapid stirring for 2 min at 4°. The mixture is centrifuged immediately at 16,000 g (2 min, 4°), and the pH of the supernatant is adjusted to 7 by the addition of a saturated solution of monobasic K phosphate. A second portion of silica (0.5 g) is then added to the neutralized supernatant and the mixture is stirred for 16 hr at 4°. After centrifugation, the adsorbent is washed with 40 ml of 50 mM K phosphate (pH 7.5), and the protein is eluted with 40 ml of 0.2 M Na₂CO₃–0.2% Triton X-100, as described above. The eluate is concentrated to 3–4 ml by vacuum dialysis against 500 ml of 20 mM K phosphate–100 mM KCl– 0.1% Triton X-100 (pH 7.5). Purification, 10-fold; recovery, 40%.

Selective Heat Denaturation. The concentrated solution from the previous step is incubated at 37° for 10 min and then centrifuged at 16,000 g (5 min, 4°). The supernatant fraction is retained, the pellet is washed with 0.5 ml of the dialysis buffer (see above), and the supernatants are combined. Purification, 2-fold; recovery, 95%.

Chromatography on Sephacryl S-300. The solution from the above step is applied to a 2.5 × 100-cm column of Sephacryl S-300 that has been equilibrated at 4° with 20 mM K phosphate–100 mM KCl–0.1% Triton X-100 (pH 7.5). Elution is achieved with the same buffer at a flow rate of 1 ml/min, and the eluate is monitored for radioactivity. Peak fractions containing [³H]folate (which appear at ~1.6 times the void volume) are pooled and concentrated to 3 ml by vacuum dialysis. Purification, 2-fold; recovery, 75%.

Comments

Triton X-100 and Nonidet P-40 are equally effective for extracting the folate transport protein. Solubilization can also be achieved with octylglucoside and deoxycholate, but 5- to 10-fold higher concentrations are required. Cholate and 3-(3'-cholamidopropyl)dimethylammonio-1-propane sulfonate (CHAPS) are ineffective.

Bound [³H]folate provides a convenient means for following the folate transport protein during purification and also for quantitating the yield and degree of purification achieved in each step. The protein can also be extracted from membranes in the absence of folate, as judged by positive ELISA and Western blot responses using a polyclonal rabbit antibody directed against the purified antigen (L. Pope, unpublished results). However, a procedure for further purification of the folate-free transport protein has not yet been developed.

The yield of folate transport protein from 100 g of cells is 4–5 mg. Concentrations of the protein in the presence of Triton X-100 are determined by the following procedure. An aliquot of the solution is treated with an equal volume of acetone and, after 1 hr at −20°, the precipitated protein is recovered by centrifugation, washed with 2 ml of 50% acetone, and dissolved in 1 ml of 10 m*M* Na phosphate–0.5% SDS (pH 7.0). The solution is then analyzed for protein content by a modified Lowry procedure in the presence of SDS[13]; 1.0 mg of transport protein is equivalent to 0.56 mg of bovine serum albumin (BSA). The absolute amount of transport protein, used to determine the protein/BSA ratio in the Lowry assay, is obtained from amino acid analysis[4] performed on acetone-precipitated protein.

The folate transport protein obtained by the above procedure contains approximately 30 nmol bound folate/mg protein.[4] It is homogeneous, as judged by SDS–PAGE, and the molecular weight is 20,000 ± 4000 (depending on the electrophoretic system and the standards employed[14]). Heating the binding protein at 100° in SDS should be avoided in the electrophoresis procedure, since it can result in the formation of high molecular weight aggregates. Amino acid analysis[4] reveals that the pro-

---

[13] M. A. K. Markwell, S. M. Haas, L. L. Bieber, and N. E. Tolbert, *Anal. Biochem.* **87**, 206 (1978).

[14] M. Ananthanarayanan, J. M. Kojima, and G. B. Henderson, *J. Bacteriol.* **158**, 202 (1984).

tein lacks cysteine, but it has an abundance of methionine and other hydrophobic amino acids; the calculated polarity of the protein is 32%.

Folate Transport Protein from a Transport-Defective Subline of *L. casei*

Background

Sublines of *L. casei*, characterized by alterations in the folate transport system, have provided additional proof that the folate-binding protein is a component of the transport system.[5] One of these mutants[14] is of considerable interest, since it has an affinity for folate substantially lower than that observed in wild-type cells. Treatment with mercaptoethanol or other reducing agents restores the affinity to normal levels; removal of mercaptoethanol, however, causes a rapid reversal to the low-affinity state.

Procedures

Folate Binding by Intact Cells. Binding characteristics of the mutant cells are measured by the same procedures employed for wild-type *L. casei*, except that the cells in 50 mM K HEPES–5 mM MgCl$_2$ (pH 7.5) are first treated for 5 min at 30° with 20 mM mercaptoethanol. Binding (at 4°) is then measured in the presence of the mercaptoethanol to retain the maximum response.

Purification of the Transport Protein. Purification of the protein in its dimeric form (see below) is accomplished[14] by the same procedure employed for the wild-type protein, except for the following changes: (1) the protein must be labeled covalently with EDC-activated [^{3}H]folate prior to extraction from the membrane; (b) the heat-treatment step is omitted; and (c) final purification is achieved using preparative SDS–gel electrophoresis. In the latter procedure, protein recovered from the Sephacryl column is precipitated by the addition of acetone (final concentration, 50%), recovered by centrifugation, dissolved in 62.5 mM Tris–HCl (pH 6.8) containing 2% SDS and 10% glycerol, and heated at 80° for 3 min. The sample is then applied to a preparative SDS slab gel[15] in which the well-forming comb for the stacking gel has been omitted, and subjected to electrophoresis at 140 V until the tracking dye reaches the bottom of the gel. The binding protein is located by staining a vertical strip of gel (0.5 cm) with Coomassie blue for 5 min. The horizontal strip of the remaining gel that

[15] U. K. Laemmli, *Nature (London)* **227**, 680 (1970).

contains the binding protein is then excised, and the protein is eluted by overnight incubation in 4 ml of 10 mM Na phosphate (pH 7.0) containing 0.1% SDS.

Comments

The mutant transport protein exhibits a K_D value for folate of 280 nM in the absence of mercaptoethanol, but it is transformed to a high-affinity state (K_D 1.2 nM) after treatment with the reducing agent at pH 7.5. The apparent molecular weight of the mutant protein is 33,000 ± 5,000, but it decreases to 20,000 ± 4,000 upon treatment with mercaptoethanol. The amino acid compositions of the mutant and wild-type proteins are indistinguishable, except that the former has a single cysteine residue. These results indicate that the mutant protein ordinarily exists as a homodimer, formed via an S–S linkage between cysteine residues. Conversion of the low-affinity dimeric form to the high-affinity monomeric form is achieved by reduction of the S–S linkage.

In order to purify the mutant transport protein, it is necessary to modify the procedure used for the wild-type counterpart. The heat treatment step is eliminated because it results in precipitation of the protein. Covalent labeling with [³H]folate is also essential because noncovalently bound folate dissociates from the protein, both in the presence or absence of mercaptoethanol.

Folate Transport Protein from L1210 Cells

Background

L1210 mouse leukemia cells contain a membrane-associated protein that mediates the transport of folate compounds. The substrate specificity, however, is somewhat different from that of the *L. casei* transporter.[8,16] A relatively high affinity is observed in anion-deficient buffers for 5-methyltetrahydrofolate (K_D 0.11 μM) and methotrexate (K_D 0.35 μM), while folate is bound with a much lower affinity; as a competitive inhibitor of 5-methyltetrahydrofolate binding, folate has a K_i value of 10 μM. The amount of the protein in wild-type cells is 1 pmol/mg protein, which corresponds to 80,000 copies/cell. Genetic variants of L1210 cells, however, contain up to 14-fold higher amounts of the binding protein.[17]

[16] G. B. Henderson, B. Grzelakowska-Sztabert, E. M. Zevely, and F. M. Huennekens, *Arch. Biochem. Biophys.* **202**, 144 (1980).
[17] F. M. Sirotnak, D. M. Moccio, and C-H. Yang, *J. Biol. Chem.* **259**, 13139 (1984).

Procedures

Growth of Cells. L1210 cells are grown in RPMI 1640 medium supplemented with 2.5% fetal bovine serum and 100 units each of penicillin and streptomycin. Capped culture flasks (2 liter) containing 1 liter of medium and an inoculum of 10^8 cells are incubated with gentle shaking (90 rpm) for 48 hr at 37°. Cells are harvested by centrifugation at 500 g (5 min, 4°), washed with 150 ml of assay buffer (20 mM HEPES–225 mM sucrose, pH 6.8 with MgO), and suspended to a density of 5 × 10^7/ml.

Folate Binding by Intact Cells. Assay mixtures, prepared at 4° in triplicate, consist of freshly grown cells (4 × 10^7), [^{3}H]methotrexate (specific activity, 250 cpm/pmol) or *dl*-5-[^{14}C]methyltetrahydrofolate (specific activity, 110 cpm/pmol) in concentrations ranging from 0.05 to 2.0 μM, and sufficient assay buffer to give a final volume of 1.0 ml. After incubation for 5 min at 4°, the cells are pelleted by centrifugation at 1000 g (5 min, 4°), the supernatant withdrawn by aspiration, and residual fluid around the pellet removed with a cotton swab. The pellets are dispersed in 500 μl of 0.9% NaCl, transferred to counting vials with the aid of two 5-ml aliquots of scintillation fluid, and monitored for radioactivity. Nonspecific binding is determined in control samples which are prepared identically, except that excess unlabeled methotrexate (200 μM) is added prior to the labeled substrate.

Partial Purification of the Transport Protein. Covalent labeling of the protein is achieved using an *N*-hydroxysuccinimide (NHS) ester of [^{3}H]methotrexate.[18,19] [^{3}H]Methotrexate (4 nmol; 10,000 cpm/pmol) is acidified by the addition of HCl (10 nmol) and dried *in vacuo* over P_2O_5. The sample is dissolved in 1.0 ml of anhydrous dimethyl sulfoxide containing 20 mM EDC and 20 mM NHS, and the mixture is incubated for 1 hr at 23°. The reagent is added to cells from a 4-liter culture (1.6–2.1 × 10^6 cells/ml) that have been centrifuged (5 min, 500 g), washed with 200 ml of assay buffer, and resuspended in the same volume of buffer. After 10 min at 23°, the labeled cells are lysed by repeated passage through a Dounce homogenizer, and the plasma membranes are isolated by the procedure of Koizumi *et al.*[20] Extraction of the binding protein is achieved by adding 2 ml of 20% Triton X-100 to the isolated membranes in 18 ml of 50 mM K HEPES (pH 7.5), and incubating the mixture for 10 min at 23°. After centrifugation at 40,000 g (15 min, 4°), the transport protein is precipitated by adding acetone to 50% and incubating for 1 hr at −20°. The sample is

[18] G. B. Henderson and B. Montague-Wilkie, *Biochim. Biophys. Acta* **735**, 123 (1983).
[19] G. B. Henderson and E. M. Zevely, *J. Biol. Chem.* **259**, 4558 (1984).
[20] K. Koizumi, S. Shimizu, K. Koizumi, K. Nishida, C. Sato, K. Ota, and N. Yamanaka, *Biochim. Biophys. Acta* **649**, 393 (1981).

recovered by centrifugation, dissolved in 1 ml of 10 mM Na phosphate–2% SDS–50 mM mercaptoethanol (pH 7.0), and dialyzed overnight against 1 liter 10 mM Na phosphate–0.1% SDS–200 mM NaCl (pH 7.0). Fractionation of the transport protein (at 23°) on a column (1.8 × 65 cm) of Sephacryl S-300 that has been equilibrated with the dialysis buffer gives a single, labeled component which is purified 50-fold relative to intact cells.

Comments

Covalent labeling of the transport protein via the NHS ester of [³H]methotrexate shows a relatively high specificity; only 20–40% of the label is distributed among other membrane proteins. Noncovalently bound methotrexate cannot be used to label the protein during purification, since it dissociates during solubilization of the protein from the membrane. The molecular weight of the protein is 36,000, as judged by its mobility in SDS–PAGE. When various detergents are compared for their ability to extract the labeled protein,[18] Nonidet P-40 was found to be equivalent to Triton X-100, while 5- to 10-fold higher concentrations of CHAPS, cholate, and octylglucoside are required to achieve the same degree of solubilization.

Acknowledgments

Publication BCR-3753 from the Research Institute of Scripps Clinic. The work was supported by grants from the National Cancer Institute (CA06522, CA32261, CA23970) and the American Cancer Society (CH-31 and CH-229). The authors are indebted to Karin Vitols for assistance in preparation of the manuscript.

[43] High-Performance Liquid Chromatographic Separation of the Naturally Occurring Folic Acid Derivatives[1]

By Susan D. Wilson *and* Donald W. Horne

The high resolution inherent in high-performance liquid chromatography (HPLC) along with the speed and reproducibility of this technique has led to its application to the task of separation and quantitation of the

[1] Supported by the Veterans Administration and by NIH Grants AM15289 and AM32189.

naturally occurring folate derivatives.[2-6] An equally important aspect of this analytical methodology is a tissue extraction procedure which protects easily oxidized folates. Recently, we discovered that using ascorbate solutions at elevated temperature to extract tissue folates results in chemical interconversion of several of these derivatives.[7] Procedures have now been developed which allow tissue extraction without these interconversions, baseline separation of naturally occurring folate monoglutamates by HPLC, and quantitation via *Lactobacillus casei* microbiological assay. These procedures have been described elsewhere.[8]

Standard Folic Acid Derivatives. Pteroylglutamic acid (PteGlu, folic acid), H_4PteGlu, 5-CH_3-H_4PteGlu, 5-HCO-H_4PteGlu, and H_2PteGlu are available from Sigma. 10-HCO-H_4PteGlu and 5,10-CH_2-H_4PteGlu may be synthesized by published procedures as outlined by Wilson and Horne.[7] These derivatives may be purified as described below using HPLC. The purified standards should be stored in 10% (w/v) sodium ascorbate at $-20°$ (or below) under a N_2 atmosphere in Thunberg tubes (Kontes) at reduced pressure.

Microbiological Assay of Folates. Assays were performed as described by Wilson and Horne[8,9] with the following modifications. The glycerol-cryoprotected *L. casei* inoculum cultures were prepared by directly suspending the lyophilized culture from the American Type Culture Collection (*L. casei* subspecies *rhamnosus,* ATCC 7469) in 1 ml of single-strength Folic Acid Casei Medium (Difco) supplemented with 0.25 mg sodium ascorbate and 0.30 ng PteGlu per milliliter. This suspension (250 μl) was inoculated into 250 ml of the same, sterile medium and incubated at 37° for about 18 hr. An equal volume of sterile 80% (v/v) glycerol was added and the suspension stored at $-70°$ in 2-ml aliquots. Prior to assay, these aliquots were made up to 10 ml with sterile, 0.9% (w/v) NaCl (saline), washed twice by centrifugation in saline, and resuspended in 2 ml of saline. In our hands, a 1:13 dilution of this suspension gave an optimal standard curve using HPLC purified (6R,S)-5-HCO-H_4PteGlu from 0 to 0.5 ng per tube (calculated as the active S isomer) with an 18-hr incubation

[2] K. E. McMartin, V. Virayotha, and T. R. Tephly, *Arch. Biochem. Biophys.* **209**, 127 (1981).

[3] D. S. Duch, S. W. Bowers, and C. A. Nichols, *Anal. Biochem.* **130**, 385 (1983).

[4] I. Eto and C. L. Krumdieck, *Anal. Biochem.* **120**, 323 (1982).

[5] B. Shane, *Am. J. Clin. Nutr.* **35**, 599 (1982).

[6] J. F. Gregory, D. B. Sartian, and B. P. F. Day, *J. Nutr.* **114**, 341 (1984).

[7] S. D. Wilson and D. W. Horne, *Proc. Natl. Acad. Sci. U.S.A.* **80**, 6500–6504 (1983).

[8] S. D. Wilson and D. W. Horne, *Anal. Biochem.* **142**, 529–535 (1984).

[9] S. D. Wilson and D. W. Horne, *Clin. Chem. (Winston-Salem, N.C.)* **28**, 1198 (1982).

at 37°. This procedure represents a modification of our previous method.[8,9]

Preparation of Liver Extracts. Male Sprague–Dawley rats were anesthetized with sodium pentobarbital, the liver removed, weighed, and minced with scissors, and 10 volumes of 100° extraction buffer (2%, w/v, sodium ascorbate, 0.2 M 2-mercaptoethanol, 50 mM HEPES, 50 mM 2-(N-cyclohexyl-amino)-ethane-sulfonic acid (Ches), pH 7.85) was added. The test tube was heated at 100° for 10 min in the dark, cooled in an ice bath, and the liver was homogenized in a Teflon-glass homogenizer. The homogenate was centrifuged at 40,000 g for 20 min and the resulting supernatant again centrifuged at 40,000 g for 10 min in a Sorvall RC2-B centrifuge. The floating lipid layer was removed by aspiration and the supernatant stored at $-20°$ in Thunberg tubes under a N_2 atmosphere at reduced pressure.

Conjugase Treatment of Extracts. The source of γ-glutamyl hydrolase (conjugase) used to hydrolyze folate polyglutamates was rat plasma. Heparinized blood was centrifuged at 2,000 g for 10 min at 4° and dialyzed overnight against buffer containing charcoal to remove endogenous folates. HPLC–*L. casei* analysis of this preparation indicated that it contained 10 ng 5-CH_3-H_4PteGlu per milliliter.

Aliquots of extracts were filtered through Amicon YMT membranes to remove high molecular weight compounds (synthetic PteGlu$_6$ freely passed through the membrane). The filtrate was warmed to 37° and plasma conjugase (0.25 volumes) was added in three equal portions at 15-min intervals. The tubes were incubated an additional 1.5 hr, heated at 100° for 5 min, centrifuged to remove precipitated protein, and stored at $-20°$ at reduced pressure under N_2 until assayed.

High-Performance Liquid Chromatography. HPLC separation of folate monoglutamates was performed as described previously.[9,10] Briefly, this entailed eluting folates from a Beckman-Altex Ultrasphere I.P. column with a concave ethanol gradient (8–15% ethanol, setting #8, Waters Model 660 Solvent Programmer) using tetrabutylammonium phosphate (TBAP) as an ion-pair reagent. An alternative procedure employed a Spectra-Physics SP8700 Solvent Delivery System equipped with a 2-ml injection loop. Solvents employed were (A) water, (B) 25% ethanol (v/v), and (C) 80% methanol (v/v). Solvents A and B each contained 1 mM sodium ascorbate and 7 mM TBAP reagent. Solvent C was employed to remove tightly adhering substances and as a column storage solvent. Folates were eluted from the Ultrasphere I.P. column isocratically (70%

[10] D. W. Horne, W. T. Briggs, and C. Wagner, *Anal. Biochem.* **116**, 393 (1981).

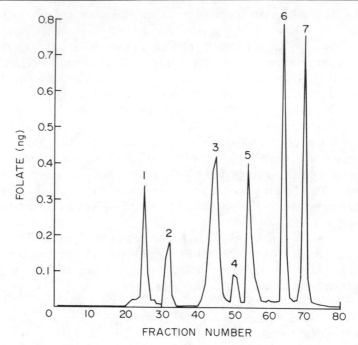

FIG. 1. High-performance liquid chromatographic separation and *L. casei* determination of standard folic acid derivatives. Elution conditions and *L. casei* assay protocol are described in the text. (1) 10-HCO-H₄PteGlu, (2) H₄PteGlu, (3) 5-HCO-H₄PteGlu, (4) H₂PteGlu, (5) 5,10-CH₂-H₄PteGlu, (6) 5-CH₃-H₄PteGlu, and (7) PteGlu; 5–10 ng of each derivative was applied to the column. From Wilson and Horne.[7]

DISTRIBUTION OF RAT LIVER FOLATES[a,b]

Cofactor	Distribution	
	(μg/g)	(%)[c]
10-HCO-H₄PteGlu	1.42 ± 0.13	22.6 ± 1.1
H₄PteGlu	2.06 ± 0.16	32.7 ± 1.4
5-HCO-H₄PteGlu	0.49 ± 0.04	7.7 ± 0.5
5-CH₃-H₄PteGlu	2.35 ± 0.22	37.3 ± 2.6

[a] From Wilson and Horne.[8]

[b] The distribution of folate derivatives in rat liver was determined by HPLC–*L. casei* analysis of conjugase-treated extracts prepared as described in the text. The values reported represent the mean ± SE of 15 experiments.

[c] Percentage of total folate eluted from the column.

A, 30% B) for the first 25 min, followed by a linear gradient ending at 75 min (40% A, 60% B). Fractions of 1 ml were collected into tubes containing 0.1 ml of 10% (w/v) sodium ascorbate. Appropriate aliquots were chromatographed via HPLC and folates were quantitated using the *L. casei* microbiological assay.

Figure 1 shows the elution profile when standard folates were analyzed via the HPLC–*L. casei* procedure described above. All standard folates were baseline resolved, a fact which makes quantitation of each derivative unambiguous. The elution position of each standard was confirmed by UV spectroscopy. The data in the table were obtained by applying our procedures (extraction, conjugase treatment, and HPLC–*L. casei* assay) to determining the distribution of folates in rat liver.

[44] Separation of Pteridines from Blood Cells and Plasma by Reverse-Phase High-Performance Liquid Chromatography

By HANS-JÖRG ZEITLER and BERTA ANDONDONSKAJA-RENZ

Reduced pterins (e.g., biopterin, folates) are intermediates in anabolic and catabolic reactions in pteridine metabolism and serve as coenzymes for different enzymatic reactions. In the biosynthesis of melanins, neurotransmitters, and prostaglandins mixed-function oxygenases need reduced biopterins, whereas the conversion of some amino acids and the biosynthesis of purines and pyrimidines (transfer of one-carbon units) are mediated by several folate derivatives. These reactions are important in various tissues[1] and in the development, differentiation, and maturation of blood cells.[2] The folate coenzymes in nonanimal cells are synthesized from guanosine triphosphate and are one of the systems of self-regulation that may get out of control in malignant conditions.[3] The importance of various pteridines in cell metabolism of higher mammals was recognized by Kaufman in 1963.[4] The identification of 6-hydroxymethylpterin[5] as a

[1] C. Kutzbach and E. L. R. Stokstad, *Biochim. Biophys. Acta* **139**, 217 (1967).

[2] D. Watkins, T. E. Shapiro, and B. A. Cooper, *in* "Biochemical and Clinical Aspects of Pteridines" (H. C. Curtius, W. Pfleiderer, and W. Wachter, eds.), Vol. 2, p. 351. de Gruyter, Berlin, 1983.

[3] A. Albert, *in* "Chemistry and Biology of Pteridines" (W. Pfleiderer, ed.) p. 1. de Gruyter, Berlin, 1975.

[4] S. Kaufman, *Proc. Natl. Acad. Sci. U.S.A.* **50**, 1085 (1963).

[5] B. Stea, P. S. Backlund, P. B. Berkey, A. K. Cho, B. C. Halpern, R. M. Halpern, and R. A. Smith, *Cancer Res.* **38**, 2378 (1978).

folate catabolite and its isolation from cultured cancer cells as their biopterin content increased[6] led to the understanding of an altered metabolism of various pteridines and folates in tumor cells.

The distribution of pteridines in various tissues was analyzed using HPLC systems by several laboratories.[7–10] Most of the reported methods, utilizing fluorometric or electrochemical (reduced pteridines[11]) detection techniques, concern themselves with the determination of one pteridine or with the separation of a few unconjugated pteridines from biological materials (mostly urine). Although HPLC methods for the separation of various pterins have been developed, these systems do not separate lumazine from biopterin. Moreover, lymphocytes and the buffy coat fraction of white cells cannot be analyzed by ion-exchange techniques because of small extraction volumes and the variable recovery[12] of pteridines. To study the pteridine pattern in body fluids, tissues, and cellular material, a method was developed and optimized for the separation and quantitation of up to 13 pteridines (pterin-6-carboxylic acid, xanthopterin, neopterin, monapterin, isoxanthopterin, lumazine, biopterin, 6-hydroxymethylpterin, pterin, 6-methylpterin, sepiapterin, deoxysepiapterin, 3'-hydroxysepiapterin).[13] This method was applied successfully to different material from patients with leukemia and tumors.[14–16]

Principle. The method is based on the extraction of free and conjugated pteridines from plasma and blood cells (peripheral or bone marrow) without or after acidic (alkaline) iodine oxidation. A pteridine-releasing procedure or a deproteinization employing both strong acidic pH and

[6] A. M. Albrecht, J. L. Biedler, H. Baker, O. Frank, and S. H. Hutner, *Res. Commun. Chem. Pathol. Pharmacol.* **19,** 377 (1978).

[7] T. Fukushima and J. C. Nixon, *Anal. Biochem.* **102,** 176 (1980).

[8] B. Stea, R. M. Halpern, and R. A. Smith, *J. Chromatogr.* **188,** 363 (1980).

[9] J. H. Woolf, C. A. Nichol, and D. S. Duch, *J. Chromatogr.* **274,** 398 (1983).

[10] S. W. Bailey and J. E. Ayling, *in* "Chemistry and Biology of Pteridines" (W. Pfleiderer, ed.), p. 633. de Gruyter, Berlin, 1975.

[11] M. Bräutigam, R. Dreesen, and H. Herken, *Hoppe-Seyler's Z. Physiol. Chem.* **363,** 341 (1982).

[12] B. Stea, R. M. Halpern, and R. A. Smith, *J. Chromatogr.* **168,** 385 (1979).

[13] B. Andondonskaja-Renz and H. J. Zeitler, *Anal. Biochem.* **133,** 68 (1983).

[14] H. J. Zeitler, B. Andondonskaja-Renz, M. Fink, and W. Wilmanns, *in* "Biochemical and Clinical Aspects of Pteridines" (H. C. Curtius, W. Pfleiderer, and W. Wachter, eds.), Vol. 2, p. 89. de Gruyter, Berlin, 1983.

[15] H. J. Zeitler and B. Andondonskaja-Renz, *in* "Biochemical and Clinical Aspects of Pteridines" (W. Pfleiderer, H. Wachter, and H. C. Curtius, eds.), Vol. 3, p. 313. de Gruyter, Berlin, 1984.

[16] B. Andondonskaja-Renz and H. J. Zeitler, *in* "Biochemical and Clinincal Aspects of Pteridines" (W. Pfleiderer, H. Wachter, and H. C. Curtius, eds.), Vol. 3, p. 295. de Gruyter, Berlin, 1984.

elevated temperature is followed by separation of the pteridines using reverse-phase high-performance liquid chromatography (rp-HPLC). The pteridines are determined quantitatively (femtomolar range) by either fluorescence detection at excitation/emission of 360/460 nm (blue fluorescence) and 425/530 nm (yellow fluorescence), or ultraviolet detection (sepiapterins) at 410 nm.

Reagents and Buffer Solutions

Pteridines (Dr. B. Schircks, Jona, Switzerland)
Trifluoroacetic acid, sequanal grade (Fluka, Neu-Ulm, West Germany), 8 M
Acidic iodine solution (5.4% I_2 and 10.8% KI in 1 M trifluoroacetic acid)
Ascorbic acid, aqueous 10% solution (fresh prepared)
Starch solution, 2% in water
Methanol (HPLC grade), 15% solution in water
Ammonium chloride, 0.83% solution in redistilled water
Dipotassium hydrogenphosphate, 3.1–3.5 mM, adjusted to pH 7.0–7.8 with orthophosphoric acid
Potassium dihydrogenphosphate (cleavage buffer), 70 mM, pH 4.45
Potassium triphosphate, 2 M, 0.25 M, 0.12 M
PBS Dulbecco's buffer (without Ca^{2+} and Mg^{2+}) (Seromed, Munich, West Germany)
Ficoll-Paque (Pharmacia)
Neo-Plasmagel (Braun Melsungen, Melsungen, West Germany)
Mono-Poly Resolving Medium (M-PRM) (Flow Laboratories)
Silicon solution (Serva, Heidelberg, West Germany)

For the preparation of all solutions including buffers, double-glass-distilled water (Schott Bidistillator) or HPLC water (CAP type I/ASTM type I; Milli-Q system, Millipore) must be used.

Reverse-Phase High-Performance Liquid Chromatography

Technical Description of the HPLC System

A variety of suitable instruments is available. We have used Altex pumps, model 110 A, and Beckman 112 Solvent Delivery Modules. Rheodyne 7125 and Negretti and Zambra Model 190 valve injectors were used with 20-μl loops. The fluorometric detection was accomplished with Shimadzu fluorometers, Model RF 530. The ultraviolet detection of sepiapterins was performed with a Kontron spectrophotometer, Uvikon

LCD 725. The fluorometers and the UV spectrophotometer were equipped with electronic integrators, Spectra-Physics, Model 4270.

When using other fluorometers and integrators, their compatibility as well as background noise level must be examined to get satisfactory signal-to-noise ratios in the range of the detection limit.

All reverse-phase HPLC analyses were carried out at first with Kontron ready-to-use columns (250 × 5 mm; 5-μm C_{18}-ODS), and later, the best results were obtained with self-packed columns (Shandon Hyperchrome, 250 × 4.6 mm, and Du Pont, 250 × 4.6 mm; separation material, Spherisorb 5μ C_{18}-ODS, Phase Separation Ltd., Queensferry, United Kingdom). All column end fittings were equipped with porous metal frits rather than glass fiber filters (filter sandwich), which were not suited for HPLC analysis of biological materials. The columns were packed according to a procedure optimized in our laboratory. Following sonication (ultrasonic bath, e.g., Bransonic 221) of the packing slurry in 2-propanol for about 15 min the columns were filled upward at ~8600 psi for 5 min using n-heptane as the packing solvent (Shandon 9000 psi pressure intensifier packing pump). They were conditioned on the packing pump by passing 2-propanol, in the downward direction, for 10 min followed by methanol for an additional 10 min. After washing with HPLC water and equilibrating with the mobile phase (degassed by sonication or helium, and filtered through a Millipore 0.45-μm HA-type filter), the column was ready to use.

rp-HPLC Chromatography (Isocratic System)

In studying the metabolism of pteridines in pathological cells, we needed an effective HPLC system for the separation and the quantitation of pteridines. We studied the dependence of the chromatographic behavior (mobility and retention) of pteridines on the ionic strength of the elution buffer at constant pH. For the mobile phase the range of 3–4 mM hydrogen phosphate ions offers a good compromise between buffer capacity and separation of the pteridines. We also studied the mobility of the pteridines as a function of pH of the mobile phase at constant ionic strength (3.1 mM HPO_4^{2-}) with the resulting optimum pH of 7.3–7.8 (see Fig. 1). However, elution of the pairs monapterin/isoxanthopterin and lumazine/biopterin is particularly sensitive to pH variations in the eluant under the conditions described above (isocratic elution). An increase in hydrogen ion concentration improves the separation of the monapterin/isoxanthopterin and pterin/6-hydroxymethylpterin pairs without considerable influence on retention times of the other pteridines (see Fig. 2). Shifting the pH toward the more alkaline values favors the separation of lumazine and biopterin.[13] The optimum pH for each column, therefore, has to be determined by a test chromatogram.

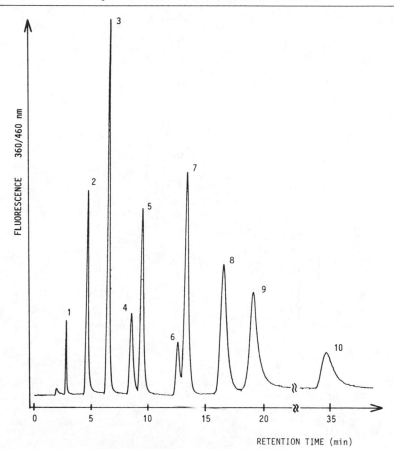

FIG. 1. rp-HPLC separation of pteridines on 5-μm C_{18}-ODS (Spherisorb). Chromatographic conditions: self-packed column, 250×4.6 mm, mobile phase, 3.5 mM dipotassium hydrogen phosphate, adjusted to pH 7.45; flow rate, 1 ml/min; column temperature, ambient; fluorescence detection, excitation/emission = 360/460 nm. 1, Pterin-6-carboxylic acid; 2, xanthopterin; 3, D-*erythro*-neopterin; 4, L-*threo*-neopterin (monapterin); 5, isoxanthopterin; 6, lumazine; 7, L-biopterin; 8, 6-hydroxymethylpterin; 9, pterin; 10, 6-methylpterin.

Because of the participation of sepiapterins in the biosynthesis of tetrahydrobiopterin[17-21] and the oxidative conversion of reduced forms of

[17] T. Fukushima and T. Shiota, *J. Biol. Chem.* **249**, 4445 (1974).

[18] M. Matsubara, M. Tsutsue, and M. Akino, *Nature (London)* **199**, 908 (1963).

[19] S. Katoh, *Arch. Biochem. Biophys.* **146**, 202 (1971).

[20] S. Katoh, M. Nagai, Y. Nagai, T. Fukushima, and M. Akino, *in* "Chemistry and Biology of Pteridines" (K. Iwai, M. Akino, M. Goto, and Y. Iwanami, eds.), p. 225. International Academic Press, Tokyo, 1970.

[21] M. Tsutsue, *J. Biochem. (Tokyo)* **69**, 781 (1971).

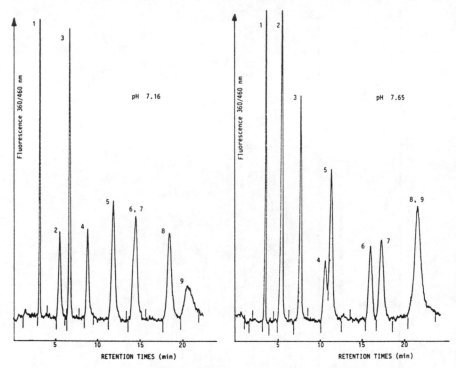

FIG. 2. rp-HPLC of pteridines at different pH values on Spherisorb 5-μm C_{18}-ODS (column, 250 × 4.6 mm); fluorescence detection at 360/460 nm (excitation/emission); Spectra-Physics integrator SP 4270; mobile phase, 3.5 mM K_2HPO_4 (adjusted with H_3PO_4 to pH 7.16 and 7.65); temperature, 20°. For numbering of peaks, see Fig. 1. (From Andondonskaja-Renz and Zeitler.[13])

biopterin to sepiapterins,[22] we have also studied some sepiapterins. On C_{18}-ODS the yellow fluorescent pteridines (xanthopterin, 3'-hydroxy-sepiapterin, sepiapterin, deoxysepiapterin) can be separated with 15% aqueous methanol as mobile phase (see Fig. 3).

Sensitivity and Calibration

The sensitivity of the described rp-HPLC system (7–10 pg biopterin at a signal-to-noise ratio $\geq$4) is comparable to the bioassay of biopterin[23]

[22] T. Fukushima and J. C. Nixon, in "Chemistry and Biology of Pteridines" (R. L. Kisliuk and G. M. Brown, eds.), p. 31. Elsevier/North-Holland Biomedical Press, Amsterdam, 1979.

[23] J. R. Leeming and J. A. Blair, Clin. Chim. Acta 108, 103 (1980).

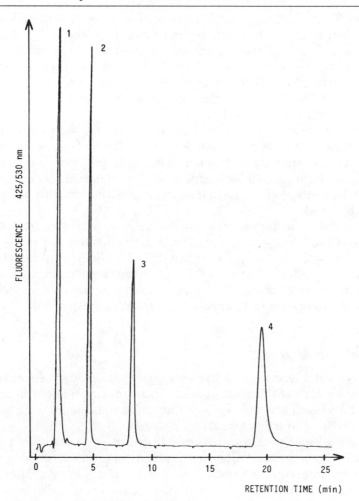

FIG. 3. rp-HPLC separation of yellow fluorescing pteridines on 5-μm C$_{18}$-ODS (Spherisorb). Chromatographic conditions: self-packed column, 250 × 4.6 mm; mobile phase, methanol/water (15/85); flow rate, 1 ml/min; column temperature, ambient; fluorescence detection at 425/530 nm (excitation/emission). 1, Xanthopterin; 2, 3'-hydroxysepiapterin; 3, sepiapterin; 4, deoxysepiapterin.

with *Crithidia fasciculata* (coefficient of variation 8%), but the HPLC methods show better specificity for the many pteridines present in biological materials.

For the quantitation of the identified pteridines in the biological samples the integration was performed with an *external standard*. Each peak of interest in the sample is compared with the same peak in the calibration solution. Because the amount of each component in the calibration solution is known, a ratio between that component and the analyzed sample can be calculated, giving the amount of the component in the sample (concentration = area/response factor). Because the response factor is determined by peak area (calibration substance) per unit of concentration, the uniformity of the sample injection size determines the accuracy of the calibration and the analytical values.

The *internal standard method* was not suited for the analysis of biological samples with numerous fluorescing compounds owing to the prerequisite of an internal standard which has to be positioned in the middle part of the chromatogram as a large peak. The standard chosen must not be identical with components in the sample material and must not overlap with other peaks.

To calibrate the integrator, a standard solution of all investigated pteridines was used with a concentration in approximately the range of the analysis sample (50–200 pg/20 μl and 2.5–10 μg/liter, respectively; solvent, HPLC buffer). Because of a decreasing fluorescence yield due to lamp deterioration the correlation between concentration and peak area (response factor) has to be checked routinely using standards.

Biological Applications

Since pteridines are destroyed by light, the standard solutions and preparation of blood samples (bone marrow) are carried out in minimal light, and vessels are covered with aluminum foil. Figure 4 outlines the procedure for extraction and HPLC analysis.

Preparation of Glassware. To avoid adsorption on glass and polymer surfaces, all glassware, Pasteur pipets, and centrifuging tubes were siliconized by immersing in a silicone solution for about 10 min. After allowing excess silicone solution to run off, the glassware is dried in an oven for 1 hr at 100–150°. After drying, the siliconized vessels should be washed with redistilled water.

Blood Fractionation. Anticoagulant-treated blood (fresh peripheral blood, bone marrow, blood from the blood bank) was separated into plasma and cell fractions by various centrifugation techniques.

Normal Centrifugation. The separation of the blood into plasma, erythrocytes, and the white cell fraction (buffy coat) was performed by centrifuging the samples (blood, 10–12 ml; bone marrow, 1–4 ml) in glass tubes at ~1000 g for 20 min. After pipetting off the plasma (Pasteur pipet) the buffy coat cell layer was drawn into a shortened Pasteur pipet, from the erythrocytes. To remove any remaining red cells by lysis, the white cell fraction was suspended in a 0.83% solution of ammonium chloride at

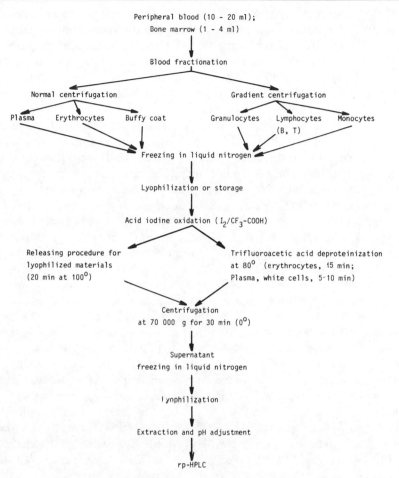

FIG. 4. Outline of the multiple-assay procedure for the HPLC analysis of pteridines in fractions of peripheral blood and bone marrow.

37° for 5 min by gently drawing it in and out of a Pasteur pipet. After centrifuging at 500 g for 10 min, the supernatant was discarded. This procedure was repeated if necessary. The white cell fraction was washed with a balanced salt solution (e.g., PBS Dulbecco without Ca^{2+} and Mg^{2+}) by gently drawing it in and out of a Pasteur pipet. From the suspension the differential blood count and the viability (trypan blue dye exclusion method) was determined and the cells were isolated by centrifugation at 500 g for 10 min. All blood fractions were frozen immediately in liquid nitrogen and stored at temperatures below −20° or lyophilized (Modulyo; Edwards-Kniese, Marburg, West Germany).

Ficoll Centrifugation. Blood was diluted with an equal volume of a balanced salt solution, and 8 ml of this mixture was carefully pipetted onto 6 ml of Ficoll-Paque (density, 1.077 g/ml). After centrifugation at 1 600 *g* for 30–40 min, the fractions (plasma, lymphocytes/monocytes, erythrocytes) were isolated. The mononuclear cells and the erythrocytes were washed with the balanced salt solution and treated as above.

Isolation of Granulocytes, Lymphocytes (B, T), and Monocytes. Ten milliliters of fresh (within 2 hr after collecting, if possible), anticoagulant-treated blood was mixed with 10 ml of Neo-Plasmagel (polymerized fragments of succinylated gelatin, NaCl, CaCl$_2$, and sterile water) and incubated in a vessel at a 45° angle and at 37° until the erythrocytes have settled to one-quarter of the initial blood volume. The supernatant containing plasma and white cells (including some remaining red cells) was removed and centrifuged at 300 *g* for 10 min. The cell pellet was carefully resuspended in 6 ml of a balanced salt solution with a Pasteur pipet and layered onto 6 ml of Ficoll-Paque. After centrifugation at 1 600 *g* for 30 min, the layer of mononuclear cells and platelets (plasma–Ficoll interface) was isolated as above. Remaining erythrocytes in the granulocyte pellet were lysed with ammonium chloride as described above and the granulocytes centrifuged at 500 *g* for 10 min. The washed granulocytes (suspension in a balanced salt solution, after differential blood count and viability test) were centrifuged at 500 *g* for 10 min, resuspended in 1–2 ml of redistilled water, and lyophilized after freezing in liquid nitrogen. Another suitable method is the single-step separation of erythrocytes, granulocytes, and mononuclear leukocytes on discontinuous gradients of Ficoll-Hypaque[24] (e.g., Mono-Poly Resolving Medium). Seven milliliters of fresh heparinized blood (3.5 ml of bone marrow) was carefully layered onto 6 ml (3 ml) of Mono-Poly Resolving Medium and centrifuged at 300 *g* for 30 min in a swinging bucket rotor at ambient temperature. The red cells (pellet) were separated from the mononuclear leukocytes (at plasma–medium interface; viability, >98%; recovery of lymphocytes and monocytes, 20–75 and >90%, respectively; remaining granulocytes, 0–15%) and the polymorphonuclear leukocytes (PMN band below the interface; viability, >98%; recovery, 30–70%; remaining mononuclear cells and eosinophils, 0–15%), which were transferred to individual tubes and washed twice with a balanced salt solution. The cells were then isolated and worked up as described before.

For the separation of blood from patients with various diseases (e.g., severe anemia, infections, chronic granulomatosis) the centrifugation time may be increased to 50–60 min. Some medications also affect the

[24] D. English and B. R. Andersen, *J. Immunol. Methods* **5**, 249 (1974).

separation of white cells, as the blood of patients with juvenile rheumatoid arthritis or microcytic hypochromic anemia may fail to separate by the above-mentioned M-PRM procedure.

The separation of monocytes and the purification of human B and T cells (lymphocytes) on density gradients of Percoll can be achieved according to methods reported previously.[25–28]

Oxidation of Reduced Pteridines

Some methods for the oxidation of the unstable,[7] reduced forms of pteridines in tissues and body fluids have been previously described.[8,22,29] We adapted these oxidation procedures by employing iodine at strong acid pH. Acid oxidations with manganese dioxide were found to be unsuitable for plasma and cell materials. Because of the possibility of the formation of pterin from reduced forms of biopterin or neopterin in phosphoric acid-containing solutions,[14] we used as a strong acid medium an aqueous solution of trifluoroacetic acid which had none of the side reactions with any reduced form of biopterin and neopterin. In addition, trifluoroacetic acid was used successfully for the deproteinization of biological samples (see Deproteinization). The native and the lyophilized blood fractions (plasma, cell fractions) were weighed and diluted (with the exception of native plasma) with redistilled water as shown in Table I (for the lyophilized and the native fractions of blood or bone marrow, the following data are appropriate mean values: plasma, 90 mg/ml; erythrocytes, 340 mg/ml; lymphocytes, 200 mg/ml; buffy coat, 1 ml of whole blood yields approximately 0.5–1 mg of lyophilized white cells).

The magnetically stirred suspensions were acidified dropwise with 8 M trifluoroacetic acid (TFA) and then oxidized with the appropriate amount of a solution of 5.4% I_2 and 10.8% KI in 1 M TFA at ambient temperature for 60 min or at 37° for 30 min. The final concentration of TFA was 0.1 M. The excess iodine was reduced by addition of a fresh 10% aqueous solution of ascorbic acid (see Table I) until the starch test was negative.

Extraction of Pteridines from Biological Materials

For extraction of pteridines from the different biological materials two methods were used.

[25] H. Pertoft, A. Johnsson, B. Wärmegård, and R. Seljelid, J. Immunol. Methods 33, 221 (1980).
[26] C. Gutierrez, R. R. Bernabe, J. Vega, and M. Kreisler, J. Immunol. Methods 29, 57 (1979).
[27] A. J. Ulmer and H. D. Flad, J. Immunol. Methods 30, 1 (1979).
[28] F. Gmelig-Meyling and T. A. Waldmann, J. Immunol. Methods 33, 1 (1980).
[29] T. Fukushima, K. J. Kobayashi, I. Eto, and T. Shiota, Anal. Biochem. 89, 71 (1978).

TABLE I

SCHEME FOR ACIDIC IODINE OXIDATION OF BLOOD AND BONE MARROW SAMPLES

Sample	Weight (mg)	Dilution with water (ml)	Addition of 8 M TFA (μl)	Oxidation with I_2/KI/1 M TFA[a] (μl)	Incubation time	Ascorbic acid, 10% (μl)[a]
Plasma						
Native	10,000	—	75	600		225
Lyophilized	100	1.2	9	68	1 hr	25
Erythrocytes					at 20°	
Native	10,000	40	370	2300	or	500
Lyophilized	100	1.2	11	68	30 min	15
Thrombocytes, white cells					at 37°	
Native	t.a.[b]	1.0	12	10		3
Lyophilized	100	1.5	10	60–70		10–15

[a] Mean values for samples from healthy donors.
[b] t.a., Total amount from 10 ml of whole blood.

1. Pteridine-Releasing Procedure. To differentiate in tissues and protein-containing body fluids between free and protein-bound (conjugated) pteridines, an effective releasing procedure for the determination of the conjugated form was needed. A suitable method was described by Baker *et al.*[30] in which the release of conjugated pteridines could be achieved by irreversible heat denaturation (autoclaving at 120°). We have modified the above-mentioned procedure based on our experimental results. The aconitate buffer was replaced by a phosphate buffer (70 mM KH_2PO_4, pH 4.45) because of the phosphate-containing mobile phase of the HPLC system used. Also, reproducible cleavage times are obtained when performed at 100° without excess pressure. Comparing the results with both methods, we have found better yields of pteridines when the extraction was performed at the lower temperature.

To release pteridines, nonoxidized native samples were suspended in eight times their volume of cleavage buffer (pH 4.45), whereas for lyophilized erythrocytes, plasma, and other cells a 4-fold, 1- to 2-fold, and 0.2- to 0.5-fold volume of the initial volume of the whole blood (bone marrow) sample had to be used. These suspensions have a pH 5.3–5.4 for plasma and 6.0 for erythrocytes. The pH of the mixture with plasma was adjusted to pH 4.45 with orthophosphoric acid. Iodine-oxidized samples were ad-

[30] H. Baker, O. Frank, A. Shapiro, and S. H. Hutner, this series, Vol. 66, p. 490.

justed to pH 4.45 with 0.1 M potassium triphosphate prior to their addition to the cleavage buffer. The suspensions were then put in a preheated (100°) autoclave (Labor Autoklav M-4-20, filled with the necessary amount of distilled water; Sasse, Weinheim, West Germany) and heated for 20 min (with opened steam valve). The less acid suspensions of erythrocytes or of their lyophilizates in the cleavage buffer should not be adjusted from pH 6 to 4.45, owing to the formation of a deep red gelatin after the releasing treatment.

Subsequently, the samples were centrifuged (0°) at 70,000 g for 20 min and the clear supernatants (those from erythrocytes are slightly red) were lyophilized after freezing in liquid nitrogen. This is a necessary step for concentrating the pteridines because of their low concentration in biological materials (except urine). Samples could be either stored at -20 to $-80°$ (protected from light) for several weeks without measurable alterations in the pteridine content, or extracted with a minimum amount of HPLC buffer (mobile phase) by sonication (ultrasonic bath, Bransonic 221) for several seconds or by vortexing (Vortex Genie Mixer). The clear supernatant of the centrifuged extracts could then be immediately submitted to HPLC analysis.

However, according to Fukushima et al.[22] the reduced forms of biopterin (dihydro-, quinoid dihydro-, tetrahydrobiopterin) are air-oxidized and converted to biopterin, pterin, xanthopterin, sepiapterin, and deoxysepiapterin in different buffers and at elevated temperatures.[13,14] In solutions of reduced forms of biopterin and neopterin (under conditions described for pteridine release at 120 or 100°) we found a relatively constant conversion, depicted in Fig. 5. When using acid phosphate buffer (pH 4.5) at elevated temperatures only low yields of deoxysepiapterin from 7,8-dihydrobiopterin and low yields of xanthopterin from deoxysepiapterin and sepiapterin were obtained (see Fig. 5). The additional pathway from tetrahydrobiopterin to deoxysepiapterin as reported occurring in buffers with other ionic compositions[22] could be excluded here as well as the formation of sepiapterin. Moreover, 6-hydroxymethylpterin formed from dihydrobiopterin and pterin formed from tetrahydrobiopterin are produced via the 7,8-dihydro form (commercially available) and the quinoid form, respectively. In analogous experiments with the reduced forms of neopterin (see Fig. 6), 6-hydroxymethylpterin was produced only from the 7,8-dihydro form, whereas pterin formation from tetrahydroneopterin requires a quinoid intermediate as reported for the biopterin conversion. Isoxanthopterin and lumazine are not produced from any reduced form of biopterin and neopterin under these conditions. At 120°, however, the described conversion reactions were found to be increased to some extent. Because most of the biopterin in the fluids and tissues of mammals is

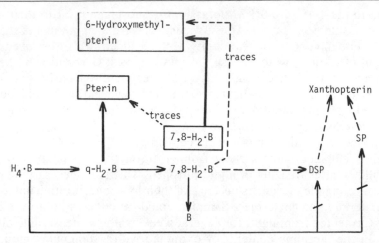

FIG. 5. Oxidative conversion of reduced forms of biopterin under the conditions of the "releasing procedure" (autoclaving at 120° for 5 min or heating at 100° for 20 min). $H_4 \cdot B$, Tetrahydrobiopterin; $q\text{-}H_2 \cdot B$, quinoid dihydrobiopterin; B, biopterin; SP, sepiapterin; DSP, deoxysepiapterin. The dashed lines refer to reactions with low yields.

present in its reduced forms, the possibility of chemical conversion has to be taken in consideration when interpretating analyses of pteridines, and extraction techniques which do not involve oxidation of the reduced part of the pteridines must be carefully examined.

2. *Deproteinization with Trifluoroacetic Acid.* In addition to the afore-mentioned cleavage method at 100° for an extraction of free and conjugated pteridines from biological materials, we have developed a deproteinization procedure at acidic pH with a subsequent incubation at elevated temperature. This method is easy to apply to analyses with or without iodine oxidation (either acidic or alkaline). Of several deproteinizing acids which could also be used as the acid medium in the iodine oxidation procedure, hydrochloric acid and orthophosphoric acid were unsuited because of insufficient protein precipitation. The hygroscopic and nonvolatile trichloroacetic acid and perchloric acid also were not suitable for the deproteinization of biological samples concentrated by lyophilization. Because of a low concentration of pteridines and higher concentrations of these unsuitable acids in the supernatants of the centrifuged samples, sufficient separation of the pteridines does not occur in the HPLC analysis. Moreover, following deproteinization with perchloric acid the supernatants of the centrifuged samples were destroyed totally during lyophilization. For this reason we substituted successfully the vol-

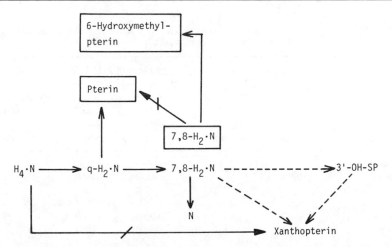

Fig. 6. Oxidative conversion of reduced forms of neopterin under the conditions of the "releasing procedure" (see Fig. 5). $H_4 \cdot N$, Tetrahydroneopterin; $q\text{-}H_2 \cdot N$, quinoid dihydroneopterin; N, neopterin; $3'\text{-}OH\text{-}SP$, $3'$-hydroxysepiapterin. The dashed lines refer to low yield reactions.

atile *trifluoroacetic acid* for these deproteinizing agents. This procedure can be used before any oxidation of the biological sample to estimate the oxidized portion of pteridines. When used after an acidic (or alkaline) iodine oxidation,[7] the deproteinization procedure permits a simple determination of the total (or alkaline stable) amount of each pteridine present in the investigated sample.

To the intensely stirred and light-protected suspension (of the weighed native biological fraction in water, or after acid/alkaline iodine oxidation), the calculated amount (according to Table II) of an aqueous solution of 8 *M* trifluoroacetic acid was dropped slowly into the suspension until a molarity of 0.6 (cell material) to 0.8 (plasma) was reached. The acidified suspensions were then magnetically stirred for 5–10 (plasma, tissue, cell material except red cells) or 15 min (erythrocytes) in a metal rack positioned in a water bath (bath temperature, 80°) or in a metal block with corresponding bore holes (ReactiBlock, heated to 80°; Pierce Chemical Company, Rockford, Illinois). From the tested temperatures (40–80°) and incubation times (5–25 min), the best results were obtained with the conditions given in Table II.

Subsequent to heating, each suspension together with washes of the vessel was centrifuged at 0° at 70,000 *g* (or 93,000 *g*) for 30 (or 20) min and the clear supernatants were lyophilized to dryness after freezing in liquid nitrogen.

TABLE II
DEPROTEINIZATION OF BIOLOGICAL MATERIALS WITH AQUEOUS TRIFLUOROACETIC
ACID (TFA)[a]

Sample	Weight (mg)	Dilution with water (ml)	Addition of 8 M TFA		Incubation time at 80° (min)
			Without acid oxidation (μl)	With acid oxidation (μl)	
Plasma					
Native	10,000	—	5540	4950	5
Lyophilized	100	1.2	690	545	5
Erythrocytes					
Native	10,000	40	13600	13600	15
Lyophilized	100	1.2	400	400	15
Thrombocytes, white cells					
Native	t.a.[b]	1.0	90	95	10
Lyophilized	100	1.5	130	140	10

[a] Followed by incubation at elevated temperature (80°).
[b] t.a., Total amount from 10 ml of whole blood.

Similarly, directly after the releasing procedure the hot suspensions from the biological fractions may be acidified to a final concentration of ~0.7 M by the dropwise addition of 8 M TFA, centrifuged, frozen, and lyophilized as described before.

The lyophilizates may be stored below −20° or used immediately for HPLC analysis. A further advantage of the described deproteinization procedure with trifluoroacetic acid is the nearly complete solubility of the lyophilizates in water or aqueous buffer systems.

High-Performance Liquid Chromatography of Extracts of Biological Fractions

In contrast to standard solutions, optimum HPLC separation and quantitation (at least with rp—HPLC systems and aqueous buffers as mobile phases) of pteridines in extracts of biological fractions are very dependent on the pH of the biological sample. Advantageous hydrogen ion concentrations were found at pH 4 for cell extracts (pH >5.2 was unsuited for extracts from red cells because of turbidity or precipitate formation) and at pH 5 for extracts from plasma. It was observed that under less acidic conditions than those recommended (see Table III), the amount of a few fluorescing substances eluted was increased (especially

TABLE III
OPTIMUM ADJUSTMENT OF BIOLOGICAL SAMPLES PRIOR TO HPLC ANALYSIS

Sample	Releasing procedure: extraction volume (μl) of HPLC buffer	pH adjustment to	Deproteinization procedure with TFA: dissolving in aqueous potassium triphosphate	
			Molarity	Volume
Plasma	~2000[a]; 1000[b]	5.0	0.25	1000
Erythrocytes	~4000[a]; 1000[b]	3.8–4.2	0.12	1000
White cells (buffy coat); thrombocytes	500	3.8–4.2	0.12	100–500

[a] Refers to a releasing procedure without subsequent acidification with TFA.
[b] Refers to a releasing procedure with subsequent acidification with TFA.

from erythrocytes), resulting in a poorer separation of the pteridines. Moreover, in extracts from both red and white blood cells, strong fluorescence peaks appeared at pH >4.5, which influenced the detection and the quantitation of pteridines such as xanthopterin and neopterin. On the other hand, in more acidic extracts (pH ≈2–3) of biological samples, the fluorescence of the investigated pteridines was partly decreased, resulting in an incorrect peak area.

The lyophilizates from the releasing procedure with or without treatment with trifluoroacetic acid were extracted with a minimum amount of water or HPLC buffer (mobile phase) according to Table III. The resulting suspension of each extract was adjusted to the appropriate pH (using a microelectrode; Pierce Reacti-Block; Reacti-Vials with Reacti-Vial Magnetic Stirrer) by adding a few microliters of an appropriate solution (e.g., concentrated phosphoric acid; 2 M potassium triphosphate). This adjustment was especially important for extracts of TFA-treated samples because their acid milieu (plasma, pH 2.2; erythrocytes, pH 2.3) interfered with the separation of the pteridines in the HPLC system. The supernatant of each sample (2-ml Eppendorf tubes) was submitted to HPLC analysis or stored below −20° *separately* from the precipitate. The given extraction and dissolution volumes shown in Table III refer to the lyophilized residues from plasma (native, 10 ml; lyophilized, 900 mg), erythrocytes (native, 4 ml; lyophilized, 1360 mg), and the white cell fraction (buffy coat; from 20 ml of whole blood).

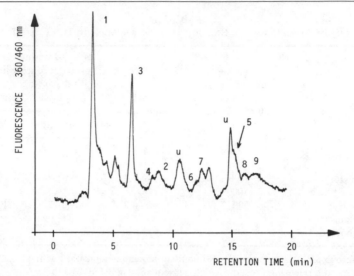

FIG. 7. rp-HPLC elution profile of pteridines extracted from plasma of healthy donors (fresh whole blood) following acid iodine oxidation with subsequent deproteinization (trifluoroacetic acid). Chromatographic conditions: column, 250×4.6 mm; stationary phase, Spherisorb 5-μm C_{18}-ODS; mobile phase, 3.5 mM K_2HPO_4, pH 6.5 (adjusted with H_3PO_4); flow rate, 1 ml/min; column temperature, ambient; fluorescence detection at 360/460 nm. 1, Among others, pterin-6-carboxylic acid; u, unknown; for numbering of other peaks, see Fig. 1.

The lyophilizates which were obtained by the deproteinization procedure with trifluoroacetic acid were highly soluble in aqueous buffers with a strong acid pH, as described before. To achieve suitable hydrogen ion concentrations, the lyophilized samples were dissolved in the appropriate solution of K_3PO_4 according to Table III, and pH alterations were corrected as above. To prevent both overloading of the HPLC column and deterioration of the separation, samples having too high concentrations of fluorescing substances were diluted with water. To separate pteridines having major differences in their concentrations, a change to other attenuations of the integrator unit is recommended. Representative HPLC profiles after acid iodine oxidation and deproteinization treatment are presented in Figs. 7–9.

Compared to standard solutions the retention times of pteridines in biological fractions could be shifted somewhat. The identification of the peaks in the chromatograms were performed by cochromatography with known pteridines at different hydrogen ion concentrations in the mobile phase (e.g., pH 7.4, 7.6, and 6.5), which give rise to characteristic shifts of the retention times of pteridines. For this purpose 3 volumes of the

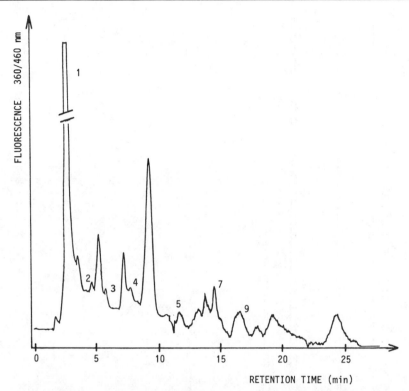

Fig. 8. rp-HPLC elution profile of pteridines extracted from erythrocytes of healthy donors after acid iodine oxidation with subsequent deproteinization (trifluoroacetic acid). For chromatographic conditions, see Fig. 1. Integration, Spectra-Physics SP 4270; the attenuation was changed at retention time of 10 min from 16 to 4. 1, Among others, pterin-6-carboxylic acid; u, unknown; for numbering of other peaks, see Fig. 1.

analysis sample were mixed with 1 volume of a standard solution of nonoverlapping pteridines (pterin-6-carboxylic acid, neopterin, biopterin; or monapterin, biopterin, 6-hydroxymethylpterin; or xanthopterin, isoxanthopterin, lumazine, pterin) having a concentration approximately 5–10 times that present in the sample to be analyzed.

After about 20 analyses (or at increasing pressure), it was found best to wash the HPLC column with slightly acidified water (adjusted with orthophosphoric acid to pH 3–5) for about 30 min and follow with a methanol wash for 30 min, and then water. Prior to the next series of analyses, the HPLC column should be equilibrated with the mobile phase.

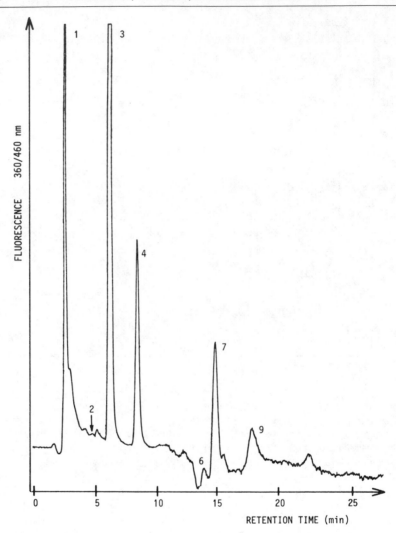

FIG. 9. rp-HPLC elution profile of pteridines extracted from the white cell fraction (buffy coat) of healthy donors after acid iodine oxidation with subsequent deproteinization (trifluoroacetic acid). For chromatographic conditions, see Fig. 1. At retention time of 10 min the attenuation was changed from 16 to 4. 1, Among others, pterin-6-carboxylic acid; u, unknown; for numbering of other peaks, see Fig. 1.

Rapid Screening (Microscreening) for Pteridines in Plasma/Serum Samples

This simple and unexpensive procedure is suited to screening daily pteridine levels (e.g., neopterin, biopterin) in body fluids and is of special interest for an early diagnosis of reactions with a preceding rise of neopterin (e.g., graft-versus-host disease; allograft rejection in transplant operations) or of another pteridine.

1. Procedure without Iodine Oxidation. (Test time, approximately 30–40 min.) Two hundred and fifty microliters of fresh plasma (serum) (1.5 ml test tube) were mixed thoroughly on a Vortex with 25 μl of 8 M trifluoroacetic acid or of 50% aqueous trichloroacetic acid and centrifuged at 8800 g for 15 min. To 100 μl of the acid supernatant was added 3–5 μl (serum, 0.5–2 μl) of 2 M aqueous potassium triphosphate, and after mixing the clear solution (pH ~3) was directly applied for rp-HPLC analysis.

2. Procedure with Iodine Oxidation. (Test time, approximately 90–100 min.) Two hundred and fifty microliters of fresh plasma (serum) (siliconized glass tube) was vortexed for 30 sec with a mixture of 20 μl of acid iodine solution (5.4% I_2/10.8% KI [see p. 275] in 1 M trifluoroacetic acid) and of 25 μl of 8 M trifluoroacetic acid (or of 50% aqueous trichloroacetic acid). After oxidation for 60 min (protected from light; room temperature) excess iodine was reduced with a few crystals of ascorbic acid (vortexing) and the mixture centrifuged as described before. To 100 μl of the supernatant was added 4–6 μl (serum, 2–3 μl) of 2 M aqueous potassium triphosphate and after mixing the clear solution (pH, 2–3) was analyzed directly for pteridines (e.g., neopterin).

[45] Separation of Methotrexate Analogs Containing Terminal Lysine or Ornithine and Their Dansyl Derivatives by High-Performance Liquid Chromatography[1]

By JAMES H. FREISHEIM and A. ASHOK KUMAR

High-performance liquid chromatography (HPLC) has been increasingly used for the rapid detection and preparative separation of several pteridines,[2] folic acid derivatives, and pterins[3] and pteroyloligoglutamates.[4] Methotrexate (MTX), a very potent antineoplastic drug, and its

[1] Supported by Grant CA11666 from the NCI, USPHS, DHHS and by Grant PCM 821-5886 from the NSF.

[2] T. Fukushima and J. C. Nixon, this series, Vol. 66, Part E, p. 429.
[3] M. C. Archer and L. S. Reed, this series, Vol. 66, Part E, p. 452.
[4] A. R. Cashmore, R. N. Dreyer, C. Horváth, J. O. Knipe, J. K. Coward, and J. R. Bertino, this series, Vol. 66, Part E, p. 459.

metabolites 7-hydroxy-MTX and 4-amino-4-deoxy-N^{10}-methylpteroic acid (APA) have been subjected to HPLC analysis by a number of investigators.[5,6] A variety of packing materials and solvent conditions have been employed to separate these folate antagonist compounds. Sample identification has been achieved by monitoring the absorption at 303, 305, and 315 nm, as well as by detecting the fluorescence of MTX derivatives obtained by permanganate oxidation[7] or dithionite reduction.[8] The emphasis of most of these studies was on improving the analytical sensitivity, ease of sample preparation, and rapid quantitation of the folate antagonist MTX and its metabolites in biological fluids such as plasma, serum, saliva, and urine. High-performance liquid chromatography (HPLC) has also been employed in the analysis of amides, esters, and poly(γ-glutamyl) derivatives of MTX[9-11] and to demonstrate the cellular uptake of radioactive impurities found in commercial [^{3}H]MTX.[12] A procedure utilizing a reverse-phase, semipreparative HPLC column and a binary solvent system consisting of trifluoroacetic acid and 1-propanol has been developed for the semipreparative-scale purification and analytical identification of four analogs of methotrexate. The methotrexate analogs containing a lysine or ornithine residue in place of a terminal glutamate residue together with their respective dansyl derivatives were purified in milligram quantities by the procedures described herein.

Materials. Methotrexate was obtained from Lederle Laboratories (Pearl River, NY). 4-Amino-4-deoxy-N^{10}-methylpteroic acid (APA) was the product of carboxypeptidase G_1 cleavage of MTX.[13] CM-Sephadex C-25 was purchased from Pharmacia Fine Chemicals (Piscataway, NJ). Trifluoroacetic acid (TFA) was the sequanal grade from Pierce Chemical Co. (Rockford, IL). 5-Dimethylaminonaphthalene-1-sulfonyl chloride (DNS-Cl) was purchased from Sigma Chemical Co. (St. Louis, MO). Dimethyl sulfoxide (DMSO), acetonitrile (CH_3CN), methanol, and 1-propanol were obtained from Burdick and Jackson Laboratories (Muskegon, MI). All

[5] M.-L. Chen and W. L. Chiou, *J. Chromatogr.* **226**, 125 (1981).
[6] H. Breithaupt, E. Kuenzlen, and G. Goebel, *Anal. Biochem.* **121**, 103 (1982).
[7] J. A. Nelson, B. A. Harris, W. J. Decker, and D. Farquhar, *Cancer Res.* **37**, 3970 (1977).
[8] W. M. Deen, P. F. Levy, J. Wei, and R. D. Partridge, *Anal. Biochem.* **114**, 355 (1981).
[9] C. Temple, Jr., R. D. Elliott, J. R. Piper, J. D. Rose, A. T. Shortmacy, and J. A. Montgomery, *in* "Biological/Biomedical Applications of Liquid Chromatography" (G. Hawk, ed.), Vol. 10, p. 345. Dekker, New York, 1979.
[10] A. Rosowsky, G. P. Beardsley, W. D. Ensinger, H. Lazarus, and C.-S. Yu, *J. Med. Chem.* **21**, 380 (1978).
[11] J. Jolivet and R. L. Schilsky, *Biochem. Pharmacol.* **30**, 1387 (1981).
[12] B. A. Kamen, A. R. Cashmore, R. N. Dreyer, B. A. Moroson, P. Hsieh, and J. R. Bertino, *J. Biol. Chem.* **255**, 3254 (1980).
[13] J. L. McCullough, B. A. Chabner, and J. R. Bertino, *J. Biol. Chem.* **246**, 7207 (1971).

analytical and preparative TLC studies were done on silica gel G plates purchased from Analtech, Inc. (Newark, DE).

Equipment and Synthetic Procedures

Preparative and analytical-scale HPLC were performed using a Waters Associates (Milford, MA) system consisting of a Model 720 system controller, an M45 solvent delivery system, and a Model 450 variable-wavelength detector. The samples were loaded onto the column in a minimal volume of DMSO (200–500 μl) using a Waters U6K universal sample injector with a 2.0-ml sample loop. The elution profiles were recorded with a Houston Instruments Omniscribe recorder. The solvents used in HPLC were degassed for 15 min before the gradient system was started. All the compounds were monitored at 302 nm (AUFS 2.0) during the purification. The elution from the column was achieved with a solvent flow rate of 1.0 ml/min (initial pressure 660 psi). The ornithine analog N^α-(4-amino-4-deoxy-N^{10}-methylpteroyl)-L-ornithine (APA-Orn) and the lysine analog N^α-(4-amino-4-deoxy-N^{10}-methylpteroyl)-L-lysine (APA-Lys) were synthesized employing a modification of a procedure published previously from our laboratories.[14] Diethylphosphorocyanidate was used to couple appropriately blocked ornithine and lysine amino acids to APA.[15] The synthesis of N^α-(4-amino-4-deoxy-N^{10}-methylpteroyl)-N^δ-(5-[N,N-dimethylamino]-1-naphthalenesulfonyl)-L-ornithine (APA-Orn-DNS) was carried out using a modification of the procedure followed in the synthesis of N^α-(4-amino-4-deoxy-N^{10}-methylpteroyl-N^ϵ-(5-[N,N-dimethylamino]-11-naphthalenesulfonyl)-L-lysine (APA-Lys-DNS).[16] The ornithine analog APA-Orn (34.5 mg) in 40 mM lithium carbonate buffer (pH 9.5), was allowed to react with 5-dimethylaminonaphthalene-1-sulfonyl chloride (21.6 mg) in acetonitrile for 35 min. The advantages of using lithium carbonate buffer and acetonitrile in the dansylation of amino acids have been documented.[17] The excess unreacted dansyl chloride was removed by triturating with benzene. The concentrated residue from the reaction was redissolved in a minimal volume of $CH_3OH : CH_3CN$ mixture (1 : 1, v/v), and chromatographed on preparative TLC using $CHCl_3 : CH_3OH$ (3 : 2, v/v) as the solvent. The major yellow fluorescent band (R_f 0.38) was eluted with methanol (100%) and stored in dry form at $-20°$.

[14] R. J. Kempton, A. M. Black, G. M. Anstead, A. A. Kumar, D. T. Blankenship, and J. H. Freisheim, *J. Med. Chem.* **25,** 475 (1982).

[15] T. Shiori, Y. Yokoyama, Y. Kasai, and S. Yamada, *Tetrahedron* **32,** 2211 (1976).

[16] A. A. Kumar, J. H. Freisheim, R. J. Kempton, G. M. Anstead, A. M. Black, and L. Judge, *J. Med. Chem.* **26,** 111 (1983).

[17] Y. Tapuhi, D. E. Schmidt, W. Lindner, and B. L. Karger, *Anal. Biochem.* **115,** 123 (1981).

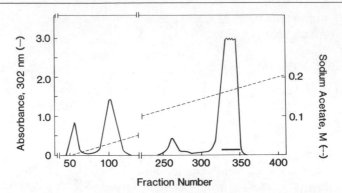

FIG. 1. Chromatography of APA-Lys on CM-Sephadex C-25. A sample of impure APA-Lys (24.0 mg) in 0.05 M sodium acetate (2.5 ml), pH 4.5, was loaded onto a CM-Sephadex column equilibrated with 0.05 M sodium acetate, pH 4.5. Following a 100-ml wash with the same buffer, the compound was eluted with a gradient of sodium acetate (dashed line) (volume, 800 ml; 0.05–0.2 M; and pH 4.5–7.8). Fractions (2.0 ml) were collected at a flow rate of 60 ml/hr and monitored at 302 nm. The material in the fractions identified by the horizontal bar was pooled.

HPLC PURIFICATION OF MTX, APA, AND
MTX ANALOGS[a]

| | Retention times (min) | |
Compound	Preparative scale	Analytical scale
MTX	38	38
APA	43	46
APA-Orn	37[b]	—
	40[c]	—
	—	29
APA-Orn-DNS	58	60
APA-Lys	29	33
APA-Lys-DNS	66	66

[a] Unless otherwise indicated, the gradient conditions were 0–30% B (linear) in 100 min. Solvents: A, 0.1% TFA in water; B, 0.1% FTA in 1-propanol; and C, 0.1% TFA in methanol.

[b] The column was washed with 100% A for 10 min prior to starting the gradient.

[c] The isocratic conditions, 90% A, 2% B, and 8% C, were achieved in 10 min following 100% A wash for 10 min. The amount of material loaded onto the semipreparative HPLC column was in the range of 2–4 mg per trial.

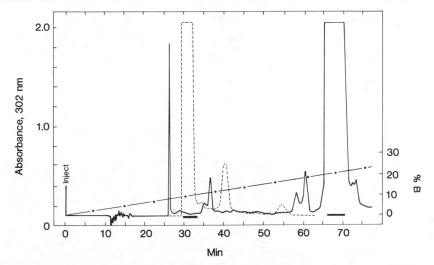

FIG. 2. HPLC of APA-Lys and APA-Lys-DNS. The compounds were purified on a Waters μBondapak C_{18} column (7.8 mm i.d. × 300 mm; 10 μm; 80–100 Å). Samples of APA-Lys (2.0 mg) and APA-Lys-DNS (3.0 mg) in DMSO (400 μl) were injected into the column. The solvent consisted of a linear gradient of 0–30% B in 100 min (●). The profiles shown are a composite of two individual HPLC runs representing the purification of APA-Lys (---) and APA-Lys-DNS (——). In both cases, 1.0-ml fractions were collected at a flow rate of 1.0 ml/min and monitored at 302 nm (AUFS 2.0). Pooled fractions are represented by solid horizontal bars.

Separation of Lysine and Ornithine Analogs of Methotrexate

The lysine analog (APA-Lys) was subjected to initial purification on a CM-Sephadex C-25 column (1.5 × 20 cm) equilibrated with 0.05 M sodium acetate, pH 4.5 (Fig. 1). The compound was eluted from the column using a linear gradient of 0.05 M sodium acetate (pH 4.5) to 0.2 M sodium acetate (pH 7.8) in a total volume of 800 ml. The fractions were monitored at 302 nm and the major component eluting in fractions 326–353 was pooled and subjected to rotary evaporation. This orange-colored material was stored in the dark at −20°.

In order to purify the MTX analogs in milligram quantities, an HPLC method was developed using a Waters μBondapak C_{18} column (7.8 × 300 mm) and 1-propanol gradients (0–30% B in 100 min, see table). Figure 2 represents a typical example of one such purification. The MTX analogs APA-Lys and APA-Lys-DNS elute from the column with retention times (Rt) of 29 and 66 min, respectively. The purity of the analogs was confirmed by rechromatography of the major components on an analytical

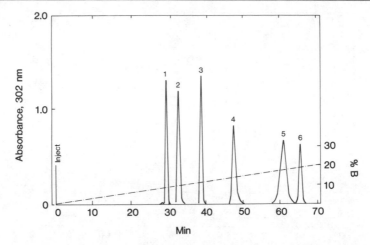

Fɪɢ. 3. HPLC of MTX, APA, and MTX analogs. Column: μBondapak C_{18} (7.8 × 300 mm); elution with 0–30% B in 100 min (dashed lines); flow rate, 1.0 ml/min; detection, absorbance at 302 nm (AUFS 2.0). Peaks: 1, APA-Orn (23 μg); 2, APA-Lys (27 μg); 3, MTX (18 μg); 4, APA (48 μg); 5, APA-Orn-DNS (42 μg); and 6, APA-Lys-DNS (20 μg). The peaks shown are a composite of six individual HPLC runs under identical conditions.

scale using the same column and the same gradient system (Fig. 3) as well as by TLC.

The purification of APA-Orn, its dansyl derivative (APA-Orn-DNS), MTX, and APA was also carried out under conditions identical to those described for the analogs APA-Lys and APA-Lys-DNS (cf. Fig. 2). HPLC of the ornithine analog APA-Orn without a preliminary purification on the CM-Sephadex C-25 column using the above gradient system indicated the presence of contaminants closely associated with the major component. As indicated in the table, an isocratic solvent system was developed as a second step in the HPLC purification of the ornithine analog APA-Orn. The table also gives the retention times of all the analogs examined including MTX and APA. Figure 3 depicts the analytical elution profile of purified MTX, APA, as well as lysine and ornithine analogs when rechromatographed on the same column under identical conditions. The structures of the four analogs of MTX are indicated in Fig. 4.

The results presented herein demonstrate the utility of HPLC on a C_{18} reverse-phase column in the purification of MTX, APA, and analogs derived therefrom. APA-Lys purification on HPLC was preceded by cation-exchange chromatography on a CM-Sephadex C-25 column and indicated the presence of three impurities in addition to the main component APA-

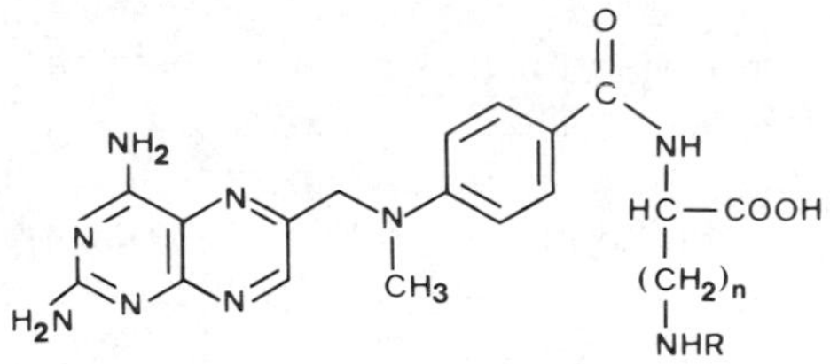

COMPOUND	n =	R =
APA–Orn	3	H
APA–Orn–DNS	3	DNS
APA–Lys	4	H
APA–Lys–DNS	4	DNS

FIG. 4. Analogs of methotrexate.

Lys (cf. Fig. 1). Further purification of APA-Lys by means of HPLC (cf. Fig. 2) resulted in a singly migrating component as indicated by analytical HPLC (Fig. 3) and by TLC. APA-Orn, following synthesis, was loaded directly onto the HPLC column and the results indicated the presence of a total of six components (results not shown). The major component (Rt = 37 min; table) was rechromatographed on the same column but with different elution conditions. The APA-Orn obtained during this chromatography (Rt – 40 min; table) was demonstrated to be pure by analytical scale HPLC (Fig. 3). Purification of the fluorescent analogs APA-Lys-DNS (Fig. 2; Rt = 66 min) and APA-Orn-DNS (Rt = 58 min; table) was achieved in a single HPLC run. The purity of these samples was also established by rechromatography on the same column (Fig. 3) and from mass spectra.[16] The results shown in Fig. 3 indicate that a baseline separation of all the folate analog compounds can be achieved within 70 min using the solvent conditions described.

In terms of order of elution, the retention times were APA-Orn < APA-Lys < MTX < APA < APA-Orn-DNS < APA-Lys-DNS (table and Fig. 3). The difference between APA-Orn and APA-Lys retention times most probably originates from the difference of one methylene group between the two molecules. A similar explanation may account for differences in the retention time of APA-Orn-DNS and APA-Lys-DNS (table). The large variation in retention times of APA-Orn between the preparative-scale and analytical-scale HPLC might have been influenced by the contaminating components (~50%) in the APA-Orn sample. The prolonged retention of APA on the C_{18} column compared to MTX, APA-Orn, and APA-Lys is due to the lack of a terminal amino acid. At pH 2.0, APA-Orn and APA-Lys would carry an extra positive charge as compared to MTX and hence would be more hydrophilic, which may explain, in part, their lower retention times. HPLC purification of these MTX analogs in milligram quantities has greatly facilitated the study of these compounds as inhibitors of dihydrofolate reductase and, in the case of dansyl analogs, as fluorescent probes.

[46] Determination of Pterins by Liquid Chromatography/Electrochemistry

By Craig E. Lunte and Peter T. Kissinger

Pterins have attracted much attention in recent years. Tetrahydrobiopterin has been shown to be a required cofactor and probable mediating agent for several mixed-function oxidases,[1] and several pterin species have been identified in a variety of biological fluids and tissues. These pterins are of interest because of their possible involvement in a number of diseases. One obstacle to a better understanding of the biological role of pterins has been the lack of a general analytical method.

A general method should be able to individually determine the various oxidation states of several pterin species in a wide variety of sample types. In this chapter, an analytical method based on liquid chromatography/electrochemistry (LCEC)[2] which does meet these requirements is presented. Previously, several approaches have been taken to the determination of pterins; these include a bioassay using Crithidia fasciculata,[3] a radioenzymatic assay,[4] an immunoassay,[5] and liquid chromatography with fluorescence detection.[6-8] However, none of these methods meet all

[1] S. Kaufman and D. B. Fisher, in "Molecular Mechanisms of Oxygen Activation" (O. Hayaishi, ed.), p. 285. Academic Press, New York, 1974.

[2] R. E. Shoup, ed., "Recent Reports on Liquid Chromatography/Electrochemistry." BAS Press, West Lafayette, Indiana, 1982.

[3] H. Baker, O. Frank, A. Shapiro, and S. Hutner, this series, Vol. 66, p. 490.

[4] G. Guroff, C. Rhoads, and A. Abramowitz, Anal. Biochem. 21, 273 (1967).

[5] T. Nagatsu, T. Yamaguchi, T. Kato, T. Sugimoto, S. Matsuura, M. Akino, S. Tsushima, N. Nakazawa, and H. Ogawa, Anal. Biochem. 110, 182 (1981).

[6] T. Fukushima and J. Nixon, Anal. Biochem. 102, 332 (1980).

[7] B. Stea, R. Halpern, B. Halpern, and R. Smith, J. Chromatogr. 188, 363 (1980).

[8] J. H. Woolf, C. A. Nichol, and D. S. Dutch, J. Chromatogr. 274, 398 (1983).

of the requirements of a general analytical method for the determination of pterins in biological samples.

Materials. Pterins were obtained from the following sources: pterin, pterin-6-carboxylic acid, 6-methylpterin, and xanthopterin (Sigma, St. Louis, MO), biopterin (Calbiochem-Behring, La Jolla, CA), and neopterin (Fluka, Basle, Switzerland). Pterin-6-aldehyde and 6-hydroxymethylpterin were prepared by the method of Thijssen.[9] The reduced pterins were prepared from their corresponding oxidized forms as previously described.[10] Octylsodium sulfate was purchased from Kodak (Rochester, NY). All other chemicals were reagent grade or better and used without purification.

Chromatography

Apparatus. The chromatography system consisted of an Altex 110 pump and a Rheodyne 7125 injection valve with a 20-μl sample loop. Biophase ODS 5-μm columns of either 25 or 10 cm lengths were used depending upon the application (see specific section). The column was thermostatted with an LC-23 column heater and an LC-22 temperature controller (Bioanalytical Systems, West Lafayette, IN). Detection was with dual BAS LC-4B amperometric detectors. A schematic diagram of the LCEC system is shown in Fig. 1.

Mobile Phase. A reverse-phase "ion-pair" retention mechanism was used to separate the pterins. The mobile phase consisted of 3 mM octylsodium sulfate in a 0.1 M sodium phosphate buffer (pH 2.5). Depending upon the sample type, various amounts of methanol were added to the mobile phase as an organic modifier (see specific section). The mobile phase was prepared from distilled, deionized water and glass-distilled methanol. The mobile phase was filtered through a 0.22-μm filter (Millipore, Milford, MA) prior to use. Oxygen was removed by continuous purging with nitrogen and maintaining the mobile-phase reservoir at a temperature of 40°.

Electrochemistry

A property of the pterins essential to their biological role as enzyme cofactors is their electrochemistry. Fully oxidized pterins can be reversibly reduced to 5,8-dihydropterins which rapidly tautomerize to 7,8-dihydropterins. These 7,8-dihydropterins can either be further reduced to 5,6,7,8-tetrahydropterins or be oxidized back to the fully oxidized state.

[9] H. Thijssen, *Anal. Biochem.* **54,** 609 (1973).
[10] C. E. Lunte and P. T. Kissinger, *Anal. Chem.* **55,** 1458 (1983).

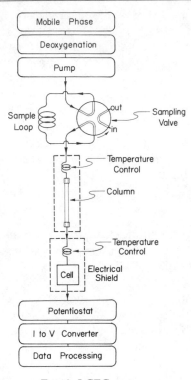

FIG. 1. LCEC system.

Tetrahydropterins are reversibly oxidized to quinonoid dihydropterins, which again tautomerize to 7,8-dihydropterins. This scheme is illustrated in Fig. 2. An exception to this pathway is xanthopterin, in which an enolic hydroxyl group at the 6 position blocks reduction to the tetrahydro state.

From an analytical perspective, the pterin redox reactions which are important are the reduction of the fully oxidized pterins (A), the oxidation of tetrahydropterins (B), and the oxidation of 7,8-dihydropterins (C). The redox potentials of these reactions, as determined by hydrodynamic voltammetry, are listed in the table for several pterins. These potentials are used to determine the detector operating conditions for LCEC.

Electrochemical Detection of Pterins

The three stable oxidation states of the pterins cannot be detected simultaneously using only one working electrode because two of the oxidation states are oxidizable while the other is reducible. This problem can

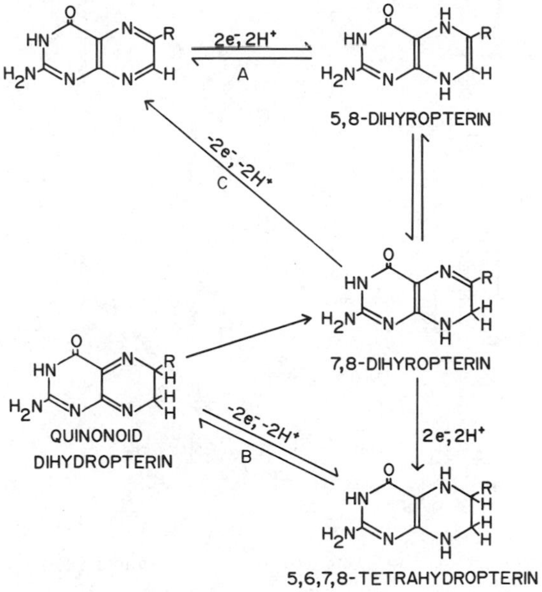

FIG. 2. Reaction scheme for pterin electrochemistry.

be overcome by using a dual-electrode detector which uses two working electrodes placed adjacent to each other and normal to the direction of flow. This is known as the parallel–adjacent configuration (Fig. 3). Using this detector, one electrode is operated at an oxidizing potential to detect

OXIDATION AND REDUCTION POTENTIALS
OF PTERINS

Pterin	Potential[a] (V)		
	A	B	C
Biopterin	−0.58	0.19	0.64
6-Hydroxymethylpterin	−0.50	0.19	0.64
6-Methylpterin	−0.40	0.25	0.75
Neopterin	−0.52	0.21	0.65
Pterin	−0.45	0.20	0.64
Pterin-6-carboxylic acid	−0.40	0.30	0.51
Xanthopterin	−0.58	—	0.51

[a] A, Reduction of oxidized pterins; B, oxidation of tetrahydropterins; C, oxidation of 7,8-dihydropterins.

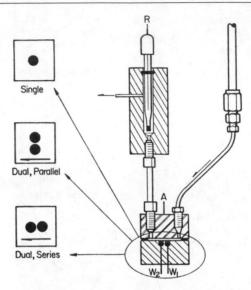

FIG. 3. Electrode configurations with a thin-layer amperometric transducer.

the dihydro- and tetrahydropterins while the other electrode is operated at a reducing potential to detect the oxidized pterins.

The oxidized forms of the pterins can be reduced at a glassy carbon electrode. Although the maximum sensitivity for electrochemical detection is found by operating on the limiting-current plateau of the hydrodynamic voltammogram, when operating at high potentials the background and resulting noise can become limiting. Under these conditions, operating at a slightly lower potential can often result in better detection limits. For this reason, a potential of −0.70 V versus the Ag/AgCl electrode was chosen as a compromise between maximizing the current response and minimizing the background current for the detection of the oxidized pterins. Although the 7,8-dihydropterins can be reduced, the redox potential of these reactions is sufficiently negative to be inaccessible by LCEC.

Because the tetrahydropterins are easily oxidized, background considerations are unimportant and a potential on the limiting-current plateau can be used. In this case, a potential of +0.50 V is sufficient for the detection of tetrahydropterins. A more positive potential is required to oxidize the dihydropterins, but again background considerations are not significant. A potential of +0.80 V versus Ag/AgCl is used to detect the 7,8-dihydropterins. In the overall dual-electrode scheme, one electrode is operated at +0.80 V and the other electrode at −0.70 V versus Ag/AgCl.

Determination of Pterins in Urine

Sample Preparation. A series of cation-exchange clean-up columns were used for urine samples. Urine (5.0 ml) was acidified to pH 2 with phosphoric acid. This sample was applied to a 0.5-ml SP-Sephadex C-25 column. The eluant of this column was applied to a 0.5-ml Dowex 50-X8 column (H^+, 200–400 mesh). Both columns were washed with 10 ml of water. The Dowex 50 column was then developed with 5 ml of 0.1 M NaOH which was 5 mM in ascorbic acid (the ascorbic acid is necessary to inhibit oxidation of the reduced pterins in the high-pH eluant). The eluant from this column was applied directly to the SP-Sephadex column, and the eluant collected and acidified to pH 2 with phosphoric acid. The reproducibility of this procedure is excellent if the columns are prepared in a single batch.

Chromatography. For the determination of pterins in urine, a 25-cm chromatography column thermostatted at 30° was used. The mobile phase contained 5% methanol (v/v). A flow rate of 1.0 ml/min was employed.

Results. Typical chromatograms obtained from a urine sample are shown in Fig. 4. Peaks were identified based on both their retention time and voltammetric characterization. One advantage of electrochemical detection is the large range of concentrations over which response is linear. With this dual-electrode method, the current response was found to be linear over four orders of magnitude. Detection limits of less than a picomole at signal-to-noise ratios greater than 3 were achieved for all of the pterin species studied.

Determination of Pterins in Mouse Tissue

Sample Preparation. These samples are much less complex than urine samples, and therefore less sample preparation is necessary. The animal was sacrificed by decapitation and the organ of interest removed. The tissue sample was then homogenized in 0.1 M phosphoric acid containing 2 mM ascorbic acid. Approximately 1 ml of acid solution was used per gram of tissue. The homogenized sample was centrifuged for 15 min at 15,000 g. The supernatant was saved and the pellet resuspended in a second volume of acid solution. This was recentrifuged and the two supernatants combined. The combined supernatants were filtered through a 0.22-μm filter before injection onto the analytical column.

Chromatography. For tissue samples, a 25-cm chromatography column thermostatted at 30° was employed. The mobile phase contained 15% methanol (v/v). A flow rate of 1.0 ml/minute was used.

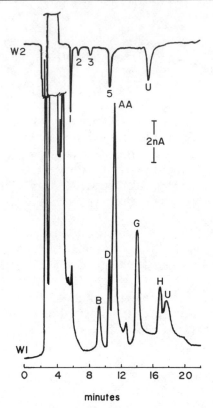

FIG. 4. Dual-electrode detection of reduced and oxidized pterins in urine. Chromatographic conditions: mobile phase, 3 mM octylsodium sulfate in 0.1 M phosphate buffer (pH 2.5), 5% MeOH, 30°, W1 = +800 mV, W2 = −700 mV. Peak identities: 1, *erythro*-neopterin; 2, *threo*-neopterin; 3, xanthopterin; 5, biopterin; B, tetrahydroneopterin; D, dihydroxanthopterin; G, tetrahydrobiopterin; H, dihydrobiopterin; AA, ascorbic acid; U, unidentified.

Results. The only pterin identified in either mouse brain or liver samples was biopterin. This biopterin occurred in all three of its oxidation states, with tetrahydrobiopterin being the most concentrated. Typical chromatograms from a liver sample and a brain sample are shown in Figs. 5 and 6, respectively. Similar to the urine samples, detection limits of less than a picomole were achieved.

Detection of Quinonoid Dihydropterins

The intermediate in the oxidation of tetrahydropterins is a quinonoid dihydropterin. As mentioned earlier, quinonoid dihydropterins rapidly

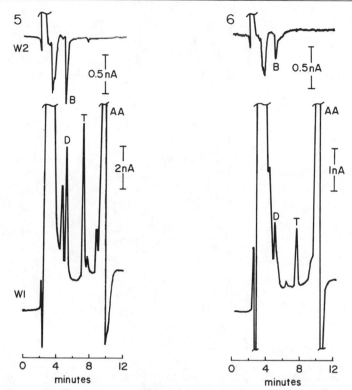

FIG. 5. Chromatogram of biopterin in mouse liver tissue. Chromatographic conditions: Mobile phase, 3 mM octylsodium sulfate in 0.1 M phosphate buffer (pH 2.5), 15% MeOH, 30°, W1 = +600 mV, W2 = −700 mV. Peak identities: B, biopterin; D, 7,8-dihydrobiopterin; T, 5,6,7,8-tetrahydrobiopterin; AA, ascorbic acid.

FIG. 6. Chromatogram of biopterin in mouse brain tissue. Chromatographic conditions and peak identities as in Fig. 5.

tautomerize to the corresponding 7,8-dihydropterin. LCEC is capable of detecting quinonoid dihydropterins in both synthetic and enzymatic samples.

Sample Preparation. Mice were killed by decapitation and their livers removed and homogenized. Phenylalanine 4-monooxygenase (phenylalanine hydroxylase) was isolated by precipitation between 35 and 45% saturation with ammonium sulfate. The ammonium sulfate precipitate was dissolved in 0.01 M sodium phosphate buffer (pH 7.5), and purified on a DEAE-Sephadex column as described by Shiman *et al.*[11]

[11] R. Shiman, D. W. Gray, and A. Pater, *J. Biol. Chem.* **254,** 11300 (1979).

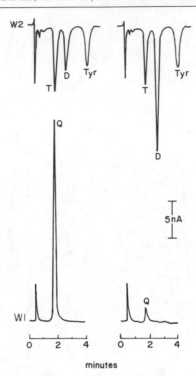

Fig. 7. Chromatograms of quinonoid dihydrobiopterin in a phenylalanine hydroxylase incubation. Chromatographic conditions: mobile phase, 3 mM octylsodium sulfate in 0.1 M phosphate buffer (pH 2.5), 15% MeOH. Peak identities: T, 5,6,7,8-tetrahydrobiopterin; D, 7,8-dihydrobiopterin; Q, quinonoid dihydrobiopterin; Tyr, tyrosine.

The incubation mixture was as described previously[12] except that tetrahydrobiopterin was employed as the pterin cofactor. The incubation was allowed to proceed for 2 min, after which a 200-μl aliquot was removed and diluted to 2.0 ml with 0.1 M sodium phosphate buffer (pH 2.5). This sample was deoxygenated for 2 min and then 20 μl was injected onto the chromatography column. It is vital to keep sample handling time to a minimum to avoid excessive tautomerization.

Chromatography. A 10-cm chromatography column at room temperature was used for this section. The mobile phase contained 15% methanol (v/v). A flow rate of 2.0 ml/min was employed.

Results. Due to similarity in structure, tetrahydrobiopterin and quinonoid dihydrobiopterin are not resolved chromatographically. However,

[12] C. E. Lunte and P. T. Kissinger, *Anal. Biochem.* **139,** 139 (1984).

the two compounds can be resolved electrochemically because quinonoid dihydrobiopterin is easily reduced while tetrahydrobiopterin is easily oxidized. Chromatograms obtained 2 and 15 min after incubation of tetrahydrobiopterin with phenylalanine and phenylalanine hydroxylase are shown in Fig. 7. These chromatograms show an initial oxidation product which is easily reduced (quinonoid dihydrobiopterin) which is converted into an oxidizable product (7,8-dihydrobiopterin). This method should be capable of detecting quinonoid dihydrobiopterin in biological samples, providing the sample preparation time can be kept to a minimum to avoid tautomerization.

[47] Preparation of (6S)-5-Formyltetrahydrofolate Labeled at High Specific Activity with ^{14}C and ^{3}H[1]

By RICHARD G. MORAN, KHANDAN KEYOMARSI, and PAUL D. COLMAN

The 6R,S mixture of diastereoisomers of 5-formyl-5,6,7,8-tetrahydrofolate (5-CHO-H_4PteGlu) has been called folinic acid or leucovorin, and the calcium salt of this mixture is used therapeutically to reverse the toxicity of large doses of methotrexate. Although some studies suggest that the "unnatural" or 6R diastereoisomer is not biochemically inert,[2–4] the majority of the growth-promoting and antimetabolite-reversing effects of leucovorin seem due to the 6S diastereoisomer. In this report, we describe a facile preparation of (6S)-5-CHO-H_4PteGlu labeled either with ^{14}C at position 5 or with ^{3}H at positions 3′,5′,7, and 9 at specific activities equivalent to those of the labeled starting materials (10–60 Ci/mmol and 50–60 mCi/mmol, respectively). The procedure takes advantage of the reactivity of the 5 position of tetrahydrofolate toward attack by carboxylic acids in the presence of carbodiimides.[5] The mild conditions used allow good yields of both labeled compounds.

[1] This research was supported in part by Grants CA27146 and CA36054 from the National Institutes of Health, DHHS.
[2] R. P. Leary, Y. Gaumont, and R. L. Kisliuk, *Biochem. Biophys. Res. Commun.* **56**, 484 (1974).
[3] J. C. White, B. D. Bailey, and I. D. Goldman, *J. Biol. Chem.* **254**, 242 (1978).
[4] J. K. Sato and R. G. Moran, *Proc. Am. Assoc. Cancer Res.* **25**, 312 (1984).
[5] R. G. Moran and P. D. Colman, *Anal. Biochem.* **122**, 70 (1982).

Preparation of (6S)-5-CHO-[3′,5′,7,9-³H]H₄PteGlu

Reagents. [3′,5′,7,9-³H]Folic acid (PteGlu) is available from either Amersham (Arlington Heights, IL) or Moravek Biochemicals (Brea, CA). 1-Ethyl-3-(3-dimethylaminopropyl)carbodiimide-HCl (EDC) and NADPH were purchased from Sigma Chemical (St. Louis, MO), DEAE-cellulose from Eastman Chemical (Rochester, NY), and ammonium acetate (NH₄Ac) from Mallinkrodt Chemical (St. Louis, MO). Dihydrofolate reductase (4.1 IU/mg protein) was purified from *Lactobacillus casei* resistant to methotrexate.

Procedure. [³H]PteGlu (0.1 mCi, approximately 2.5 nmol) was incubated with dihydrofolate reductase (0.04 IU) and NADPH (15 nmol) in 0.1 ml of 50 m*M* phosphate buffer (pH 7.0), containing 50 m*M* 2-mercaptoethanol (2-ME). After 10 min at 37°, 900 μl of 100 m*M* formic acid containing 50 m*M* phosphate and 150 m*M* 2-ME was added. The pH of the solution was 4 at this point and was gently adjusted if necessary. EDC (5 mg) was added as a solid or as a *freshly* prepared concentrated solution and incubation was continued for 10 min at room temperature. The reaction mixture was immediately purified by chromatography on a 0.9 × 50-cm column of DEAE-cellulose equilibrated with NH₄Ac.[6] Under these chromatographic conditions,[5,7] 5-CHO-H₄PteGlu elutes with 0.48 *M* NH₄Ac at a position intermediate between *p*-aminobenzoylglutamate (a breakdown product of tetrahydrofolate, H₄PteGlu) and any unreacted H₄PteGlu, which elute with 0.40 and 0.54 *M* NH₄Ac, respectively.[7] Yields of 50–75% have been obtained by this method with purities >95%, as judged by high-performance liquid chromatography (HPLC). Tubes containing product are usually pooled and lyophilized to dryness. NH₄Ac is removed by this process, but substantial (25–50%) losses of tritiated product often occur during lyophilization. Alternatively, product can be purified by HPLC using reverse-phase columns with paired-ion chromatography.[8]

---

[6] This column had been prepared by swelling 5 g of DEAE-cellulose in 300 ml of 0.1 *N* NaOH over 15 min, then washing the DEAE-cellulose with water to neutrality on a Büchner funnel, followed by washing with 1 liter of 2.5 *M* NH₄Ac. The washed DEAE-cellulose was packed into a 0.9 × 50-cm column under 10 psi of N₂ and equilibrated with 50 m*M* NH₄Ac at a flow of 0.8 ml/min. The column was eluted with a gradient of NH₄Ac produced by pumping 1.5 *M* NH₄Ac into 475 ml of stirring 0.01 *M* NH₄Ac, using a peristaltic pump and by pumping the contents of this mixer directly onto the column. All solutions contained 150 m*M* (1%) 2-ME. The rate of addition of concentrated NH₄Ac to the mixer and rate of removal of NH₄Ac solution from the mixer onto the column were held equal by using the same-diameter tubing and the same peristaltic pump, thus generating a convex gradient.

[7] R. G. Moran, S. F. Zakrzewski, and W. C. Werkheiser, *J. Biol. Chem.* **251,** 3569 (1975).

[8] R. G. Moran and P. D. Colman, *Biochemistry* **23,** 4580 (1984).

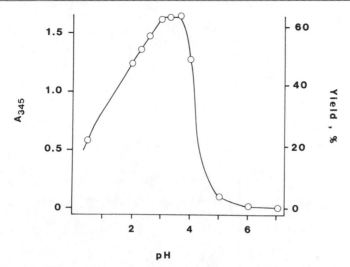

Fig. 1. pH dependency of carbodiimide-promoted formylation of $H_4PteGlu$. $(6S)$-H_4-PteGlu (0.1 μmol) was added to 100 mM formic acid and 0.5 mg of EDC in a total volume of 100 μl. After 10 min, 900 μl of 1 N HCl containing 150 mM 2-ME was added to convert any 5-CHO-H_4PteGlu to 5,10-methenyltetrahydrofolate. After 30 min at room temperature the latter compound was quantitated by its unique absorption maximum at 345 nm.

The formylation of H_4PteGlu promoted by carbodiimides is very sensitive to pH (Fig. 1). Inefficient synthesis of product by this procedure is most often traced to incorrect pH. This preparation can be scaled up by proportionately increasing [^{3}H]PteGlu, NADPH, and dihydrofolate reductase while holding reaction times and volumes constant.

Preparation of (6S)-5-CHO-[5-^{14}C]H_4PteGlu

Reagents. Sodium [^{14}C]formate is available from either Amersham or New England Nuclear (Boston, MA) and is supplied as a solid. It is dissolved in 50 mM phosphate (pH 3.6), at 10 mCi/ml. EDC is best made up as a *fresh* solution at 50 mg/ml. (6S)-H_4PteGlu is prepared enzymatically from dihydrofolate using *L. casei* dihydrofolate reductase and is purified by column chromatography as previously described[7] or by any similar method.[9] Chromatographically purified (6S)-H_4PteGlu is lyophilized in 2-ml ampules in aliquots of 4 μmol while the ampules are attached to the exterior of a lyophilizer. Dry N_2 is introduced into the lyophilizer with care to avoid introducing air, and the ampules are sealed with a torch.

[9] F. M. Huennekens, P. P. K. Ho, and K. G. Scrimgeour, this series, Vol. 6 [114].

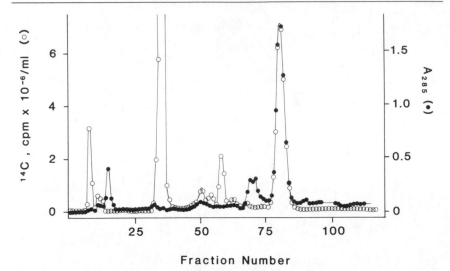

FIG. 2. Purification of (6*S*)-5-CHO[5-^{14}C]H$_4$PteGlu by DEAE-cellulose chromatography. [^{14}C]Formate, (6*S*)-H$_4$PteGlu, and EDC were incubated as described in the text and the reaction mixture was applied to a 0.9 × 50-cm column of DEAE-cellulose. The column was eluted with a convex gradient of NH$_4$Ac (0.01–1.5 *M*). Fractions (3 ml) were collected and monitored for folates (A_{285}, ●) and ^{14}C (○). Markers were not added to the sample.

When the ampules are wrapped in foil and stored at $-25°$, this material is stable for >2 years. Immediately prior to use for this procedure, an ampule of (6*S*)-H$_4$PteGlu is quickly rehydrated with 100 µl of 150 m*M* 2-ME.

Procedure. (6*S*)-H$_4$PteGlu (1.6 µmol, 40 µl) is added to sodium [^{14}C]formate (5.4 µmol, 300 µCi, 30 µl), and phosphate buffer (75 µl of 50 m*M*, pH 7), and acidified to the correct pH with 30 µl of 0.5 *N* HCl. EDC is added (1 mg, 20 µl). After 15 min at room temperature, the reaction mixture is applied to a 0.9 × 50-cm column of DEAE-cellulose equilibrated with NH$_4$Ac. A convex gradient of 0.01–1.5 *M* NH$_4$Ac (mixer volume, 475 ml) is used to elute the column at a flow rate of 0.8 ml/min (see above).

(6*S*)-5-CHO-[5-^{14}C]H$_4$PteGlu is virtually the only ultraviolet-absorbing compound eluted from this column (Fig. 2). This suggests a yield (relative to the H$_4$PteGlu starting material) comparable to those found in 5-CHO-[^{3}H]H$_4$PteGlu syntheses (50–75%). Excess [^{14}C]formate is used in this procedure to avoid inter- or intramolecular reactions of the 5-nitrogen with the α- or γ-carboxyl groups of the glutamate side chain. Hence, relative to the ^{14}C-labeled starting material, the yields of purified product are on the range of 7–12%. The low cost of [^{14}C]formate makes this approach economically acceptable.

[48] Tissue Folate Polyglutamate Chain Length Determination by Electrophoresis as Thymidylate Synthase–Fluorodeoxyuridylate Ternary Complexes[1]

By D. G. PRIEST and M. T. DOIG

The selective determination of the polyglutamate status of individual folate cofactors has been difficult because of the complex mixture of closely related structures present in most tissues. In addition, these cofactors are present in low concentrations and are chemically labile. To obviate these analytical problems we have developed an enzyme radioassay specific for methylenetetrahydrofolate polyglutamates (CH_2H_4-$PteGlu_n$).[2,3] Other folate cofactors, which can be chemically or enzymatically converted to the methylene form, can also be examined using this technique.

The assay is based upon the stoichiometric entrapment of methylenetetrahydrofolate into a covalent, ternary complex (**I**) with thymidylate

I

synthase and tritiated 5-fluorodeoxyuridylate ([^{3}H]FdUMP). The ternary complex is exceptionally stable[4] and will not dissociate during polyacrylamide gel electrophoresis.[5] The electrophoretic mobility of the complex

[1] Supported by PHS Grant Number CA22754 awarded by the National Cancer Institute, DHHS.

[2] D. G. Priest, K. K. Happel, M. Mangum, J. M. Bednarek, M. T. Doig, and C. M. Baugh, *Anal. Biochem.* **115**, 163 (1981).
[3] D. G. Priest, K. K. Happel, and M. T. Doig, *Biochem. Biophys. Methods* **3**, 201 (1980).
[4] D. V. Santi, C. S. McHenry, and H. H. Sommer, *Biochemistry* **13**, 471 (1974).
[5] J. L. Aull, J. A. Lyon, and R. B. Dunlap, *Microchem. J.* **19**, 210 (1974).

is a linear function of increasing charge associated with the poly-γ-gluta-mate portion of the folate cofactor. The method has the following advantages: the specificity of an enzyme assay, the sensitivity of a radioassay, and stabilization of the labile, reduced folate and its polyglutamates once it is incorporated into the ternary complex.

Reagents and Enzymes. [³H]FdUMP (18–20 Ci/mmol) was purchased from Moravek Biochemicals, Brea, CA. Folypolyglutamate standards with two to seven residues were obtained from C. M. Baugh, University of South Alabama, Mobile, AL, and reduced to the tetrahydro state with *Lactobacillus casei* dihydrofolate reductase and NADPH.[6] Typically 0.3 units of dihydrofolate reductase is added to 1.0 ml of a 100 μM solution of each standard containing 400 μM NADPH, 50 mM Tris-HCl, 50 mM sodium ascorbate, and 1 mM formaldehyde. Reactions are allowed to proceed for 2 hr at 25°, after which the solution is flushed with nitrogen and stored at −70°. Thymidylate synthase was either purchased from the New England Enzyme Center, Medford, MA, or purified from methotrexate-resistant *L. casei* cells by the method of Galivan *et al.*[7] In this method, *L. casei* cells are lysed by sonication in 7 volumes of a 50 mM Tris–HCl, 50 mM KCl, 1 mM EDTA, and 20 mM 2-mercaptoethanol buffer (pH 7.4). After centrifugation at 40,000 g for 20 min, the supernatant is brought to 75% saturation with ammonium sulfate and the precipitated thymidylate synthase collected, dissolved in 10 mM potassium phosphate, 20 mM 2-mercaptoethanol (pH 7.1), and dialyzed against this buffer. The dialyzed enzyme solution is applied to a phosphocellulose column and the enzyme eluted with 100 mM potassium phosphate buffer after thorough washing of the column with 50 mM potassium phosphate buffer. The purified thymidylate synthase may be concentrated by precipitation with 85% ammonium sulfate and stored at −70° after dissolving in a 50 mM Tris–HCl buffer (pH 7.4) containing 1 mM EDTA, 10 mM 2-mercaptoethanol, and 15% sucrose. Folic acid, NADPH, 2-mercaptoethanol, ascorbic acid, buffers, and other reagents were purchased from Sigma Chemical Company, St. Louis, MO.

Instruments. Slab gel electrophoresis is typically conducted with a Bio-Rad (Richmond, CA) apparatus using glass plates which have a 32-cm pathlength in a water-cooled bath. A shorter pathlength apparatus can also yield satisfactory results but greater care must be taken with evaluation of bands. A Joyce-Loebl microdensitometer was used to quantitate individual bands from developed X-ray films.

---

[6] D. G. Priest and M. Mangum, *Arch. Biochem. Biophys.* **210,** 118 (1981).
[7] J. H. Galivan, G. F. Maley, and F. Maley, *Biochemistry* **14,** 3338 (1975).

Ternary Complex Preparation

Ternary complexes are typically prepared by incubation of 1.28×10^{-3} units of thymidylate synthase, 125 nM [^{3}H]FdUMP, and CH$_2$H$_4$-PteGlu$_n$ for 30 min at 25° in 100 μl of a pH 7.4 buffer containing 50 mM Tris–HCl, 50 mM sodium ascorbate, 1 mM EDTA, 0.25 M sucrose, and 6.5 mM formaldehyde.

Ternary Complex Estimation. In order to assess the amount of tritiated ternary complex formed, aliquots are withdrawn and complexes denatured by boiling for 10 min after the introduction of 1% SDS. No detectable dissociation of bound [^{3}H]FdUMP occurs during this denaturation process. Bound [^{3}H]FdUMP is determined by application to minicolumns of Sephadex G-25 which have been equilibrated with the denaturing buffer. Typically, 25 μl of denatured complex is applied to a column prepared in a 400-μl plastic tube with a syringe needle puncture in the bottom. These columns are suspended on the upper rim of larger collection tubes and eluted by centrifugation at 1000 g in a tabletop centrifuge for 3 min. Columns should be preequilibrated by introduction of 25 μl of buffer and centrifugation at the same speed and for the same time used for elution. Over 95% of bound [^{3}H]FdUMP is obtained in the 25-μl eluate volume with negligible contamination from free [^{3}H]FdUMP.

Electrophoresis

In order to evaluate the extent of polyglutamylation of complexed CH$_2$H$_4$PteGlu$_n$, ternary complexes are electrophoresed according to the method of Laemmli[8] except in the absence of SDS. A 1.5-mm-thick slab gel of 9% polyacrylamide with a 4.5% polyacrylamide stacking gel is poured. The maximum volume applied to each sample well is 40 μl. Tris/glycine (25 mM Trisma base and 0.2 M glycine) is used at pH 7.8 for running buffer. Circulation of cold water and operation at reasonably low amperage, to prevent overheating, is helpful in obtaining well-resolved bands. Approximately 16 hr is required for a complete electrophoretic run on a 32-cm gel with an initial current setting of 25 mA. When the tracking dye, bromphenol blue, is permitted to reach the bottom of the gel, labeled ternary complex bands are typically located 50–85% of the distance from the origin. Subsequent to electrophoresis the gel is removed and fixed in 12.5% trichloroacetic acid for 1 hr.

[8] U. K. Laemmli, *Nature (London)* **227**, 680 (1970).

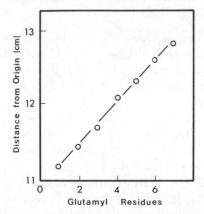

FIG. 1. Linearity between electrophoretic mobility of thymidylate synthase–[³H]-FdUMP-CH₂H₄PteGluₙ complexes and number of glutamyl residues. Complexes were prepared from thymidylate synthase, [³H]FdUMP, and CH₂H₄PteGluₙ standards with one through seven glutamyl residues. Reaction mixtures (200 μl) contained 1.28×10^{-3} units of *L. Casei* thymidylate synthase, 125 n*M* [³H]FdUMP (20 Ci/mmol), and 40 pmol of each CH₂H₄PteGluₙ standard. Forty microliters of each solution was electrophoresed at 4° on 1.5-mm slab gels of 9% polyacrylamide. Mobilities of bands resulting from 1 : 1 : 1 complexes are shown.

Fluorography

Gels are fluorographed according to the method of Laskey and Mills.[9] Briefly, fixed gels are treated with DMSO in three consecutive baths to remove the aqueous phase and then treated with DMSO containing a fluor (typically 22.2% PPO, w/v) to impregnate the gel. The fluor is precipitated in a water bath, and the gel placed on a card of heavy filter paper and dried under vacuum. The dried gel is exposed to Kodak X-Omat AR X-ray film at −70°. When 10,000 cpm of ³H is applied to the original electrophoretic gel, exposure for 24 hr is required to obtain easily visualized bands. Approximately 400 cpm, equivalent to about 0.03 pmol of CH₂H₄PteGluₙ, is the practical lower limit.

Analysis of Results

Examination of the ternary complex structure (**I**) shows that the form with a single glutamyl residue has a charge of −4 (in addition to the charge associated with the protein). Each additional glutamyl residue contributes one additional negative charge and therefore increases the electrophoretic mobility. Figure 1 shows the linear relationship between mobility and the

[9] R. A. Laskey and A. D. Mills, *Eur. J. Biochem.* **56,** 335 (1975).

number of glutamyl residues for ternary complexes prepared from $CH_2H_4PteGlu_n$ polyglutamate standards with one to seven residues. Thus, mobility can be used to directly evaluate the number of glutamyl residues present.

In order to assure accurate evaluation of polyglutamate chain length, it is essential to maintain limiting concentrations of $CH_2H_4PteGlu_n$. Thymidylate synthase has a dimeric structure and each subunit can form a ternary complex with FdUMP and $CH_2H_4PteGlu_n$. This can lead to more than one type of complex with the same total charge and therefore identical mobilities. However, if the concentration of $CH_2H_4PteGlu_n$ presented to the enzyme is sufficiently low, only a single subunit will be occupied and charge redundancy cannot occur. A second potential problem, which is also avoided by low concentrations of $CH_2H_4PteGlu_n$, is the significantly tighter binding by longer chain length polyglutamate forms.[6]

Applications

Although mobility is a linear function of the number of glutamyl residues, assessment of unknown polyglutamates can best be accomplished by direct comparison with standards. Coelectrophoresis of standards eliminates experimental fluctuations such as amperage, degree of acrylamide polymerization, and temperature. A typical application of the method is the estimation of mouse liver $CH_2H_4PteGlu_n$ chain length shown in Fig. 2. It can be seen that the predominant chain length form is a pentaglutamate. When gels were prepared in a similar fashion from other tissue sources, and the resultant fluorographs densitometrically scanned to quantitate individual oligomers, the polyglutamate distributions shown in the table were obtained. It can be seen that under normal conditions methylenetetrahydrofolate pools in liver from diverse sources all contain four to seven glutamyl residues, with penta- and hexaglutamates predominant. Other chain length distributions have been observed in perturbed systems such as folate-starved rats[10] and hepatoma cells.[11] Again, it should be emphasized that care should be exercised during sample preparation to ensure that tissue $CH_2H_4PteGlu_n$ concentration is limiting, to prevent formation of dimeric complexes and so that total incorporation of all forms of $CH_2H_4PteGlu_n$ is assured. In addition, it is at times necessary to consider degradation of polyglutamates even during the relatively short time required for extract preparation. Placing complexation components

[10] I. A. Cassady, M. M. Budge, M. J. Healy, and P. F. Nixon, *Biochim. Biophys. Acta* **633**, 258 (1980).

[11] D. G. Priest, M. T. Doig, and M. Mangum, *Biochim. Biophys. Acta* **756**, 253 (1983).

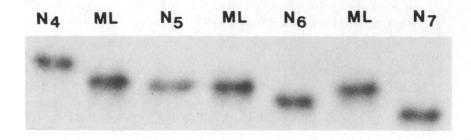

FIG. 2. Comparison of mouse liver $CH_2H_4PteGlu_n$ (ML) with standards containing four, five, six, and seven glutamyl residues, N_4, N_5, N_6, and N_7, respectively. Ternary complexes of thymidylate synthase–[^{3}H]FdUMP-$CH_2H_4PteGlu_n$ were prepared in 200-μl reaction mixtures which contained 1.28×10^{-3} units of *L. casei* thymidylate synthase, 125 n*M* [^{3}H]FdUMP (20 Ci/mmol), and mouse liver extract which contained 5 μg of protein. [Reproduced with permission from D. G. Priest, K. K. Happel, M. Mangum, J. M. Bednarek, M. T. Doig, and C. M. Baugh, *Anal. Biochem.* **115**, 163 (1981).]

(i.e., thymidylate synthase and [^{3}H]FdUMP) directly in extraction buffers can be used to test for and prevent such difficulties. We have experienced instances in which polyglutamate hydrolase (conjugase) in extracts is so active that this immediate complexation technique was essential for accurate evaluation of endogenous polyglutamates.

DISTRIBUTION OF LIVER
METHYLENETETRAHYDROFOLATE
POLYGLUTAMATE CHAIN LENGTH[a]

Animal	Distribution (%) for various n			
	4	5	6	7
Mouse	—	85	15	—
Rat	—	40	60	—
Rabbit	—	20	60	—
Pig	—	15	80	—
Monkey	—	20	80	—
Fish	—	10	75	15
Chicken	5	80	15	—

[a] Ternary complexes were prepared from liver samples as described in Fig. 2. Electrophoretic bands were evaluated quantitatively from densitometric scans of fluorographs.

The method described here is directly applicable to tissue extract $CH_2H_4PteGlu_n$ and can also be used, indirectly, to assess the polyglutamate chain length of other reduced folates, provided they can be chemically or enzymatically converted to the methylene form. For example, the addition of formaldehyde to extracts converts tetrahydrofolate to methylenetetrahydrofolate, which can then be incorporated into the ternary complex and polyglutamate chain length evaluated as before.[12] Enzymatic conversion has been useful with methyltetrahydrofolate.[13] Methylenetetrahydrofolate reductase, forced to function in what would be physiologically the reverse reaction with menadione as an artificial oxidant, has been used successfully to evaluate the polyglutamate chain length of this pool. If necessary, preexisting $CH_2H_4PteGlu_n$, which can interfere with evaluation of the desired methyltetrahydrofolate pool, can be removed by titration with unlabeled FdUMP. It is also potentially possible to evaluate the polyglutamate chain length of other reduced folate pools with a similar approach.

Finally, the method can be used to estimate reaction products of folylpolyglutamate synthetase *in vivo*[11] and *in vitro*[14] as well as folylpolyglutamate hydrolase *in vitro*[12] and thus is a suitable basis for the assay of these enzymes.

[12] D. G. Priest, C. D. Veronee, M. Mangum, J. M. Bednarek, and M. T. Doig, *Mol. Cell. Biochem.* **43,** 81 (1982).
[13] D. G. Priest, M. T. Doig, M. Dang, J. R. Peters, and P. Sur, *Proc. Fed. Am. Soc. Exp. Biol.* **42,** 2103 (1983).
[14] D. G. Priest, M. T. Doig, J. M. Bednarek, J. J. McGuire, and J. R. Bertino, *Biochem. Biophys. Methods* **5,** 273 (1981).

[49] Preparative Electrochemical Reduction of Pteroylpentaglutamate

By PETER B. ROWE and GARRY P. LEWIS

In recent years there has been a renewed interest in the enzymology of reactions involving reduced pteroylglutamate derivatives as cofactors.[1,2] Although it has been clearly shown that the natural polyglutamate derivatives generally increase the V/K_m values for these enzymatic reactions,

[1] S. Benkovic, *Annu. Rev. Biochem.* **49,** 227 (1980).
[2] S. Benkovic, *Trends Biochem. Sci.* **320** (1984).

the availability of these compounds is still quite limited. Tetrahydro-pteroylglutamate is the starting compound for the synthesis of the reduced one-carbon derivatives. Because it is extremely sensitive to air oxidation it is either usually prepared immediately prior to use or purified if supplied from a commercial source.

The methods used for the reduction of pteroylglutamic acid include catalytic hydrogenation over platinum, chemical reduction with agents such as dithionite, or enzymatic reduction with dihydrofolate reductase.[3] Although the last method produces the biologically active diastereoisomer, these methods generally have certain disadvantages including (1) the inability to effectively monitor the course of the reduction reaction which may result in low yields either from incomplete reduction or from the breakdown of tetrahydrofolic acid and (2) the contamination of the end product by a variety of salts. These objections are particularly relevant when reducing the polyglutamate derivatives, which are extremely expensive and not available in large quantities. The reduction technique described here obviates many of these problems and gives a high yield of fully reduced derivatives.

Synthesis of Pteroylpentaglutamate

Pteroylpentaglutamate was synthesized by a modification of the solid-phase method previously described.[4] This method was itself based on that originally described by Krumdieck and Baugh.[5]

These modifications were essential to achieve a reproducibly high product yield. Strict attention was paid to temperature control in the esterification of the first glutamate residue to the Merrifield resin. The oil bath temperature was slowly increased to 82° at a rate of 10°/hr and never exceeded 86°. This resulted in a 60% substitution, i.e., 1.2 mEq glutamate/g of resin. Coupling efficiency was significantly improved by increasing the molar excess of N-methylmorpholine from 2.5- to 2.75-fold in the formation of the substituted glutamate anhydride and from 1.8- to 2.1-fold in the formation of the trifluoroacetylpteroate anhydride. In addition, the reaction vessel was designed to allow smooth efficient mixing for long time periods. With these modifications and precautions a typical synthesis of pteroylpentaglutamate gave a 55% yield (expressed in terms of the amount of glutamate originally substituted onto the resin) of a product which was 92% pure.

[3] R. L. Blakley, "The Biochemistry of Folic Acid and Related Pteridines," p. 78. North-Holland Publ., Amsterdam, 1969.
[4] M. Silink, R. R. Reddel, M. Bethel, and P. B. Rowe, J. Biol. Chem. 250, 5982 (1975).
[5] C. Krumdieck and C. M. Baugh, Biochemistry 8, 1568 (1969).

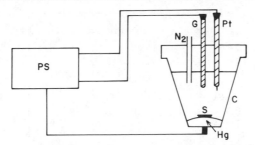

Fig. 1. Apparatus for electrochemical reduction: PS, potentiostat-regulated power supply; C, polarographic reaction vessel; Hg, mercury pool working electrode; G, graphite auxiliary (output) electrode; Pt, platinum reference (input) electrode; N_2, nitrogen gas inlet; S, stirring bar.

Reduction of Pteroylpentagluamate

This method involves the electrochemical reduction of pteroylpentaglutamate using a voltage stepping procedure in a relatively volatile buffer which can be readily removed at the completion of the reaction.

Triethylamine bicarbonate buffer (1.0 M) was prepared by adding 140 ml of freshly distilled triethylamine to 860 ml of distilled, degassed water and titrating to pH 9.25 with carbon dioxide at 4°.

Pteroylpentaglutamate (50 μmol) was dissolved in 15 ml of 0.5 M triethylamine bicarbonate (pH 9.25), in a polarographic reaction vessel (Metrohm EA 996-20). A magnetically stirred mercury pool constituted the working electrode with additional reference (platinum) and auxiliary (graphite) electrodes (Fig. 1). Pure nitrogen gas was gently bubbled through the solution for 10 min prior to the application of current and maintained over the surface of the solution throughout the reaction. Voltage was applied via a potentiostat-regulated power supply (McKee-Pederson Model MP-1026 A) in a stepwise manner as illustrated in Fig. 2 until the maximum (−950 mV) was attained after 3.5 hr. This was maintained for a further 20 hr at which time current flow was terminated and 2-mercaptoethanol added to a final concentration of 0.2 M. The gradual increase in voltage was essential to prevent cleavage of the pteroylglutamate at the C-9–N-10 bond, producing p-aminobenzoylglutamate and presumably 6-methyl-7,8-dihydropterin. While any one of a variety of stepping techniques could be used, the point to be made is that the voltage must be increased slowly. Samples were removed throughout the reduction process for analysis by ultraviolet spectroscopy, anion-exchange HPLC,[6] and polarography (Metrohm Polarecord E 506).

[6] G. P. Lewis and P. B. Rowe, *Anal. Biochem.* **93**, 91 (1979).

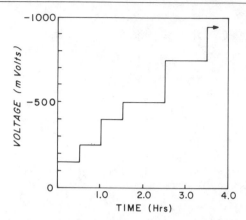

Fig. 2. Graphical representation of time-related voltage steps during reduction.

The solution was rapidly freeze-dried, which removed the relatively volatile buffer and yielded a fine white powder. Based on spectroscopic and HPLC data the final yield was of the order of 96% with respect to the starting material with no detectable breakdown of the polyglutamate chain.

Qualitative studies on the electrochemical reduction of pteroylglutamate to its dihydro- and tetrahydro derivatives and of 5,10-methenyltetrahydropteroylglutamate to 5,10-methylenepteroylglutamate and subsequently to 5-methyltetrahydropteroylglutamate have been briefly reported by Kwee and Lund.[7] They also indicated that the relative proportions of the diastereoisomers could be modified by alterations in the reducing conditions but have not revealed any details.

Studies with the enzyme formyltetrahydropteroylglutamate synthetase confirmed that the reduced pteroylglutamate produced by the synthetic procedure described here consists of an equal mixture of the two diastereoisomers.

It is worth noting that we have attempted, without success, the electrochemical reduction of 10-formylpteroylglutamate, a readily synthesized, highly stable derivative in an effort to bypass the isolation step necessary for the synthesis of 5,10-methenyltetrahydropteroylglutamate.[8] For some as-yet-unexplained reason, 10-formylpteroylglutamate breaks down under a wide range of electrochemical reducing conditions.

[7] S. Kwee and H. Lund, in "Chemistry and Biology of Pteridines" (R. L. Kisliuk and G. M. Brown, eds.), p. 247. Elsevier, Amsterdam, 1979.

[8] P. B. Rowe, this series, Vol. 18, p. 733.

[50] Identification of Folylpoly(γ-glutamate) Chain Length by Cleavage to and Separation of p-Aminobenzoylpoly(γ-glutamates)

By BARRY SHANE

The identification of naturally occurring folate coenzymes is complicated by the large number of possible derivatives arising from different combinations of oxidation state, one-carbon moiety, and polyglutamate chain length and also the presence of breakdown products which arise during extraction of folates from biological samples and their subsequent storage. Cleavage of naturally occurring folates at the C-9–N-10 bond, using a variety of conditions optimized for each one-carbon derivative, yields unsubstituted p-aminobenzoylpolyglutamates and pterin derivatives. The generation of a homologous series of compounds, differing only in glutamate chain length, greatly simplifies the determination of the glutamate chain length of folates.[1-4] p-Aminobenzoylpolyglutamates arising from the cleavage procedure and from folate breakdown can be purified as azo dyes of naphthylethylenediamine and then separated according to glutamate chain length by a variety of chromatographic techniques. This method allows the identification of labeled folate metabolites in bacteria and mammalian cells and the identification and quantitation of unlabeled endogenous folylpolyglutamates in mammalian tissues.

Extraction Procedure

Bacteria are washed several times with isotonic saline and cultured mammalian cells with Dulbecco's phosphate-buffered saline to remove contaminating medium and, in the case of labeling experiments, extracellular labeled vitamin or vitamin precursor. The washed cells are suspended in 20 mM potassium phosphate buffer containing 50 mM 2-mercaptoethanol (typically 2 ml) and the suspension is heated at 100° for 5 min. After cooling, cellular debris is removed by centrifugation and the supernatant collected. The inclusion of mercaptoethanol in the extraction buffer stabilizes some of the reduced folates and allows an aliquot of the supernatant to be set aside for subsequent γ-glutamylhydrolase treatment and identification of the folate one-carbon distribution if desired.

[1] T. Brody, B. Shane, and E. L. R. Stokstad, *Anal. Biochem.* **92,** 501 (1979).
[2] S. K. Foo, D. J. Cichowicz, and B. Shane, *Anal. Biochem.* **107,** 109 (1980).
[3] B. Shane, *Am. J. Clin. Nutr.* **35,** 599 (1982).
[4] I. Eto and C. L. Krumdieck, *Anal. Biochem.* **115,** 138 (1981).

METHODS IN ENZYMOLOGY, VOL. 122

Cleavage Procedure

The products obtained after each step of the cleavage procedure are summarized in Table I.

Step 1. Acidification. The cell extract is adjusted to approximately pH 1 with 5 N HCl and stored overnight at 4° to allow the conversion of 10-formyl- and 5-formyltetrahydrofolate to 5,10-methenyltetrahydrofolate. Under these conditions, 5,10-methylenetetrahydrofolate and tetrahydrofolate are converted to p-aminobenzoylglutamate.

Step 2. Reduction. One drop of n-octanol is added to the solution and the solution is carefully adjusted to pH 5–6 with 5 N NaOH. NaBH$_4$ (100 mg) is *immediately* added to reduce 5,10-methenyltetrahydrofolate to 5-methyltetrahydrofolate. *Caution:* If the solution is left standing at pH 6 before addition of the reducing agent, 5,10-methenyltetrahydrofolate is converted to 10-formyltetrahydrofolate which is not reduced by NaBH$_4$. After 15 min, 5 N HCl is slowly added to destroy the excess borohydride, as indicated by a cessation of bubbling. The pH of the solution should drop from about 10 to 5. Oxidized folates and dihydrofolate are partially reduced to the tetrahydro level in this step.

Step 3. Base Treatment. HgCl$_2$ (0.2 M) is added to the solution to remove mercaptoethanol. Approximately 1.0 mol HgCl$_2$ per mole mercaptoethanol are required. The mixture is centrifuged and the white precipitate discarded. One drop of 0.2 M HgCl$_2$ is added to the supernatant to check the removal of mercaptoethanol. If a white precipitate forms, additional HgCl$_2$ should be added and the centrifugation repeated until no precipitate is observed. The solution is adjusted to pH 12 with 5 N NaOH and left at room temperature for at least 4 hr with occasional mixing, or overnight at 4°, to allow the deformylation of any 10-formylfolate that might be present. Under these conditions, 5-methyltetrahydrofolate is converted to 5-methyldihydrofolate, and dihydrofolate and any tetrahydrofolate present are converted to p-aminobenzoylglutamate. Any precipitate of Hg(OH)$_2$ formed does not interfere with subsequent steps and should not be removed, as p-aminobenzoylglutamate can bind to the precipitate and the overall yield is sometimes reduced after centrifugation.

Step 4. Acidification. The solution or suspension is adjusted to approximately pH 0.5 with 5 N HCl and is stored at room temperature for 1 hr or overnight at 4° to allow the conversion of 5-methyldihydrofolate to p-aminobenzoylglutamate.

Step 5. Zinc Reduction. A zinc dust suspension (0.05 volume; 1 g Zn suspended in 4 ml 0.5% gelatin with stirring) is added and the mixture is shaken intermittently for 10 min to allow the cleavage of folic acid to p-aminobenzoylglutamate. The presence of precipitated borate does not interfere with the reduction. The mixture is centrifuged and the superna-

TABLE I

PRODUCTS OF CLEAVAGE PROCEDURE FOR DIFFERENT FOLATE ONE-CARBON FORMS[a]

Compound	0.1 N HCl and mercaptoethanol	pH 6 and NaBH$_4$	HgCl$_2$, pH 12 for 4 hr	0.1 N HCl	Zn/HCl	Recovery (%)
pABAglu	pABAglu	pABAglu	pABAglu	pABAglu	pABAglu	97
PteGlu	PteGlu	H$_4$PteGlu + PteGlu	pABAglu + PteGlu	pABAglu + PteGlu	pABAglu	102
H$_2$PteGlu	H$_2$PteGlu	H$_2$PteGlu + H$_4$PteGlu	pABAglu	pABAglu	pABAglu	93
H$_4$PteGlu	pABAglu	pABAglu	pABAglu	pABAglu	pABAglu	89
10-Formyl-PteGlu	10-Formyl-PteGlu	10-Formyl-PteGlu + 10-formyl-H$_4$PteGlu	PteGlu + pABAglu	PteGlu + pABAglu	pABAglu	99
10-Formyl-H$_4$PteGlu	5,10-Methenyl-H$_4$PteGlu	5-Methyl-H$_4$PteGlu	5-Methyl-H$_2$PteGlu	pABAglu	pABAglu	91
5-Formyl-H$_2$PteGlu	5,10-Methenyl-H$_4$PteGlu	5-Methyl-H$_4$PteGlu	5-Methyl-H$_2$PteGlu	pABAglu	pABAglu	95
5,10-Methylene-H$_4$PteGlu	pABAglu	pABAglu	pABAglu	pABAglu	pABAglu	93
5-Methyl-H$_4$PteGlu	5-Methyl-H$_4$PteGlu	5-Methyl-H$_4$PteGlu	5-Methyl-H$_2$PteGlu	pABAglu	pABAglu	98
PteGlu$_7$	PteGlu$_7$	H$_4$PteGlu$_7$ + PteGlu$_7$	pABAglu$_7$ + PteGlu$_7$	pABAglu$_7$ + PteGlu$_7$	pABAglu$_7$	96

[a] Abbreviations: pABAglu, p-aminobenzoylglutamate; PteGlu, pteroylglutamate, folic acid; H$_4$PteGlu$_n$, tetrahydropteroylpoly(γ-glutamate), in which n indicates glutamate chain length.

tant carefully removed with a Pasteur pipet. Any zinc particles remaining in the supernatant should be removed by centrifugation. The supernatant can be stored for several months at $-20°$. *Caution:* In the absence of gelatin, the added zinc forms clumps and a variable proportion of the p-aminobenzoylglutamate is trapped in these clumps and removed in the centrifugation step. Any elemental zinc remaining in the extract will interfere with the conversion of p-aminobenzoylglutamate to an azo dye derivative (described below).

Alternate Extraction and Cleavage Procedure for Tissue Samples

Mammalian tissues are homogenized in 3 volumes of 6.7% trichloroacetic acid using a Polytron homogenizer and the precipitate removed by centrifugation. The supernatant is extracted three times with 4 volumes peroxide-free diethyl ether to remove the bulk of the trichloroacetic acid and the aqueous extract is stored at $-20°$ until used.

Folates in the extract are cleaved to p-aminobenzoylpolyglutamates as described above except the borohydride reduction procedure (Step 2) can be omitted and the addition of $HgCl_2$ is unnecessary as no mercaptoethanol is present. Under the described conditions, oxidation of formyl derivatives occurs and the base treatment results in deformylation of these derivatives.[3]

Modifications of the Cleavage Procedure for Individual
Folate Derivatives

In cases in which the sample to be analyzed contains a mixture of polyglutamate forms of a single one-carbon derivative, such as might be found in the analysis of products of an enzyme reaction or after chemical syntheses of folate derivatives, the cleavage procedure can be modified along the lines suggested in Table I. For instance, 5,10-methylenetetrahydrofolate and tetrahydrofolate are rapidly converted to p-aminobenzoylglutamate by acid treatment alone (Step 1), while dihydrofolate is cleaved by base treatment alone (Step 3) and folate by reduction with zinc under acid conditions (Step 5).

Formation and Purification of Azo Dyes of
p-Aminobenzoylpolyglutamates

p-Aminobenzoylpolyglutamates are converted to azo dye derivatives by the procedure of Bratton and Marshall.[1,5] The solution obtained after

[5] A. C. Bratton and E. R. Marshall, Jr., *J. Biol. Chem.* **128**, 537 (1939).

the cleavage procedure (pH < 1.5, adjusted if necessary) is mixed with 0.5% $NaNO_2$ (0.3 ml) to generate the diazonium salts. After 2–3 min, 2.5% ammonium sulfamate (0.3 ml) is added with mixing to destroy excess HNO_2. After a further 2–3 min, 0.1% naphthylethylenediamine (0.5 ml) is added and the solution left for at least 30 min to ensure complete conversion of the diazonium salts to the purple-colored azo dye derivatives ($A_{max (556 nm)} = 5.04 \times 10^4 M^{-1} cm^{-1}$ at pH 1). The naphthylethylenediamine solution should be stored in the dark at 4° and discarded if a brown coloration is observed. The overall recoveries of various folate standards, after cleavage and conversion to azo dyes, are indicated in Table I. Azo dye formation is not affected by the presence of competing aromatic amines in biological extracts as $NaNO_2$ is added in excess. Dye formation is inhibited by 6 M urea and by reducing agents. The azo dyes are fairly stable when stored for several weeks at 4° in the dark although the presence of $HgCl_2$ in the solutions causes some breakdown of these derivatives. Under acid conditions, zinc reduces the azo dyes to the parent unsubstituted amines.

The azo dye solution is applied to a 4 × 0.7-cm BioGel P2 (200–400 mesh) column equilibrated with 0.1 N HCl. Columns are conveniently prepared using Bio-Rad econocolumns. The surface of the gel should be just below the lip of the column reservoir. The column is washed with 0.1 N HCl (6 × 1 ml) to remove pterin cleavage products, any incomplete folate cleavage products, and nonspecific UV-absorbing material from tissue extracts, and the azo dyes are eluted with 0.1 M potassium phosphate buffer pH 7 (1 ml) and water (5 × 1 ml). The azo dyes are an orange color at neutral pH. Some separation of azo dyes according to glutamate chain length occurs on the column. The heptaglutamate is eluted by the phosphate or first H_2O wash while the monoglutamate is eluted by the third and fourth H_2O washes. Derivatives longer than the heptaglutamate are partially eluted at the end of the 0.1 N HCl wash while the azo dye of p-aminobenzoic acid is not eluted from the gel under the described conditions. Normally, the phosphate and H_2O washes are combined with the final acid wash. *Caution:* As the azo dyes bind tightly to the gel at pH 1 and are eluted at neutral pH, care should be taken to ensure that the final acid wash drains completely into the column, that the column sides are free of drops of acid before the buffer is added, and that the buffer addition causes minimal disturbance of the gel surface.

Separation of Individual p-Aminobenzoylpolyglutamates

The azo dye derivatives can be separated directly according to glutamate chain length by chromatography on BioGel P4[1] or by reverse-phase

HPLC.[6] Alternatively, p-aminobenzoylpolyglutamates can be regenerated from the azo dye derivatives by zinc reduction prior to chromatographic separation. In this procedure, 0.02 volume of 5 N HCl is added to the azo dye solution, 0.05 volume of a zinc dust suspension (1 g Zn suspended in 50 ml 0.5% gelatin with stirring) is then added, and the mixture is shaken intermittently for 20 min. In cases in which standards are added, reduction can be followed by the loss of the purple color. The solution is centrifuged to remove elemental zinc and then reduced to dryness by lyophilization or rotary evaporation at 45°. The residue is resuspended in 0.5 ml H_2O and adjusted to pH 6.5–7 by the careful addition of 2 N NaOH. The precipitate of $Zn(OH)_2$ is removed by centrifugation. When large amounts of p-aminobenzoylpolyglutamates are present, the lyophilization step can be omitted. *Caution:* If excess zinc is used to reduce the azo dyes, the large $Zn(OH)_2$ precipitate obtained after neutralization can trap some of the p-aminobenzoylglutamate, necessitating extraction of the pellet to ensure complete recovery.

Individual p-aminobenzoylpolyglutamates can be separated by HPLC using a 250 × 4.6-mm strong anionic exchanger (Partisil 10 SAX, Whatman).[3] The column is eluted isocratically for 10 min with 25 mM ammonium phosphate buffer (pH 6.5), followed by a buffer gradient from 25 to 275 mM over 30 min, and from 275 to 500 mM over 45 min. The column is eluted at a flow rate of 1 ml/min at 35°. One-milliliter (1 min) fractions are collected if labeled folate derivatives are under study and unlabeled endogenous vitamin derivatives or standards are detected by A_{280}. Individual peaks containing 10 pmol p-aminobenzoylpolyglutamate have been detected and quantitated by this method. The polyglutamate distributions of labeled and unlabeled folates in a variety of bacterial and mammalian sources[2,3,7–9] are indicated in Table II. No hydrolysis of standard pteroylheptaglutamate was detected when the standard was subjected to the complete cleavage, derivatization, and separation procedures described above.

Assessment of Overall Recovery of the Procedure

Because different folate one-carbon forms are converted to p-aminobenzoylglutamate by different steps in the cleavage procedure, there is is no standard folate derivative that can be added to the biological extract to

[6] I. Eto and C. L. Krumdieck, *Anal. Biochem.* **120,** 323 (1982).
[7] S. K. Foo and B. Shane, *J. Biol. Chem.* **257,** 13587 (1982).
[8] S. K. Foo, R. M. McSloy, C. Rousseau, and B. Shane, *J. Nutri.* **112,** 1600 (1982).
[9] B. Shane, A. L. Bognar, R. D. Goldfarb, and J. H. LeBowitz, *J. Bacteriol.* **153,** 316 (1983).

TABLE II
FOLYLPOLYGLUTAMATES IN BACTERIA AND MAMMALIAN TISSUES

	Distribution (in %) of glutamate, by chain length										
	1	2	3	4	5	6	7	8	9	10	11
Corynebacterium sp.[a]	0	0	4.0	96.0	0	0	0	0	0	0	0
Lactobacillus casei[b]	4.4	2.6	0.5	1.4	1.0	3.6	5.6	37.6	36.5	6.3	0.5
Streptococcus faecium[b]	0.9	0.5	6.0	77.3	15.3	0	0	0	0	0	0
Rat liver[c]	1.0	0.3	1.9	11.0	68.8	16.5	0.6	0	0	0	0
Hog liver[c]	0	0.4	0.4	1.8	16.3	49.6	24.1	6.3	1.1	0	0
Chinese hamster ovary cells[d]	6.4	0.3	0.7	1.1	8.7	11.1	35.2	28.4	7.5	0.6	0
Human fibroblasts[d]	2.8	1.5	0.9	1.1	4.2	18.9	30.1	28.5	10.4	1.7	0

[a] Labeled distribution after culturing in medium containing [^{14}C]pABA.
[b] Labeled distribution after culturing in medium containing [^{3}H]PteGlu.
[c] Endogenous distribution of unlabeled vitamin.
[d] Labeled distribution after culturing in medium containing $(6S)$-5-methyl-H$_4$[^{3}H]PteGlu.

assess the quantitation of each step in the procedure. When the biological extract contains labeled folate derivatives, the overall recovery of the procedure can be assessed by the recovery of the radioactive label. It should be noted that commercially available [^{3}H]PteGlu is labeled in the 3', 5', 7, and 9 positions and that only the 3' and 5' tritium atoms are recovered in p-aminobenzoylglutamate. The proportion of label in these positions can be easily assessed by Zn/HCl reduction of the [^{3}H]PteGlu followed by separation of the azo dye of the labeled p-aminobenzoylglutamate product from the labeled pterin product by chromatography on BioGel P2. The commercially available labeled vitamin contains between 45 and 65% of its label in the 3' and 5' positions.

The recovery of labeled derivatives derived from folates in rat liver and bacteria was 97% after cleavage and purification on BioGel P2 and 95% after HPLC analysis.[3,9] Overall recoveries are somewhat lower (~70%) with labeled tissue culture cells.[7,8] The lower recoveries with cultured cells may be artifactual, as some breakdown of labeled folate in the medium cannot be avoided, and the labeled pterin breakdown product obtained is actively transported by mammalian cells. The presence of labeled pterins in the cell will lead to an overestimation of intracellular labeled vitamin levels and an underestimate of the recovery after the cleavage procedure.

Small amounts of any labeled folate derivative or p-aminobenzoylglutamate can be added to extracts to assess recoveries when the endogenous unlabeled folate distribution is of interest. Although this is useful as a measure of any losses that may occur during the various manipulations, as indicated above, the recovery obtained is not necessarily an accurate indicator of the effectiveness of every step in the procedure.

Acknowledgment

Supported in part by PHS Grant CA 22717 and Research Career Development Award CA 00697 from the National Cancer Institute, Department of Health and Human Services.

[51] Detection of p-Aminobenzoylpoly(γ-glutamates) Using Fluorescamine

By PETER C. LOEWEN

The analysis of natural folate pools has been complicated by both the multiplicity of oxidation states and the multiplicity of polyglutamate lengths. Methods have been developed for the cleavage of the folate mixture to create a p-aminobenzoylpoly(γ-glutamate), pABAglu$_n$, fraction in which the polyglutamate chain lengths have been characterized by ion-exchange[1-5] and polyacrylamide gel[6] chromatography. In order to increase the sensitivity of the assay of these column eluates without resorting to radioactive precursors, fluorescamine, which forms fluorescent adducts with amines, can be used.[7] The fluorescamine adduct of p-aminobenzoic acid (pABA) and its polyglutamate derivatives exhibit 50- to 100-fold greater fluorescence than similar adducts of other amines such as amino acids, thus allowing the detection of picomole amounts of pABA derivatives.[7]

[1] J. P. Brown, F. Dobbs, G. E. Davidson, and J. M. Scott, J. Gen. Microbiol. 84, 163 (1974).

[2] C. M. Baugh, E. Braverman, and M. G. Nair, Biochemistry 13, 4952 (1974).

[3] K. U. Buehring, T. Tamura, and E. L. R. Stokstad, J. Biol. Chem. 249, 1081 (1974).

[4] R. Bassett, D. G. Weir, and J. M. Scott, J. Gen. Microbiol. 93, 169 (1976).

[5] M. J. Tyerman, J. E. Watson, B. Shane, D. E. Schutz, and E. L. R. Stokstad, Biochim. Biophys. Acta 497, 234 (1977).

[6] T. Brody, B. Shane, and E. L. R. Stokstad, Anal. Biochem. 92, 501 (1979).

[7] R. A. H. Furness and P. C. Loewen, Anal. Biochem. 117, 126 (1981).

Preparation and Chromatography of the pABAglu$_n$ Fraction

Principle. The folate pool is extracted and the pABAglu$_n$ portion is removed by reductive cleavage. The pABAglu$_n$ mixture can be directly fractionated on DEAE-Sephadex A-25 or partially purified by reversible conversion to an azo dye[6,8,9] prior to ion-exchange separation.

Procedure. Bacteria and fungi were collected by centrifugation and washed once in 0.1 M potassium phosphate (pH 7.6). Pellets containing 1–5 g of cells were weighed, resuspended in 10 ml of the same buffer, and boiled for 5 min. Tissue extracts were prepared by homogenization in 0.2 M HCl. The debris was removed by centrifugation, and the supernatant brought to pH 1.0 using concentrated HCl and stored overnight at 4°. After readjusting the pH to 6.0 with 2 M NaOH, 0.5 g of NaBH$_4$ was added and the mixture was shaken for 15 min. Excess NaBH$_4$ was destroyed by acidification and 1 g of powdered zinc was added, followed by shaking at 20° for 5 min. After the zinc was removed by centrifugation the solution was neutralized and immediately charged onto a 0.7 × 50-cm DEAE-Sephadex A-25 column equilibrated with 0.15 M NaCl in 0.04 M potassium phosphate (pH 7.6). In order to ensure reproducibility and to eliminate the possibility of contaminating peaks, all resin was pretreated first with 0.5 M HCl and then with 0.5 M KOH. The resin was washed with distilled water after each treatment and then equilibrated with buffer. A new column was prepared for each determination. After the sample was charged, the column was washed with 50 ml of equilibrating buffer and a linear gradient was applied by mixing 250 ml of 0.15 M NaCl–0.04 M potassium phosphate (pH 7.6) with 250 ml of 0.7 M NaCl in the same buffer. Fractions of 2.5 ml were collected for reaction with fluorescamine.

Alternatively, if partial purification of the pABAglu$_n$ fraction was required, the solution from zinc cleavage could be treated to form diazo derivatives of the pABAglu$_n$ component by the sequential addition at 2-min intervals of 4 ml of 0.5% sodium nitrite, 4 ml of 2.5% ammonium sulfamate, and 2 ml of 0.1% naphthylethylenediamine. After 20 min at 20°, a 4-ml suspension of Dowex AG-50 (200–400 mesh) in 0.2 M HCl was added and shaken for 5 min. The suspension was loaded into a 1.3 × 3.0-cm column and washed with 40 ml of 0.2 M HCl followed by 30 ml of 0.1 M potassium phosphate (pH 7.6) and 30 ml of H$_2$O. The purple diazo dye was reduced and the pABAglu$_n$ was eluted from the column with 10 ml of 0.3 M sodium hyposulfite. Fractions of 1 ml were collected from which 10

[8] A. C. Bratton and E. R. Marshall, Jr., *J. Biol. Chem.* **128,** 537 (1939).
[9] B. Shane, *J. Biol. Chem.* **255,** 5649 (1980).

μl were taken and added to 1 ml of 0.1 M potassium phosphate (pH 7.6) for reaction with fluorescamine. The material that reacted with fluorescamine eluted as a single peak and these fractions were pooled and concentrated to dryness. After resuspension in 10 ml of 0.15 M NaCl in 0.04 M potassium phosphate (pH 7.6), the solution was charged onto a column of DEAE-Sephadex A-25 as described above.

Comments. The amount of cell paste or tissue required for a satisfactory elution profile depended on the organism, but routinely 1 g or more was a good starting point for a first analysis of any organism. The direct application of the zinc-reduced material to the ion-exchange resin avoided a 35–40% loss of material inherent in the recovery from the diazo adduct. In addition, there was usually no enhancement of background fluorescence in the elution profile, with the only difference in results between the two procedures being the greater amount of material in the nonderivatized eluates.

Detection of pABAglu$_n$ Using Fluorescamine

Principle. Fluorescamine reacts rapidly with the amino group of *p*-aminobenzoic acid and its polyglutamate derivatives to form a highly fluorescent adduct. Portions of the eluate are treated with fluorescamine to determine the location and amount of the various sizes of pABAglu$_n$ present.

Procedure. For each set of assays, a fresh solution of 3 mg/ml fluorescamine was prepared in acetone dried over sodium sulfate. To 1 ml of sample volume, 15 μl of the fluorescamine solution was added followed by gentle mixing. After 5 min at 20° the relative fluorescence of the solution was determined using an excitation wavelength of 400 nm and an emission wavelength of 500 nm. A standard value of fluorescent units per nanomole of pABA can be determined by reacting increasing amounts of pABA with fluorescamine in the same assay system. This standard value can then be used to quantitate the amount of pABAglu$_n$ in each peak in the elution profile.

Comments. Because of the intense fluorescence of the fluorescamine–pABA adduct relative to most other fluorescamine adducts, other naturally occurring amines did not interfere with the assay. The relative fluorescence of the fluorescamine–pABA adduct was unaffected by changes in the pH between 5.8 and 9.2, whereas the fluorescence of the amino acid adducts dropped as the pH was lowered.[7] In determining the standard value, which is best done for each series of quantitative assays, as little as 25 pmol/ml of pABA could be detected in a direct assay. However, dilution of the sample during chromatography made it necessary to have at least 500 pmol of a species present for accurate quantitation following elution from the ion-exchange resin. The length of the polyglutamate chain did not affect the relative fluorescence of the pABA–fluorescamine adduct.

[52] Preparation and Analysis of Pteroylpolyglutamate Substrates and Inhibitors

By Rowena G. Matthews

In recent years, there has been considerable interest in the role of folyl- and antifolylpolyglutamates in cellular metabolism.[1,2] A number of research groups are measuring K_i values associated with inhibition of folate-dependent enzymes by pteroylpolyglutamates or kinetic parameters associated with catalysis involving pteroylpolyglutamate substrates. Since the pteroylpolyglutamates typically bind more tightly to their target enzymes than do the corresponding monoglutamates, it is essential that preparations of pteroylpolyglutamate substrates and inhibitors be free of degradation products, and that they be homogeneous preparations of defined polyglutamate chain length and of defined structure at the pteridine. Tetrahydropteroylpolyglutamate derivatives should be prepared by enzymatic reduction to avoid contamination with the unnatural C-6 diastereomer, which may be a potent inhibitor. In addition, since the pteroylpolyglutamates which serve as starting materials for these preparations are expensive and available only in limited quantities, it is highly desirable that the preparation of pteroylpolyglutamate substrates and inhibitors proceed with high yields.

In this article, I shall describe methods of preparation and analysis which we have developed for the production of $PteGlu_n$, $H_2PteGlu_n$, $H_4PteGlu_n$, $5,10-CH_2-H_4PteGlu_n$, and $5-CH_3-H_4PteGlu_n$ derivatives of the requisite purity, using as starting materials the appropriate $PteGlu_n$ derivatives prepared by solid-phase synthesis.[3] Pteroylpolyglutamates ($PteGlu_n$)

[1] I. D. Goldman, B. A. Chabner, and J. R. Bertino, eds., "Folyl and Antifolyl Polyglutamates." Plenum, New York, 1983.

[2] I. D. Goldman, ed., "Folyl and Antifolyl Polyglutamates." Praeger, New York, (in press).

[3] C. L. Krumdieck and C. M. Baugh, this series, Vol. 66, p. 523.

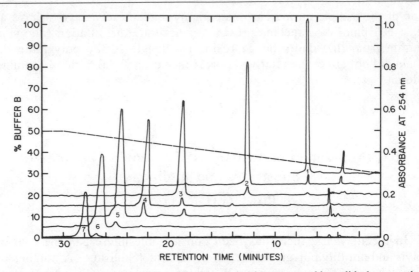

FIG. 1. Chromatographic profiles of $PteGlu_n$ derivatives prepared by solid-phase synthesis. High-performance liquid chromatography was performed as described in the text, and the absorbance at 254 nm was monitored. The solid lines represent absorbance traces associated with chromatography of pteroylpolyglutamates with one to seven glutamyl residues, and the major peak is indicated for each $PteGlu_n$ derivative. These traces have been offset for clarity with the uppermost trace being that of $PteGlu_1$. The dashed line represents the gradient profile utilized for elution. Approximately 1.5 nmol of $PteGlu_n$ was applied for each run.

containing one to seven glutamyl residues were obtained from Dr. Charles M. Baugh, South Alabama Medical Sciences Foundation, College of Medicine, University of South Alabama, Mobile, Alabama 36688.

Assessment of Purity of $PteGlu_n$ Derivatives

The purity of pteroylpolyglutamate stocks should be routinely checked. Preparation by solid-phase synthesis can result in significant contamination of a $PteGlu_n$ stock with lower chain length homologs, especially $PteGlu_{n-1}$. In the published procedure for solid-phase synthesis,[3] contamination with lower homologs is minimized by acetylating any unsubstituted resin-bound Glu-α-benzyl esters after each peptide-forming step. (Acetylated resin-bound Glu-α-benzyl esters will not undergo further chain elongation or coupling with pteroic acid, and can be separated easily from pteroylpolyglutamates after deprotection and cleavage from the resin.) We use ion-paired high-performance liquid chromatography (HPLC) on a C_{18} reversed-phase column to assess the purity of pteroylpolyglutamate stocks.[4] Approximately 1.5 nmol of $PteGlu_n$ are applied to a 5-μm Altex Ultrasphere ODS column (0.46 × 25 cm) equilibrated with 70% buffer A/30% buffer B. Buffer A consists of 5 mM tetrabutylammonium phosphate (Waters PIC A) in glass-distilled water and buffer B consists of acetonitrile (Burdick and Jackson, UV grade). The column is run at a flow rate of 1 ml per minute. Immediately upon injection of the sample a 30-min linear gradient from 70% A/30% B to 50% A/50% B is initiated and then the column is eluted isocratically with 50% A/50% B for 15 min. The eluate is monitored at 254 nm on a 0–1 absorbance range setting. Figure 1 shows the chromatographic profiles of a series of $PteGlu_n$ stocks, as purchased from the South Alabama Medical Sciences Foundation. In these samples, the major peak comprised 85–90% of all 254-nm-absorbing material, and contamination by $PteGlu_{n-1}$ was less than 10% $PteGlu_n$ as measured by peak height ratios. If required, $PteGlu_n$ stocks can be repurified as described by Krumdieck and Baugh.[3] The concentration of stock solutions is determined at 346 nm, using an extinction coefficient of 7200 M^{-1} cm^{-1},[5] and the purified stocks exhibit ultraviolet spectra with maxima at 282 and 346 nm at pH 7.

Preparation of $H_2PteGlu_n$ Derivatives

7,8-Dihydropteroylpolyglutamates are prepared by dithionite reduction of $PteGlu_n$ derivatives.[6] $PteGlu_n$ (10–20 μmol) is dissolved in 2–5 ml of 50 mM potassium phosphate buffer (pH 7.6), 0.3 mM in ethylenediaminetetraacetic acid (EDTA) and placed in a Thunberg tube. Solid dithionite ($Na_2S_2O_4$) is placed in the side arm. A 25-fold excess (mol/mol) of dithionite over $PteGlu_1$ or $PteGlu_2$ is employed, and for longer chain $PteGlu_n$ derivatives a 100-fold molar excess of dithionite is used. The Thunberg tube is alternately evacuated and equilibrated with oxygen-free nitrogen using an anaerobic train previously described.[7] Alternatively, the oxygen in solution in the Thunberg tube can be removed by passing a stream of nitrogen gas through the solution for 10 min. The $H_2PteGlu_n$ product is then purified by chromatography on DEAE-Sephadex A-25. A 0.9 × 25-cm column is equilibrated with 5 mM Tris–chloride buffer (pH 8.0), 0.2 M NaCl and 50 mM in 2-mercaptoethanol. The $H_2PteGlu_n$ is eluted with a 500-ml linear gradient of 0.2–0.7 M NaCl in the same buffer.

[4] Y.-Z. Lu, P. D. Aiello, and R. G. Matthews, *Biochemistry* **23**, 6870 (1984).
[5] J. C. Rabinowitz, *in* "The Enzymes" (P. D. Boyer, H. Lardy, and K. Myrbäck, eds.), 2nd Ed., Vol. 2, p. 185. Academic Press, New York, 1960.
[6] R. G. Matthews and C. M. Baugh, *Biochemistry* **19**, 2040 (1980).
[7] C. H. Williams, Jr., L. D. Arscott, R. G. Matthews, C. Thorpe, and K. D. Wilkinson, this series, Vol. 62, Part D, p. 185.

TABLE I
PREPARATION OF 7,8-DIHYDROPTEROYLPOLYGLUTAMATES BY
DITHIONITE REDUCTION[a]

Dihydropteroylpolyglutamate	Yield (%)	Glutamyl residues per mole[b]
$H_2PteGlu_2$	48	—
$H_2PteGlu_3$	69	2.90
$H_2PteGlu_4$	74	—
$H_2PteGlu_5$	55	5.09
$H_2PteGlu_6$	80	—
$H_2PteGlu_7$	70	—

[a] Adapted, with permission, from ref. 6. Copyright (1980) American Chemical Society.

[b] Determined by amino acid analysis following 24-hr hydrolysis. Amino acid analysis performed as described by Matthews et al.[9]

Fractions (5 ml) are collected, and those which contain $H_2PteGlu_n$ are identified by an absorbance spectrum with peaks at 224 and 282 nm and a shoulder at 304 nm. The concentration of $H_2PteGlu_n$ is determined using an extinction coefficient of 28,400 M^{-1} cm^{-1} at 282 nm, for solutions at pH 7.[8] Yields obtained in the preparation of $H_2PteGlu_n$ derivatives with one to seven glutamyl residues are summarized in Table I.[9] Amino acid analysis of tri- and pentaglutamate products indicated that little if any side-chain cleavage occurs under the described conditions of preparation. $H_2PteGlu_n$ stocks are stored at $-15°$ under nitrogen.

Preparation of $H_4PteGlu_n$ Derivatives

(6S)-$H_4PteGlu_n$ derivatives are formed by stereospecific reduction of $PteGlu_n$ precursors using dihydrofolate reductase from *Lactobacillus casei*.[10] The dihydrofolate reductase is prepared from methotrexate-resistant *L. casei* as described by Liu and Dunlap.[11] Partially purified enzyme can also be purchased from the New England Enzyme Center. It is important to use the enzyme from *L. casei* because dihydrofolate reductases from many other sources do not react with $PteGlu_n$ derivatives but only with 7,8-$H_2PteGlu_n$ derivatives. $PteGlu_n$ (20 μmol) is dissolved in 100 ml

[8] R. L. Blakley, *Nature (London)* **188**, 231 (1960).
[9] R. G. Matthews, L. D. Arscott, and C. H. Williams, Jr., *Biochim. Biophys. Acta* **370**, 6 (1974).
[10] R. G. Matthews, J. Ross, C. M. Baugh, J. D. Cook, and L. Davis, *Biochemistry* **21**, 1230 (1982).
[11] J. K. Liu and R. B. Dunlap, *Biochemistry* **13**, 1807 (1974).

of 10 mM Tris–chloride buffer (pH 7.0), 10 mM in 2-mercaptoethanol. After the buffer is equilibrated with nitrogen, 60 μmol of NADPH are added, followed by 5 units of dihydrofolate reductase. The reaction mixture is incubated overnight at 25°, under nitrogen and shielded from light. The H$_4$PteGlu$_n$ product is purified by chromatography of the reaction mixture on a 1.5 × 20-cm column of DEAE-52 previously equilibrated with 5 mM Tris–chloride buffer (pH 7.2), 0.2 M in 2-mercaptoethanol. Elution is effected with a 200-ml linear gradient of 0–1 M NaCl in the same buffer and 3-ml fractions are collected. Fractions containing H$_4$PteGlu$_n$ are identified by the presence of a symmetrical absorbance peak maximal at 297 nm. Concentrations of stock solutions are determined using an extinction coefficient of 29,100 M^{-1} cm^{-1} at 297 nm.[12] Yields for these preparations vary from 50 to 70% after purification by chromatography. The purified H$_4$PteGlu$_n$ stocks can be stored under nitrogen at −15° for several weeks.

Preparation of CH$_2$-H$_4$PteGlu$_n$ Derivatives

(6R)-5,10-CH$_2$-H$_4$PteGlu$_n$ derivatives are prepared by nonenzymatic condensation of the corresponding (6S)-H$_4$PteGlu$_n$ derivatives with formaldehyde. For assays to be conducted at neutral pH it is often convenient to generate the CH$_2$-H$_4$PteGlu$_n$ derivative in the assay mixture by addition of the H$_4$PteGlu$_n$ derivative at the desired final concentrations to an assay mixture which is 1 mM in formaldehyde and 50 mM in 2-mercaptoethanol. The assay solution should be equilibrated with nitrogen prior to addition of H$_4$PteGlu$_n$ and should then be incubated with H$_4$PteGlu$_n$ under nitrogen for 5 min prior to initiation of the assay.

Alternatively, CH$_2$-H$_4$PteGlu$_n$ can be prepared as a concentrated stock solution and purified chromatographically. H$_4$PteGlu$_n$ derivatives are prepared as described above. After overnight incubation the reaction mixture is brought to 10 mM in formaldehyde and incubated for 15 min at 25° and then the pH of the solution is raised to 9.2 by addition of 1 M ammonium carbonate. The resultant CH$_2$-H$_4$PteGlu$_n$ is purified by chromatography on a 0.9 × 15-cm column of DEAE-52, equilibrated with 0.01 M ammonium carbonate (pH 9.2), 1 mM in formaldehyde, and eluted with a 200-ml linear gradient of 0–0.5 M NaCl. Fractions of 3 ml are collected in tubes containing 10 μl of 12.3 M formaldehyde. The CH$_2$-H$_4$PteGlu$_n$ fractions are identified by the appearance of a symmetrical absorbance peak centered at 297 nm. The CH$_2$-H$_4$PteGlu$_n$ content is determined in 50 mM phosphate buffer (pH 7), 50 mM 2-mercaptoethanol, using an extinction

[12] R. G. Kallen and W. P. Jencks, *J. Biol. Chem.* **241**, 5845 (1966).

TABLE II
PREPARATION AND PROPERTIES OF
5-METHYLTETRAHYDROPTEROYLPOLYGLUTAMATES[a]

Starting pteroylpolyglutamate	Yield of purified CH_3-$H_4PteGlu_n$ (%)	Glutamyl residues per mole[b]
$PteGlu_1$	42	0.98
$PteGlu_2$	72	1.97
$PteGlu_3$	58	2.86
$PteGlu_4$	51	3.86
$PteGlu_5$	38	4.91
$PteGlu_6$	70	6.03
$PteGlu_7$	63	7.18

[a] Reprinted, with permission, from ref. 10. Copyright (1982) American Chemical Society.
[b] Determined by amino acid analysis following 24-hr hydrolysis. Aspartic acid (28 nmol) was added to each sample prior to hydrolysis as an internal standard for recovery of glutamic acid in the sample during transfer and loading on the analyser. Amino acid analysis performed as described by Matthews et al.[9]

coefficient at 297 nm of 31,700 M^{-1} cm^{-1}.[13] Purified CH_2-$H_4PteGlu_n$ stocks can be stored for several weeks under nitrogen at $-15°$. Amino acid analysis of tetra- and heptaglutamate derivatives gave 4.16 and 7.44 glutamyl residues per mole respectively.[6]

Preparation of CH_3-$H_4PteGlu_n$ Derivatives

(6S)-5-CH_3-$H_4PteGlu_n$ derivatives are prepared by reduction of the corresponding CH_2-$H_4PteGlu_n$ derivative using methylenetetrahydrofolate reductase from pig liver.[10] Methylenetetrahydrofolate reductase is prepared by published procedures[14,15] and for these purposes it is sufficient to purify the enzyme through step 3. CH_3-$H_4PteGlu_n$ is prepared in the same manner as $H_4PteGlu_n$ except that 80 rather than 60 μmol of NADPH are added to the reaction mixture and in addition 100 μmol of formaldehyde and 5 units of methylenetetrahydrofolate reductase (NADPH-menadione oxidoreductase activity) are added. After overnight incubation, the CH_3-$H_4PteGlu_n$ product is purified by chromatography of the reaction mixture on a 1.5 × 20-cm column of DEAE-52, previously

[13] R. L. Blakley, *Biochem. J.* **74,** 71 (1960).
[14] S. C. Daubner and R. G. Matthews, *J. Biol. Chem.* **257,** 140 (1982).
[15] R. G. Matthews, this volume [58].

equilibrated with 0.05 M ammonium acetate (pH 7.2), 10 mM in 2-mercaptoethanol. Elution is effected with a 100-ml linear gradient of 0–1 M NaCl in the same buffer and 3-ml fractions are collected. Fractions containing CH_3-$H_4PteGlu_n$ can be identified by the presence of a symmetrical absorbance peak, maximal at 292 nm. Table II shows the yields of purified CH_3-$H_4PteGlu_n$ and the results of amino acid analysis of hydrolyzed samples. Concentrations of CH_3-$H_4PteGlu_n$ can be determined using an extinction coefficient at 292 nm of 31,700 M^{-1} cm^{-1}.[16] Solutions of CH_3-$H_4PteGlu_n$ can be stored frozen, under nitrogen, at $-15°$.

Acknowledgements

Research from the author's laboratory was funded by Research Grants GM 24908 and GM 30885 from the National Institutes of Health. She gratefully acknowledges the contributions of many colleagues in this work, whose names are listed in the cited references.

[16] V. S. Gupta and F. M. Huennekens, *Arch. Biochem. Biophys.* **120,** 712 (1967).

[53] Analysis of Methotrexate Polyglutamate Derivatives
in Vivo

By B. A. KAMEN and N. WINICK

The discovery that methotrexate (MTX) was a substrate for folylpolyglutamate synthase approximately a decade ago spawned a surge of activity to improve upon the conventional biochemical technique of Sephadex column chromatography used for analysis of methotrexate polyglutamates, MTX(Glu$_n$).[1,2] A previous issue of Methods in Enzymology detailed the HPLC separation of pteroyloligoglutamates.[3] Subsequently a number of techniques based upon this methodology have been developed. Separations were generally done on C$_{18}$ and ODS columns using gradients of acetonitrile in 5 mM tetrabutylammonium phosphate or acetate buffer.[4,5] These techniques have been used to analyze MTX metabolites obtained from cells treated with methotrexate *in vitro*. There are

[1] V. M. Whitehead, *Cancer Res.* **37,** 408 (1977).
[2] M. Balinska, J. Galivan, and J. K. Coward, *Cancer Res.* **41,** 2751 (1981).
[3] A. R. Cashmore, R. N. Dreyer, C. Horvath, J. O. Knipe, J. K. Coward, and J. R. Bertino, this series, Vol. 66, p. 459.
[4] D. W. Fry, J. C. Yalowich, and I. D. Goldman, *J. Biol. Chem.* **257,** 1890 (1982).
[5] J. Jolivet and R. L. Schilsky, *Biochem. Pharmacol.* **30,** 1387 (1981).

fewer research results, however, concerned with MTX polyglutamation *in vivo*. Since most chemotherapeutic regimens call for MTX on a multi-dose schedule (weekly or even daily) for periods of months to years, it is important to be able to analyze $MTX(Glu_n)$ synthesized *in vivo*. This implies that there is an ability to quantitate these compounds without having the benefit of a radiolabeled precursor. The quantity of material is biopsy size samples (for example 5–25 mg of hepatic tissue which contains 0.1–1 pmol MTX/mg) necessitates the use of radioligand or radioenzymatic assays, as opposed to the usual method of spectral analysis which requires extensive sample preparation and has a lower limit of detection of 10^{-8} to 10^{-7} M.[6,7]

The material presented here details methodology for the extraction, separation, and quantitation of $MTX(Glu_n)$ formed *in vivo*. The method is based upon our previously published report using a radioligand binding assay for MTX (and polyglutamates)[6] detection and either an isocratic phosphate buffer solution with a molecular sieve column[8] or a newer methodology using a C_{18} μ-Bondapak column to effect separation of $MTX(Glu_n)$. These procedures have been used to quantitate MTX polyglutamates in clinical samples (e.g., 5–10 mg biopsy of hepatic tissue containing 0.5–1.0 pmol MTX/mg) as well as to evaluate the pharmacodynamics of MTX given to laboratory animals for periods up to 1 year.[8,9]

Materials and Methods

Equipment. Commercially available pumps/injectors and fraction collectors were used as received from the manufacturer. We have used the M45 solvent delivery system (max psi 4000) for the molecular sieve procedure. Model 510B pumps were used to generate the gradient for the reverse-phase chromatography. Both these pumps are from Waters Associates. The U6K injector (Waters) or a WISP (automatic injection system, Waters Associates) was employed for loading the columns. A Redi-rack (LKB) has proved to be a dependable, small, but high-capacity fraction collector to couple with the HPLC systems. Spectral detection of standards can be done using any high-quality UV spectrophotometer, with a small volume (10 μl) flow cell. We have used both Waters and Beckman systems.

Columns. Molecular sieve chromatography was done on an I-60 protein column (Waters Associates). When the columns are properly cared

[6] B. A. Kamen, P. L. Takach, R. V. Vatev, and J. D. Caston, *Anal. Biochem.* **70,** 54 (1976).
[7] R. Hayman, R. McGready, and M. B. Vander Weyden, *Anal. Biochem.* **87,** 460 (1978).
[8] G. R. Krakower, P. Nylen, and B. A. Kamen, *Anal. Biochem.* **122,** 412 (1982).
[9] G. R. Krakower and B. A. Kamen, *J. Pharmacol. Exp. Ther.* **227,** 633 (1983).

for and when properly prepared samples are injected we have obtained more than 200 runs/column. Reverse-phase chromatography is done on a C_{18} column. In general we have used a 15-cm C_{18} μ-Bondapak column (Waters Associates).

Guard Columns. The guard column for the I-60 column was packed with I-125 packing material and for the C_{18} column a precolumn module (Waters Associates) with a C_{18} Guard-Pak cartridge was used.

Chemicals. Sodium chloride, potassium phosphate, Tris, and other salts were obtained from J. T. Baker Company. Acetonitrile, methanol, and HPLC-grade water were obtained from Waters Associates. Methotrexate was purchased from Lederle. MTX polyglutamate standards were obtained from Dr. Charles Baugh at the University of Alabama. Tritiated MTX (10–20 Ci/mmol) was purchased from Amersham Searle and purified on the I-60 column or DEAE-cellulose.

Sample Preparation for I-60 Molecular Sieve Chromatography. For separation of methotrexate and methotrexate polyglutamates on the I-60 protein column, tissue extracts were prepared by homogenizing samples in three volumes (w/v) of ice-cold extraction buffer (0.01 M Tris, pH 8, at room temperature) containing 150 mM 2-mercaptoethanol and 5 mM EDTA. Samples (generally greater than 100 mg) were sonicated in a Brinkmann Polytron (30 sec at setting 10 using a 7-mm diameter probe, Model PT7). The homogenized sample is centrifuged 30 min at 12,000 g. The resulting supernatant is boiled for 10 min and the centrifugation step is repeated. The sample is filtered through a 0.22 μM pore filter (Millipore) prior to injection, on the I-60 column. An alternative procedure is used for the initial extraction of smaller samples (e.g., hepatic biopsy, 5–20 mg). Specifically, the tissue is put in 0.4–0.5 ml extraction buffer in a microfuge tube and then minced with a glass rod, the tip of which was fire polished to fit the bottom of the tube. A commercial microhomogenizer (Thomas or Kontes) can also be used. The sample is then freeze-thawed three times in a dry ice/acetone bath or in a $-120°$ freezer, centrifuged, and boiled as above. An analysis of more than 30 human biopsy samples revealed a reliable reproducibility. Baseline samples had total folates equivalent to published values (based upon a microbiological assay and our own work with larger quantities of tissue obtained at times of surgery or autopsy and assayed using a radioligand binding technique). Baseline samples had no detectable methotrexate either (<0.1 pmol/mg); therefore we can state that using our extraction and assay conditions the "background" in the sample accounts for an insignificant portion of the total methotrexate measured in small biopsy samples.

Samples prepared as described above are stable with respect to methotrexate; however, it is recommended that if folate is also to be analyzed that they are not stored for prolonged periods, even frozen.

Analysis of tissue samples having known quantities of methotrexate polyglutamates added at the time of homogenization has shown greater than 85% recovery in the boiled supernatant. The buffer inhibits γ-glutamylcarboxypeptidase (conjugase) for at least 1 hr at 37° when MTX(Glu$_7$) was the test substrate (plasma and liver conjugase were tested). Similarly, extraction buffer also protects folates and has proven satisfactory for allowing analysis of total folates as well as methotrexate.

Sample Preparation for C$_{18}$ Reverse-Phase Chromatography. Samples already extracted and boiled as described for the I-60 preparation are applied to a DEAE-cellulose column (0.5 × 1 cm) equilibrated in 0.05 M ammonium bicarbonate. The sample (containing 50–500 pmol MTX) is eluted from the column with 1 M ammonium bicarbonate after removing impurities with washes of 2–3 column volumes of 0.15 and then 0.3 M ammonium bicarbonate. The 1.0 M eluate contains all the methotrexate and methotrexate polyglutamates but represents a 50- to 100-fold purifica-

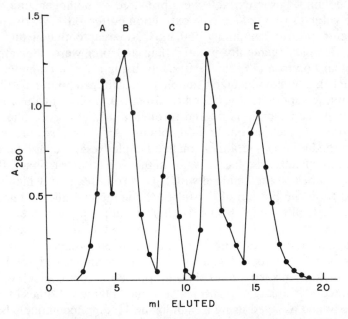

Fig. 1. Spectrally detectable quantities of MTX(Glu$_n$) standards were chromatographed on the I-60 column. The buffer was potassium phosphate 0.1 M (pH 7.0). The flow rate was 1 ml/min with 0.25-ml fractions collected. Absorbance is read at 280 nm. The concentration of the standards in the stock solution was approximately 10^{-4} M and 0.1 ml of sample was injected. A, MTX(Glu$_6$); B, MTX(Glu$_4$); C, MTX(Glu$_3$); D, MTX(Glu$_2$); E, MTX(Glu$_1$). MTX(Glu$_1$) by convention is MTX; thus the subscript (Glu$_n$) refers to the total number of glutamates in the compound.

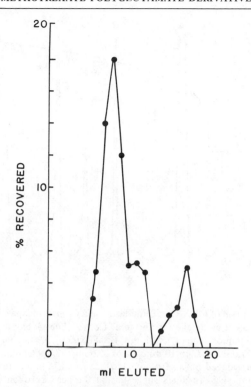

FIG. 2. A 10-mg hepatic biopsy (from a patient who had received a weekly low dose of MTX for 2 years as treatment of arthritis) containing 1.9 pmol $MTX(Glu_n)$/mg wet weight was homogenized in 0.4 ml of extraction buffer as detailed in the text and 0.2 ml (approximately 10 pmol) was chromatographed on an I-60 column. The flow rate was 1 ml/min with 0.5-ml fractions collected and assayed for MTX by radioligand binding assay. A large percentage of the MTX is in the $MTX(Glu_{3-7})$ range. The resolution is not as complete as when 0.25-ml fractions are collected, but the predominance of higher $MTX(Glu_n)$ is clearly seen. The percentage recovered is based upon the total $MTX(Glu_n)$ loaded. In general, recoveries range from 80 to 120%.

tion of these compounds as judged by loss of the A_{280} absorbance. The sample is concentrated by either lyophilization or evaporation in a partial vacuum at 50° in a nitrogen atmosphere. The concentrated, partially purified sample is dissolved in sterile water and applied to the C_{18} column.

Chromatography Procedure

Molecular Chromatography. The I-60 column is placed in line with the U6K injector and the M45 solvent delivery system with a guard column. The column is equilibrated in 0.1 M phosphate buffer (pH 7). As in all

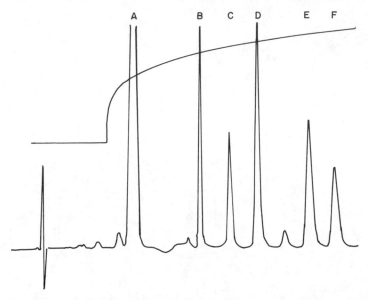

FIG. 3. Chromatogram of MTX(Glu$_n$) standards on a C$_{18}$ μ-Bondapak column using a 20–40% acetonitrile gradient as detailed in the text. Convex line plots gradient development. Samples: A, MTX, retention time (T_r) 8.01 min; B, MTX(Glu$_2$), T_r 12.33 min; C, MTX(Glu$_3$), T_r 14.26 min; D, MTX(Glu$_4$), T_r 16.10 min; E, MTX(Glu$_6$), T_r 10.60 min; F, MTX(Glu$_7$), T_r 21.30 min. MTX(Glu$_n$) loaded ranges from 0.03 to 0.55 μmol. A plot of retention time versus the number of glutamic acid residues suggests that the peak between D and E would be MTX(Glu$_5$). It represented a "contaminant" in the MTX(Glu$_6$) standard.

good standard chromatographic techniques, buffers are degassed and filtered prior to use. The flow rate is 1.0 ml/min (pressures 500–1000 psi are generated). Generally adequate resolution is obtained by collecting 0.5 min (0.5 ml) fractions, although earlier we collected 0.25-ml fractions and the flow rate was 0.5 ml/min.[8]

The separation of known methotrexate polyglutamate standards loaded in spectral quantities is shown in Fig. 1. It should be noted that the I-60 column appears to retain the lower chain length methotrexate derivatives.[8] Thus methotrexate elutes significantly later than where it should by strict molecular sieve chromatography. This allows more complete separation between the lower and higher methotrexate polyglutamates. Varying buffer concentration between 0.01 and 0.5 M has not improved resolution of the polyglutamates.

Adequate detection of the polyglutamates can be achieved with as little as 10–20 pmol of total material. This is shown in Fig. 2 (see also ref.

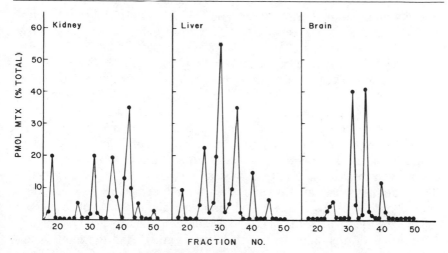

FIG. 4. Samples of liver, brain, and kidney tissue from monkeys treated with weekly IM MTX for 1 year were extracted, purified, and lyophilized as described in the text. Twenty to forty microliter samples were injected onto the C_{18} column. The flow rate was 1 ml/min with a gradient from 20–40% acetonitrile. Half-minute fractions were collected and assayed for MTX using the radioligand binding assay. MTX polyglutamates with three to five glutamic residues dominate, with an absence of the monoglutamate in the brain. The 80–100% recovery of MTX and $MTXGlu_n$ compares the total amount of drug found in the assayed column fractions to the total amount of drug found in the boiled extract samples prior to preparation for the C_{18} column.

9). Routine operating conditions are described in the legend and above. Since methotrexate will elute from the fifteenth to eighteenth ml, and the excluded volume is approximately 4 ml, we assay a total 20–30 fractions, depending on the volume of each fraction. The total run time is approximately 20 min. To analyze multiple samples the column is washed for approximately 30 min at 1–1.5 ml/min between sample runs.

It should also be noted that resolution can be improved slightly by coupling two columns in series with minimal dead-space tubing. The flow rate can be doubled, and therefore the retention time is not changed. However, since $MTX(Glu_{5-7})$ still cannot be completely resolved we do not generally use two columns.

Reverse-Phase Separation. Reverse-phase high pressure liquid chromatography yields a more complete separation of methotrexate and its polyglutamate derivatives. The DEAE-purified samples are dissolved in sterile water and applied to a μ-Bondapak C_{18} column. The two buffers used to generate the 20–40% acetonitrile gradient are buffer A (5 mM

tetrabutylammonium phosphate in water, pH 7) and buffer B (5 mM tetrabutylammonium phosphate in 40% acetonitrile, 60% water, pH 7). The flow rate is 1 ml/min, initially 50% buffer A and 50% buffer B with a convex gradient generated over 20 min to 100% buffer B (Waters Automated Gradient Controller). After the gradient is completed, buffer B continues to flow at 1 ml/min for 10 min followed by a linear return to the initial conditions in order to recycle the column. Methotrexate and polyglutamate standards (Glu$_{2,3,4,6,}$ and 7) are separated within 25 min (Fig. 3). Neither buffer A nor B in concentrations greater than ordinarily would be present in a radioligand binding assay affects the detection of known standards in this assay. Figure 4 illustrates the use of this system. Samples of monkey liver, kidney, and brain were prepared as described and injected onto the C$_{18}$ column. Half-minute fractions are collected and assayed for MTX using the sequential radiobinder assay.

DEAE and alkaline extraction are used rather than the more traditional trichloroacetic acid precipitate and/or Sep-Pak (Waters Associates) preparations because we have at times found poor and variable recovery of known methotrexate polyglutamates from tissue specimens *in vivo* as compared to cells *in vitro,* using these methods. For example, a mixture of MTX and MTX(Glu$_7$) added to a normal liver extract yielded an overall 65–70% recovery following elution on a Sep-Pak but the MTX was >90% and the MTX(Glu$_7$) was only 20–30%.

For routine, semiquantitative analysis of MTX polyglutamates the extra steps and time necessary to prepare samples for analysis on a C$_{18}$ column has generally not been necessary. As it becomes more important to determine the exact chain length, the reverse-phase column (C$_{18}$) will be used more often. When enough material is available the samples can be concentrated enough to obtain a spectral quantitation (reading at 302–304 nm).

[54] Radioenzymatic Assay for Dihydrofolate Reductase Using [^{3}H]Dihydrofolate

By M. PERWAIZ IQBAL and SHELDON P. ROTHENBERG

Assay Method

Folic acid (pteroylglutamic acid, PteGlu) or dihydrofolic acid (H$_2$PteGlu) are reduced to tetrahydrofolic acid (H$_4$PteGlu) by NADPH with the enzyme dihydrofolate reductase (DHFR) catalyzing the reaction.

The product, $H_4PteGlu$, can be separated from the substrate by precipitation of the PteGlu or $H_2PteGlu$ with zinc sulfate.[1-3] This method permits monitoring the reaction by the direct measurement of the reduction of tritium-labeled PteGlu or $H_2PteGlu$, rather than the indirect method of measuring the reduction of the substrate by monitoring the oxidation of NADPH spectrophotometrically, which has been the conventional method of determining the activity of DHFR.[4]

[³H]PteGlu was used as the substrate in the original radioenzymatic essay for DHFR.[1] However, PteGlu serves as a substrate only at an acid pH (~ pH 5.0) and it is not the physiological folate for this reaction because $H_2PteGlu$ forms in cells when 5,10-methylene-$H_4PteGlu$ donates its methyl group to dUMP for the synthesis of TMP. Thus, $H_2PteGlu$ is the more appropriate substrate to measure the activity of DHFR. Unfortunately, [³H]$H_2PteGlu$ is not available commercially and its preparation is time-consuming and cumbersome, and this folate is not stable on storage. In order to circumvent this problem, we have coupled the chemical reduction of [³H]PteGlu to [³H]$H_2PteGlu$ by dithionite with the enzymatic reduction of the $H_2PteGlu$ to $H_4PteGlu$.

Principle. [³H]PteGlu is reduced to [³H]$H_2PteGlu$ by incubating it with dithionite. The enzymatic reduction of the $H_2PteGlu$ is then initiated by the addition of NADPH and DHFR. Unlike the enzymatic reduction of PteGlu, the reduction of $H_2PteGlu$ by DHFR is optimal at neutral pH.[5]

Reagents

[³H]PteGlu, (specific activity 20–50 Ci/mmol, Amersham Corp., Arlington, IL)

Sodium citrate solution, 0.06 *M*, pH 7.2, containing 0.01 *M* 2-mercaptoethanol

PteGlu

Sodium dithionite

NADPH (Sigma Chemicals, St. Louis, MO)

$ZnSO_4$, 0.3 *N*

Dihydrofolate reductase preparation

Procedure. [³H]PteGlu (0.5–110 pmol) is preincubated with 1.5 μmol of dithionite for 10 min at 37° in 0.25 ml of 0.06 *M* sodium citrate (pH 7.2), containing 0.01 *M* 2-mercaptoethanol. Following this incubation, 60 nmol of NADPH and the preparation of DHFR are added with sufficient so-

[1] S. P. Rothenberg, *Anal. Biochem.* **6,** 176 (1966).
[2] R. Hayman, R. McGready, and M. B. Vander Weyden, *Anal. Biochem.* **87,** 460 (1978).
[3] S. P. Rothenberg, M. P. Iqbal, and M. da Costa, *Anal. Biochem* **103,** 152 (1980).
[4] M. J. Osborn and F. M. Huennekens, *J. Biol. Chem.* **233,** 969 (1958).
[5] C. K. Mathews and F. M. Huennekens, *J. Biol. Chem.* **238,** 3436 (1963).

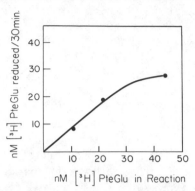

FIG. 1. The relationship between the concentration of [³H]PteGlu in the reaction mixture and the concentration of this substrate reduced to [³H]H₄PteGlu in 30 min of incubation. There was no unlabeled PteGlu in the reaction mixture.

dium citrate solution (pH 7.2) for a final volume of 0.5 ml. The enzyme reaction is terminated by the addition of 0.2 ml of a solution of PteGlu (12 mg/ml) and 0.2 ml of 0.3 N ZnSo₄. The precipitate of PteGlu, residual [³H]PteGlu, and unreduced [³H]H₂PteGlu is pelleted by centrifugation, and the radioactivity in the supernatant solution is counted in any standard scintillation cocktail for aqueous solutions. A blank which contains all the constituents of the assay mixture except the DHFR is run in parallel with the enzymatic reaction to correct for nonprecipitable radioactivity which is due either to an impurity in the original [³H]PteGlu or to some nonenzymatic reduction of the substrate to [³H]H₄PteGlu by the dithionite. The radioactivity in this blank is subtracted from the radioactivity in the enzymatic reaction to obtain the net counts of enzymatically produced [³H]H₄PteGlu. The molar concentration of H₄PteGlu produced can be computed from the percentage of the [³H]PteGlu reduced to [³H]H₄PteGlu times the initial total concentration of PteGlu in a reaction volume of 0.5 ml.

Figure 1 shows the typical curve relating the [³H]PteGlu reduced to [³H]H₄PteGlu to the concentration of [³H]PteGlu in the reaction mixture. The method is sensitive enough to detect the reduction of substrate in the nanomolar concentration range.

Figure 2 shows the competitive inhibition of the reduction of [³H]PteGlu by increasing the concentration of unlabeled PteGlu when the enzyme is saturated by substrate. This indicates that the PteGlu is first reduced to H₂PteGlu by the dithionite; otherwise there would have been no inhibition since PteGlu is not active as a substrate for DHFR at pH 7.2.

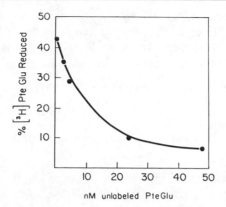

FIG. 2. The competitive inhibition of the reduction of [³H]PteGlu to [³H]H₄PteGlu by unlabeled PteGlu at pH 7.2.

Comments

There are a number of advantages of this radioenzymatic assay. First, this method measures the activity of the enzyme using $H_2PteGlu$, the more physiological substrate. Second, the coupling of the chemical reduction of [³H]PteGlu to [³H]H₂PteGlu with the enzymatic reaction eliminates the need for the complicated synthesis and storage of the labile [³H]H₂PteGlu. Third, depending on the specific activity of [³H]PteGlu, the assay is very sensitive and can monitor the reduction of as little as 0.5 nM of substrate. Finally, this radioenzymatic assay for DHFR measures directly the reduction of the substrate rather than the oxidation of NADPH, so that enzyme activity can be determined in crude tissue or cell preparations which may contain other components which can oxidize NADPH.

[55] Bacterial Folylpoly(γ-glutamate) Synthase–Dihydrofolate Synthase

By ANDREW L. BOGNAR and BARRY SHANE

Folylpolyglutamates, the major cellular forms of the vitamin, are the active coenzymes for the reactions of one-carbon metabolism. Folypolyglutamate synthase (EC 6.3.2.17), the enzyme that catalyzes the conversion of pteroylmonoglutamates to poly (γ-glutamate) derivatives, has been purified to homogeneity from *Corynebacterium*,[1] *Lactobacillus casei*,[2] and *Escherichia coli*.[3] Protein purified from bacteria that biosynthesize folate *de novo* also possesses dihydrofolate synthase activity.[1,4] The *Corynebacterium* protein catalyzes the following reactions:

Dihydropteroate + MgATP + L-glutamate → dihydrofolate + MgADP + P_i
Tetrahydrofolate + MgATP + L-glutamate
$$→ \text{tetrahydropteroyldiglutamate} + \text{MgADP} + P_i$$
5,10-Methylenetetrahydropteroyl(glutamate)$_n$ + MgATP + L-glutamate
$$→ 5,10\text{-methylenetetrahydropteroyl(glutamate)}_{n+1} + \text{MgADP} + P_i$$

where n equals 1 to 3.

Protein purified from *Lactobacillus*, an organism that requires exogenous folate for growth, lacks dihydrofolate synthase activity, and catalyzes the following reaction:

5,10-Methylenetetrahydropteroyl(glutamate)$_n$ + MgATP + L-glutamate
$$→ 5,10\text{-methylenetetrahydropteroyl(glutamate)}_{n+1} + \text{MgADP} + P_i$$

where n can be from 1 to 10 *in vivo*.

In this report, the purification and general properties of the *Corynebacterium* and *L. casei* folylpolyglutamate synthases are described.

Folylpolyglutamate Synthetase Assay

Enzyme activity is normally measured by the incorporation of [^{14}C]glutamate into folylpolyglutamates using unlabeled (6R,S)-tetrahydrofolate and (6R,S)-5,10-methylenetetrahydrofolate as the folate substrates for the *Corynebacterium* and *Lactobacillus* enzymes, respectively.[1,2,5]

Reaction mixtures contain 100 mM Tris–50 mM glycine buffer (pH 9.75 at 22°), (6R,S)-tetrahydrofolate (100 μM), formaldehyde (5 mM; *Lactobacillus* enzyme only), L-[^{14}C]glutamate (250 μM; 1.25 μCi), ATP (5 mM), MgCl$_2$ (10 mM), KCl (200 mM), dithiothreitol (5 mM), 2-mercaptoethanol (10 mM; derived from folate solution), dimethyl sulfoxide (50 μl), bovine serum albumin (50 μg), and enzyme in a total volume of 0.5 ml. The reaction tubes are capped and incubated at 37° for 2 hr.

The reaction is stopped by the addition of ice-cold 30 mM β-mercaptoethanol (1.5 ml) containing 10 mM glutamate, and the mixture is applied

[1] B. Shane, *J. Biol. Chem.* **255**, 5655 (1980).
[2] A. L. Bognar and B. Shane, *J. Biol. Chem.* **258**, 12574 (1983).
[3] A. L. Bognar, C. Osborne, B. Shane, S. C. Singer, and R. Ferone, *J. Biol. Chem.* **260**, 5625 (1985).
[4] R. Ferone and A. Warskow, *Adv. Exp. Med. Biol.* **163**, 167 (1983).
[5] J. J. McGuire, P. Hsieh, J. K. Coward, and J. R. Bertino, *J. Biol. Chem.* **255**, 5776 (1980).

to a DEAE-cellulose (DE-52, Whatman) column (2 × 0.7 cm), protected by a 3-mm layer of nonionic cellulose, that has been equilibrated with 10 mM Tris buffer (pH 7.5) containing 80 mM NaCl and 30 mM mercaptoethanol. Unreacted glutamate is eluted with the equilibration buffer (3 × 5 ml) and the labeled folate product is eluted with 0.1 N HCl (3 ml).

Dihydrofolate Synthase Assay

The dihydrofolate synthase assay is identical to the folylpolyglutamate synthase assay except dihydropteroate (50 μM) replaces the folate substrate.

Growth of Organisms

Corynebacterium is cultured in medium containing (per liter) glucose (20 g), glycine (4 g), $K_2HPO_4 \cdot 3H_2O$ (0.66 g), KH_2PO_4 (0.5 g), $MgSO_4 \cdot 7H_2O$ (200 mg), $FeSO_4 \cdot 7H_2O$ (40 mg), and thiamine $\cdot$ HCl (1 mg). The organism is grown at 30° with vigorous aeration for 2–4 days ($A_{640 \text{ nm}} \sim 15$). The pH is maintained between 6.7 and 7.0. Addition of low levels of yeast extract improves the growth rate without significantly affecting the yield of enzyme.

Lactobacillus casei (ATCC 7469) is cultured in a complex medium containing 2 nM folic acid.[6] Bacteria are grown at 37° without aeration for 22 hr. The pH of the medium drops from 6.5 to 4.8. Increased cell yield can be obtained by controlling the pH at 6.1, but the specific activity of the enzyme is reduced.

Bacteria are collected by centrifugation, washed several times with ice-cold 0.9% NaCl, and resuspended in 1 volume ice-cold 50 mM potassium phosphate buffer (pH 7), containing 50 mM KCl and the suspension is rapidly frozen as pellets and is stored at −80° until used.

Purification of the *Corynebacterium* Enzyme[1]

A typical purification of the *Corynebacterium* folylpolyglutamate synthase is shown in Table I. The purification obtained is about 7000-fold, and the purity of the final preparation ranges from 95 to 100%, as judged by sodium dodecyl sulfate–gel electrophoresis. Dihydrofolate synthase activity copurifies with folylpolyglutamate synthase at a constant ratio of specific activities.

---

[6] T. Tamura, Y. S. Shin, M. A. Williams, and E. L. R. Stokstad, *Anal. Biochem.* **49,** 517 (1972).

TABLE I
PURIFICATION OF FOLYLPOLYGLUTAMATE SYNTHASE–DIHYDROFOLATE
SYNTHASE FROM *Corynebacterium*

Fraction	Volume (ml)	Protein (mg)	Specific activity[a] (μmol/hr · mg)	Purification (-fold)	Yield (%)
1. Crude extract	420	4100	0.0049	1.0	100
2. 0–50% ammonium sulfate	150	3210	0.0058	1.2	94
3. DEAE-cellulose	10	17.5	0.846	173	74
4. Sephadex G-150	12	1.78	5.76	1176	51
5. Butyl-agarose	10	0.29	24.4	4970	35
6. AMP-agarose	11	0.088	34.3	6994	15

[a] Folylpolyglutamate synthase activity.

Step 1. Crude Extract. All operations are carried out at 0–4°. The bacterial suspension is thawed, diluted with 1 volume of 50 mM potassium phosphate buffer (pH 7) containing 50 mM KCl, and sonicated for 9 min using a Branson W-350 sonicator in the pulsed mode (30 min at 0.3 s/s). The sonicate is centrifuged and the supernatant dialyzed overnight against 50 mM potassium phosphate buffer (pH 7) containing 50 mM KCl to give a crude enzyme extract. The enzyme is unstable in the absence of KCl.

Step 2. Ammonium Sulfate Fractionation. Ammonium sulfate is slowly added to the crude extract to give a 50% saturated solution. The pH should be readjusted to 7, if necessary. After stirring for 1 hr, the suspension is centrifuged and the precipitate is redissolved in 50 mM potassium phosphate buffer (pH 7) containing 50 mM KCl and dialyzed overnight against two changes of the same buffer. A 20–50% ammonium sulfate cut gives similar results.

Step 3. DEAE-Cellulose Chromatography. Fraction 2 enzyme is applied to a DEAE-cellulose (DE-52; Whatman) column (40 × 2.5 cm) that has been equilibrated with 50 mM potassium phosphate buffer (pH 7) containing 50 mM KCl. The column is washed with the equilibration buffer (200 ml) and eluted with a linear gradient (2 liters) of KCl (50–600 mM) in 50 mM potassium phosphate buffer (pH 7). The enzyme elutes at a higher KCl concentration (~430 mM) than the bulk of the applied protein. Active fractions are combined and concentrated by ultrafiltration (PM10 membrane; Amicon) to a volume of ~5 ml. The extract is diluted with 50 mM potassium phosphate buffer (pH 7) to an approximate KCl concentration of 150 mM and reconcentrated by ultrafiltration to a volume of 5 ml.

Step 4. Sephadex Chromatography. Fraction 3 enzyme is applied to a Sephadex G-150 column (90 × 2.6 cm) equilibrated with 50 mM potassium phosphate buffer (pH 7) containing 150 mM KCl. The column is eluted with equilibration buffer and active fractions are combined and concentrated to ~8 ml by ultrafiltration.

Step 5. Butyl-Agarose Chromatography. Fraction 4 is dialyzed for 4 hr against 50 m*M* potassium phosphate buffer, pH 7 (200 volumes), and applied to a butyl-agarose (Miles Biochemicals) column (10 × 1 cm) equilibrated with the same buffer. The column is washed with the equilibration buffer (20 ml) and eluted with a linear gradient (100 ml) of KCl (0–400 m*M*) in buffer. The enzyme elutes after the majority of the protein from this column. Active fractions are combined and concentrated by ultrafiltration to about 5 ml.

Step 6. AMP Affinity Chromatography. Fraction 5 enzyme is dialyzed for 2 hr against two changes of 5 m*M* potassium phosphate buffer, (pH 7) containing 5 mM MgCl$_2$ (100 volumes). *Caution:* the enzyme is unstable under these conditions. The dialyzed extract is applied to an AMP-hexyl-agarose (coupled through N^6 of AMP; P-L Biochemicals) column (3 × 0.7 cm) equilibrated with the dialysis buffer. The column is washed with buffer (10 ml), buffer plus 2 m*M* ATP (5 ml), and buffer (10 ml), and purified enzyme is eluted with buffer plus 100 m*M* KCl (10 ml). Fractions (5 ml) are collected in tubes containing 500 m*M* potassium phosphate buffer (pH 7) plus 500 m*M* KCl (0.5 ml). Additional nonspecific protein is eluted from the column by buffer plus 500 m*M* KCl. The 60% loss of activity observed in this final purification step is due to the marked lability of the enzyme in the presence of low K$^+$ and phosphate concentrations. Additional studies indicate that most of this loss in activity can be prevented by the inclusion of 10 or 30% dimethyl sulfoxide (Me$_2$SO) in the wash and elution buffers.

Me$_2$SO is added to Fraction 6 enzyme to a final concentration of 30% and the extract is divided into 0.5-ml aliquots and stored at −20, −80, or −196°. Enzyme purified up to Step 4 is fully stable for at least 1 year when stored in 50 m*M* potassium phosphate buffer (pH 7) containing 150 m*M* KCl at −20°. The purified preparation is unstable under these conditions and over 90% of the activity is lost in 1 week at −20° when stored in 50 m*M* buffer (pH 7) containing 50 m*M* KCl. Addition of Me$_2$SO (30%) stabilizes the preparation and no loss of activity is observed in 6 months when the preparation is stored at −20 to −196°. The instability observed may be due to binding of the dilute protein preparation to glass or plastic containers.

Purification of the *Lactobacillus* Enzyme[2]

A typical purification of the *L. casei* enzyme is shown in Table II. The purification obtained ranges from 40,000- to 200,000-fold, depending on

TABLE II
PURIFICATION OF FOLYLPOLYGLUTAMATE SYNTHASE FROM *Lactobacillus casei*

Fraction	Volume (ml)	Protein (mg)	Specific activity (μmol/hr · mg)	Purification (-fold)	Yield (%)
1. Crude extract	5100	38760	0.00007	1.0	100
2. DEAE-cellulose	5700	23655	0.00013	1.9	106
3. 30–60% ammonium sulfate	600	14400	0.00018	2.6	87
4. Phosphocellulose P-11	400	330	0.012	170	81
5. Phenyl-agarose	65	9.75	0.23	3300	77
6. Sephadex G-100	14	1.36	0.82	11700	38[a]
7. AMP-agarose	17	0.17	2.9	41000	17
8. 10-Aminodecyl-Sepharose	8.5	0.014	13.8	197000	6.7

[a] A higher yield of enzyme (74%) was recovered from the column but only part of this was used in the subsequent steps.

the growth conditions of the bacterial culture, and the purified preparation is homogeneous as judged by sodium dodecyl sulfate–gel electrophoresis.

Step 1. Crude Extract. All operations are carried out at 0–4°. The bacterial suspension is thawed, diluted with 1 volume of 50 mM potassium phosphate buffer (pH 7) containing 50 mM KCl, and sonicated for 9 min using a Branson W-350 sonicator in the pulsed mode (30 min at 0.3 s/s). The sonicate is centrifuged and the precipitate resuspended in an equal volume of buffer and resonicated. This process is repeated two more times and the four supernatants obtained are combined to give a crude enzyme extract (Fraction 1).

Step 2. DEAE-Cellulose Chromatography. Fraction 1 is applied to a DEAE-cellulose (DE-52, Whatman) column (40 × 7.5 cm) that has been equilibrated with 50 mM potassium phosphate buffer (pH 7) containing 50 mM KCl. The column is washed with the equilibration buffer (700 ml) and the total flow-through is combined to give Fraction 2. This step is used primarily to remove nucleic acids and polynucleotides, but can result in up to a 10-fold purification when lower amounts of protein are applied to the column.

Step 3. Ammonium Sulfate Fractionation. Ammonium sulfate is slowly added to Fraction 2 to give a 30% saturated solution. After stirring for 2 hr, the mixture is centrifuged and the precipitate discarded. Ammonium sulfate is slowly added to the supernatant to give a 60% saturated solution and, after stirring for 2 hr, the mixture is centrifuged. The precipitate is resuspended in 50 mM potassium phosphate buffer, pH 7 (500 ml), and dialyzed overnight against the same buffer (40 volumes).

Step 4. Phosphocellulose Chromatography. The dialyzed extract (Fraction 3) is applied to a phosphocellulose (P-11, Whatman) column (11.3 × 7.5 cm) equilibrated with 50 mM potassium phosphate buffer (pH 7). The column is washed with the equilibration buffer (800 ml) and eluted with a linear gradient (2 liters) of KCl (0–250 mM) in the same buffer. Fractions containing the bulk of the enzyme activity are pooled (Fraction 4). Most of the applied protein does not bind to the phosphocellulose column.

Step 5. Phenyl-Agarose Chromatography. Ammonium sulfate is added to Fraction 4 to give a 70% saturated solution and the mixture is stirred for 90 min and centrifuged. The precipitate is redissolved in 50 mM potassium phosphate buffer (pH 7.5), containing 0.41 M ammonium sulfate (10% saturation; pH after addition of ammonium sulfate) (23 ml) and the solution is applied to a phenyl-agarose (BRL) column (9 × 1 cm) equilibrated with 50 mM potassium phosphate buffer (pH 7.5), containing 0.41 M ammonium sulfate. The column is washed with buffer/0.41 M ammonium sulfate (30 ml) and buffer/0.33 M ammonium sulfate (30 ml) and eluted with a linear gradient (100 ml) of ammonium sulfate (0.33–0 M) in buffer followed by buffer alone. Fractions containing enzyme activity, which elutes in the latter half of the gradient, are pooled.

Step 6. Sephadex Chromatography. Fraction 5 enzyme is precipitated with ammonium sulfate (70% saturation) as described above, resuspended in 50 mM potassium phosphate buffer (pH 7), and applied to a Sephadex G-100 column (115 × 1.5 cm) equilibrated with 50 mM potassium phosphate buffer (pH 7). The column is eluted with the equilibration buffer and fractions containing the majority of the enzyme activity are pooled and dialyzed for 3 hr against two changes of 5 mM potassium phosphate buffer, pH 7 (Fraction 6).

Step 7. AMP Affinity Chromatography. Fraction 6 enzyme is applied to an AMP-hexyl-agarose (coupled through N^6 of AMP; P-L Biochemicals) column (7 × 1 cm) equilibrated with 5 mM potassium phosphate buffer (pH 7), containing $MgCl_2$ (2 mM). The column is washed with the equilibration buffer (20 ml) and eluted with 50 mM potassium phosphate buffer, pH 7 (15 ml) and 50 mM potassium phosphate buffer (pH 7), containing 250 mM KCl (15 ml). The fractions with peak activity are pooled and dialyzed for 3 hr against two changes of 50 mM potassium phosphate buffer, pH 7 (Fraction 7). The enzyme is not eluted from this column by $MgATP^{2-}$, suggesting that binding is due to an ion-exchange or phosphate affinity effect. Under the described conditions, most of the applied protein binds to the column and the enzyme is eluted before the bulk of the protein. Additional studies indicate that enzyme stability and recovery from this column are increased if the enzyme is eluted with 50 mM potassium phosphate buffer (pH 7) containing 10% Me_2SO.

TABLE III
PROPERTIES OF BACTERIAL FOLYLPOLYGLUTAMATE SYNTHASES[a]

	Corynebacterium	Lactobacillus
Purification (-fold)	7000	40,000–200,000
pH optimum	9.5	9.5
Monovalent cation	200 mM K$^+$	200 mM K$^+$
Divalent cation	Mg^{2+}	Mg^{2+}, Mn^{2+}
Reducing agent	None	None
M_r		
SDS–gel electrophoresis	53,000	43,000
(Sephadex)	51,000	43,000
Dihydrofolate synthase activity	Yes	No
Monoglutamate substrate	H$_4$PteGlu	5,10-Methylene-H$_4$PteGlu
Polyglutamate substrate	5,10-Methylene-H$_4$PteGlu	5,10-Methylene-H$_4$PteGlu
Stabilizers	K$^+$, P$_i$, Me$_2$SO	K$^+$, P$_i$, Me$_2$SO

[a] PteGlu, folic acid, pteroylglutamate; H$_4$PteGlu, tetrahydropteroylpoly(γ-glutamate).

Step 8. Aminodecyl-Sepharose Chromatography. Fraction 7 is applied to a 10-aminodecyl-Sepharose column (7 × 1 cm) equilibrated with 50 mM potassium phosphate buffer (pH 7). The bulk of the protein is eluted with the equilibration buffer (30 ml) and the enzyme is specifically eluted with buffer containing 250 mM KCl. Fractions containing enzyme activity are pooled.

The purified protein is unstable, presumably due to the low protein concentration. Me$_2$SO is added to Fraction 8 enzyme to a final concentration of 20% (v/v) and the extract is divided into 1-ml aliquots and stored at −20°. Approximately 50% of the enzyme activity remains after storage for 3 months.

Properties of the Purified Proteins[1,2,7,8]

General properties of the purified proteins are listed in Table III. Both proteins are monomeric and have an absolute requirement for a monovalent cation for activity. K$^+$ (200 mM) gives maximal stimulation of activity while Rb$^+$ and NH$_4^+$ are active but less effective. The monovalent cation may cause a conformational change in the protein.[9] The Mg^{2+}

[7] B. Shane, *in* "Peptide Antibiotics—Biosynthesis and Functions" (H. Kleinkauf and H. Dohren, eds.), p. 353. de Gruyter, Berlin, 1982.

[8] B. Shane and D. J Cichowicz, *Adv. Exp. Med. Biol.* **163,** 149 (1983).

[9] B. Shane, *in* "Chemistry and Biology of Pteridines" (J. A. Blair, ed.), p. 621. de Gruyter, Berlin, 1983.

TABLE IV
FOLATE SUBSTRATES OF FOLYLPOLYGLUTAMATE SYNTHASE–DIHYDROFOLATE SYNTHASE

Values of n in Glu-$_n$[a]	H_2Pte	Relative activity (50 μM substrate)		
		(6S)-H_4PteGlu	(6R)-5,10-Methylene-H_4PteGlu	(6R)-10-Formyl-H_4PteGlu
		Corynebacterium enzyme		
0	20			
1		100	46.3	11.5
2		1.5	33.3	0.4
3		0	0.2	0
		Lactobacillus casei enzyme		
0	0.1			
1		7.3	100	0.7
2		53	246	7.8
3		6.5	79	0.4
4		0.1	2.8	0.1
5		0.1	3.7	0
6		0	2.8	0
7		0	1.7	0

[a] Subscript to Glu, n, denotes glutamate chain length in H_4PteGlu$_n$.

requirement is to generate the MgATP substrate and free ATP is a potent inhibitor of the reaction. The high pH optimum reflects the pK of the amino group of the glutamate substrate. The free amine form is the substrate for the reaction, and the reaction proceeds efficiently at pH 8 provided high levels of glutamate are provided.

One mole of ATP is hydrolyzed to ADP and P$_i$ for every mole of glutamate added to the folate or pteroate substrates.

The folate substrate specificities of the enzymes are shown in Table IV. The *Corynebacterium* protein will use dihydropteroate and tetrahydrofolate as substrates. However, 5,10-methylene derivatives are the preferred polyglutamate substrates. The purified protein will metabolize labeled 5,10-methylenetetrahydrofolate to the tetraglutamate derivative, which is the predominant polyglutamate found *in vivo*. The *Lactobacillus* enzyme will not utilize dihydropteroate as a substrate and preferentially uses 5,10-methylenetetrahydrofolate mono- and polyglutamate derivatives as its substrates. The purified *L. casei* enzyme will metabolize labeled 5,10-methylenetetrahydrofolate primarily to the tetra- and pentaglutamate derivatives and small amounts of longer glutamate chain length derivatives are formed. The major *in vivo* folates in *Lactobacillus* are octa- and nonaglutamate derivatives.

TABLE V

KINETIC CONSTANTS OF FOLYLPOLYGLUTAMATE SYNTHASE–DIHYDROFOLATE SYNTHASE

Compound	K_m (μM)	K_i (K_{is}) (μM)	(V_{max}) (μmol/hr · mg)
Corynebacterium folylpolyglutamate synthetase			
MgATP	18.0	10.0	45.1
(6R,S)-H$_4$PteGlu	2.1	12.7	45.4
L-Glutamate	160	—[a]	41.2
Phosphate	—	750	—
β,γ-Methylene-ATP	—	0.58	—
ATP-γ-S	—	3.0	—
Corynebacterium dihydrofolate synthetase[b]			
MgATP	2.9	—	—
H$_2$PteGlu	<0.4	—	—
L-Glutamate	1380	—	33.1
β,γ-Methylene-ATP	—	1.0	—
ATP-γ-S	—	0.29	—
Lactobacillus casei folylpolyglutamate synthetase			
MgATP	5600	800	64.9
(6R)-5,10-Methylene-H$_4$PteGlu$_2$	2.3	15	67.5
L-Glutamate	423	—	62.3
Phosphate	—	12,500	—
β,γ-Methylene-ATP	—	597	—
ATP-γ-S	—	105	—

[a] Not estimated or not applicable

[b] Approximate values; detailed kinetic analyses not carried out.

The kinetic mechanism for both folylpolyglutamate synthases is ordered ter ter with the nucleotide substrate binding first, followed by the folate and glutamate substrates, and the order of product release is ADP, folate product, and P_i.[2,10] This mechanism precludes the sequential addition of glutamate moieties to enzyme-bound folate. A catalytic mechanism involving the formation of a folyl-γ-glutamylphosphate intermediate followed by a nucleophilic attack by the free amine of glutamate on this mixed anhydride is suggested.

Kinetic constants for the enzymes are shown in Table V. Dihydropteroate is a noncompetitive inhibitor of the *Corynebacterium* folylpolyglutamate synthase activity while tetrahydrofolate does not significantly inhibit the dihydrofolate synthase activity, suggesting that the protein is bifunctional. The affinities of nucleotide analog inhibitors differ for the two

[10] B. Shane, *J. Biol. Chem.* **255,** 5663 (1980).

Corynebacterium synthase activities. As the nucleotide binds to free enzyme, this also suggests that the protein is bifunctional. Kinetic constants for the *Lactobacillus* folylpolyglutamate synthase are similar except for a greatly decreased affinity for the nucleotide substrate, which is also reflected by a decreased affinity for nucleotide inhibitors.

It should be noted that under the conditions of the standard assay, glutamate (250 μM) is not saturating for any of the activities and ATP (5 mM) is not saturating for the *Lactobacillus* enzyme. ATP is the preferred nucleotide substrate of both proteins although other purine triphosphates and UTP will substitute to some extent. ZTP is also a substrate for the *Lactobacillus* enzyme. The glutamate binding site is specific for L-glutamate although L-homocysteate and 4-fluoroglutamate are alternative substrates.

Other Purified Folylpolyglutamate Synthases

The *E. coli* folylpolyglutamate synthase–dihydrofolate synthase gene has recently been cloned and amplified and the protein has been purified to homogeneity.[3] Its properties are similar to the *Corynebacterium* protein. It appears to be bifunctional and utilizes 5,10-methylene derivatives as its polyglutamate substrate. The preferred monoglutamate substrate is 10-formyltetrahydrofolate.[11,12]

The hog liver folylpolyglutamate synthase has been purified to homogeneity.[13] It has a slightly higher K_{cat} value than the bacterial enzymes. In contrast to the bacterial enzymes, the mammalian enzyme has an absolute requirement for a reducing agent, it can utilize oxidized folates as substrates, and the preferred polyglutamate substrates are tetrahydrofolate derivatives.

Acknowledgment

Supported in part by PHS Grant CA 22717 and Research Career Development Award CA 00697 (B.S.) from the National Cancer Institute, Department of Health and Human Services.

[11] M. Masurekar and G. M. Brown, *Biochemistry* **14**, 2424 (1975).
[12] R. Ferone, S. C. Singer, M. H. Hanlon, and S. Roland, *in* "Chemistry and Biology of Pteridines" (J. A. Blair, ed.), p. 585. de Gruyter, Berlin, 1983.
[13] D. J. Cichowicz and B. Shane, in preparation.

[56] Dihydrofolate Reductase: A Coupled Radiometric Assay

By Philip Reyes and Pradipsinh K. Rathod

Dihydrofolate reductase (5,6,7,8-tetrahydrofolate : NADP$^+$ oxidoreductase, EC 1.5.1.3) is the target of a number of clinically useful drugs. While investigating properties of dihydrofolate reductase from malarial parasites grown *in vitro,* we sought an assay procedure for the reductase that was not only more sensitive than the standard spectrophotometric assay but also highly specific and more convenient than previously described radiometric methods. Here, we describe a coupled radiometric assay for dihydrofolate reductase based on the oxidative decarboxylation of D-6-phospho[1-^{14}C]gluconate. We also describe a simple procedure for the synthesis of D-6-phospho[1-^{14}C]gluconate from the readily available D-[1-^{14}C]glucose 6-phosphate.

Preparation of Radiolabeled 6-Phosphogluconate

Synthesis Principle. Glucose 6-phosphate labeled with carbon-14 in carbon atom 1 is readily available from commercial sources. Selective oxidation of the aldehyde function of glucose 6-phosphate to a carboxylate group by bromine[1] yields 6-phosphogluconate.

Reagents

D-[1-^{14}C]Glucose 6-phosphate, disodium salt, 57 μCi/μmol (50 μCi in 0.5 ml of ethanol : water, 7 : 3), New England Nuclear
D-Glucose 6-phosphate, 200 mM
Sodium acetate, 200 mM, pH 5.4

Synthesis Procedure. The following solutions are pipetted into a 10-ml Ehrlenmeyer flask: 175 μl of labeled glucose 6-phosphate (20 μCi), 500 μl of 200 mM glucose 6-phosphate (unlabeled), 1 ml of 200 mM sodium acetate, 825 μl of H$_2$O, and finally, 50 μl of bromine. With indicator paper, the pH of the solution is adjusted to 5 with 6 N NaOH. The progress of the reaction is followed by thin-layer chromatography of the reaction mixture on polyethyleneimine-cellulose (Brinkmann instruments). A 20% (v/v) ethanol solution containing 500 mM NaCl and 40 mM MgCl$_2$ is used as a chromatographic solvent to resolve glucose 6-phosphate from 6-phosphogluconate. Complete oxidation of the glucose 6-phosphate requires a reaction time of 3–5 hr at room temperature (Fig. 1).

[1] B. L. Horecker, this series, Vol. 3, p. 172.

METHODS IN ENZYMOLOGY, VOL. 122

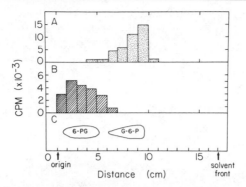

FIG. 1. Thin-layer chromatographic analysis of the conversion of [1-¹⁴C]glucose 6-phosphate to 6-phospho[1-¹⁴C]gluconate. (A) Migration of radioactivity from 5 μl of reaction mixture before the addition of bromine (see text). (B) Migration of radioactivity from 5 μl of reaction mixture after reaction with bromine. (C) Migration of authentic glucose 6-phosphate (G-6-P) and 6-phosphogluconate (6-PG). Sugar phosphates in (C) were visualized as follows: The chromatogram was first sprayed with 4.3% ammonium molybdate in 0.43 M HCl and 13% perchloric acid. The moist chromatogram was then heated for 10 min at 90° and then exposed for 45 sec to a 30-W UV lamp.

Excess bromine is removed by blowing nitrogen over the reaction mixture for 30 min.

Purification of the Radiolabeled Product. Purification of D-6-phospho[1-¹⁴C]gluconate is achieved by ion-exchange chromatography on a 0.8 × 20-cm column of Dowex-1 equilibrated with 10 mM ammonium formate (pH 7.5). The reaction mixture is diluted to 25 ml with H_2O and applied to the column. A 200-ml linear gradient (10 mM to 1 M) of ammonium formate (pH 7.5) is used to develop the column and 2-ml fractions are collected. A 2-μl aliquot from each fraction is analyzed for radioactivity. The radioactive material that elutes in fractions with a conductivity of 21–25 mmho is pooled and lyophilized. The dried product is stored at −20°.

Determination of D-6-Phospho[1-¹⁴C]gluconate Specific Activity

Principle. Because the radiolabeled 6-phosphogluconate is to be used as a substrate in the quantitation of dihydrofolate reductase activity, it is important to determine the exact specific activity of this substrate This is done by using 6-phosphogluconate dehydrogenase to catalyze the conversion of D-6-phospho[1-¹⁴C]gluconate into ribulose 5-phosphate and $^{14}CO_2$.

Decarboxylation occurs with a simultaneous reduction of NADP⁺ to NADPH. The molar stoichiometry between CO_2 release and NADP⁺ reduction is 1 : 1. Thus, one can ascertain the specific activity of the phosphogluconate simply by determining the microcuries of released CO_2 for each mole of NADPH formed. A molar extinction coefficient of 6200 M^{-1} cm^{-1} for NADPH is employed to quantitate absorbance changes at 340 nm.[2]

Reagents

D-6-Phosphogluconate dehydrogenase from baker's yeast (Sigma Chemical Co.), 200 units/ml[3]

Tris–HCl buffer, 1 M, pH 8.0

$MgCl_2$, 500 mM

D-6-Phospho[1-^{14}C]gluconate, 20 mM (see above)

Dithiothreitol, 100 mM

NADP$^+$, 2 mM

Procedure. The following reagents are pipetted into a 3-ml quartz cuvette (1-cm light path): sufficient distilled water to make the final volume 3.0 ml, 1 μl of 6-phosphogluconate dehydrogenase (0.2 unit), 100 μl of 1 M Tris–HCl (pH 8.0), 20 μl of 500 mM $MgCl_2$, 15 μl of 20 mM D-6-phospho[1-^{14}C]gluconate, 33 μl of 100 mM dithiothreitol, and 66 μl of 2 mM NADP$^+$. Immediately after the addition of NADP$^+$, an initial absorbance reading at 340 nm is taken; 150 μl of the reaction mixture is then pipetted into each of 10 flasks set up to measure the release of $^{14}CO_2$ (see below).

At each of several time points, absorbance and released CO_2 are measured in duplicate samples. Under these conditions, CO_2 release and the increase in absorbance at 340 nm show a linear relationship (Fig. 2).

Calculations. From the type of data shown in Fig. 2, one can calculate the specific activity of D-6-phospho[1-^{14}C]gluconate. To do this, two quantities are first calculated from data for a given reaction time t:

$$\frac{\text{microcuries of released } CO_2}{\text{ml of reaction volume}} = \text{cpm} \times \frac{1}{CE} \times \frac{1}{2.22 \times 10^6} \qquad (1)$$

and

$$\frac{\text{micromoles of released } CO_2}{\text{ml of reaction volume}} = \Delta A \times \frac{1}{6.2} \qquad (2)$$

where cpm indicate counts per minute of released CO_2 per ml of reaction volume; CE, liquid scintillation counting efficiency; ΔA, change in absorbance at 340 nm. Thus,

[2] C. Mathews and F. M. Huennekens, *J. Biol. Chem.* **238**, 3436 (1963).
[3] One unit of enzyme activity generates 1 μmol of product per milliliter under standard assay conditions.

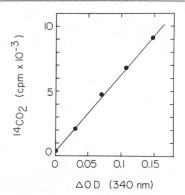

FIG. 2. Stoichiometry of radiolabeled CO_2 release and $NADP^+$ reduction. From Rathod and Reyes,[5] with permission.

$$\text{specific activity of} \atop \text{D-6-phospho[1-}^{14}\text{C]gluconate} \atop (\mu Ci/\mu mol) = \frac{cpm \times 6.2}{\Delta A \times CE \times 2.22 \times 10^6} \qquad (3)$$

For example, the data in Fig. 2 show a change in absorbance of 0.15 at 340 nm and 9000 cpm of released CO_2 per ml of reaction volume, over a period of 20 min. The counting efficiency was 80%. Therefore, the specific activity of 6-phospho[1-^{14}C]gluconate is 0.21 $\mu Ci/\mu mol$. This value agrees very well with the expected specific activity of 0.20 $\mu Ci/\mu mol$ (see synthesis procedure, above).

Comments. Bromine is a volatile and corrosive substance. All manipulations involving this reagent should be performed in a well-vented hood.

Details of the Coupled Assay for Dihydrofolate Reductase

Principle. The reduction of dihydrofolate to tetrahydrofolate by the reductase is accompanied by the oxidation of NADPH to $NADP^+$. In the coupled assay, the newly formed $NADP^+$ then serves as a cosubstrate for 6-phosphogluconate dehydrogenase. This latter enzyme regenerates NADPH as it catalyzes the oxidative decarboxylation of D-6-phospho[1-^{14}C]gluconate to ribulose 5-phosphate and $^{14}CO_2$ (Fig. 3). The labeled CO_2 is trapped in an organic base after acidifying the reaction mixture. There is a 1 : 1 molar ratio between dihydrofolate reduction and CO_2 production.

Reagents

Dihydrofolate (Sigma Chemical Co.), 20 mM in 50 mM Tris buffer (pH 7.5) and 100 mM 2-mercaptoethanol

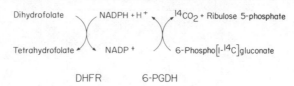

FIG. 3. Coupling of dihydrofolate reduction to the oxidative decarboxylation of 6-phospho[1-^{14}C]gluconate.

MgCl$_2$, 500 mM

Dithiothreitol, 100 mM

Tris–HCl buffer, 1 M, pH 8.0

Bovine serum albumin, 10 mg/ml

6-Phosphogluconate dehydrogenase from baker's yeast (Sigma Chemical Co.), 0.4 unit/ml (see Comments below)

D-6-Phospho[1-^{14}C]gluconate (0.21 μCi/μmol), 20 mM

NADPH, 1 mM

Sample of dihydrofolate reductase to be assayed

The reliability of the coupled assay was confirmed by using a partially purified preparation of dihydrofolate reductase from bovine liver (Sigma Chemical Co.; see Fig. 4).

Procedure. The following reagents are added to a 10-ml Ehrlenmeyer flask placed in an ice-water bath: 0.5 ml of 6-phosphogluconate dehydrogenase (0.2 units), 15 μl of 20 mM dihydrofolate, 100 μl of 500 mM MgCl$_2$, 100 μl of 100 mM dithiothreitol, 100 μl of 1 M Tris–HCl (pH 8.0), 100 μl of bovine serum albumin solution, 70 μl of 20 mM radiolabeled 6-phosphogluconate, the dihydrofolate reductase to be assayed, 1.5 ml of distilled water, and 100 μl of 1 mM NADPH. The reaction vessel is immediately stoppered with a gas-tight serum cap from which is suspended a plastic cup containing 100 μl of methylbenzethonium hydroxide (serum stoppers and plastic cups can be purchased from Kontes Glass Co.). The flask is then incubated in a 37° water bath. To terminate the reaction, the flask is returned to an ice-water bath and 0.25 ml of 6 N perchloric acid is injected through the serum stopper. After 15 min of equilibration at room temperature, each plastic cup is transferred to a glass counting vial filled with 10 ml of 0.4% Omnifluor (New England Nuclear) in toluene (w/v). The radioactivity is determined in a scintillation spectrometer. Observed counts per minute can be converted to moles of released CO$_2$ by knowing the counting efficiency of the scintillation spectrometric system and the specific activity of the D-6-phospho[1-^{14}C]gluconate. Because of the 1 : 1 molar relationship between dihydrofolate reduction and CO$_2$ release (Fig. 3), the quantity of labeled CO$_2$ is a direct

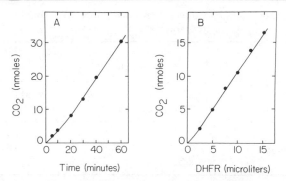

FIG. 4. Coupled dihydrofolate reductase assay. (A) $^{14}CO_2$ release as a function of time. (B) $^{14}CO_2$ release as a function of the concentration of dihydrofolate reductase. From Rathod and Reyes,[5] with permission.

measure of dihydrofolate reductase activity. Figure 4 shows typical plots of CO_2 release as a function of time and as a function of the amount of dihydrofolate reductase.

Comments. As supplied, the commercial preparation of 6-phosphogluconate dehydrogenase is usually not suitable for use in the coupled assay. In order to maximize the activity and stability of this enzyme, it is dissolved in a buffer containing 1 mg/ml of bovine serum albumin, 50 mM Tris–HCl (pH 8.0), 10 mM MgCl$_2$, and 1 mM dithiothreitol. The resulting enzyme solution is then dialyzed against a buffer containing 50 mM Tris–HCl (pH 8.0) and 1 mM dithiothreitol in order to remove ammonium sulfate. The dialyzed 6-phosphogluconate dehydrogenase solution is used directly in the coupled dihydrofolate reductase assay.

It should also be noted that the coupled reductase assay always shows a lag period prior to achieving a uniform rate of product formation (Fig. 4). The degree of this lag period is determined by, among other things, the amount of 6-phosphogluconate dehydrogenase added to the coupled assay system.[4] When 0.5 unit (or more) of the dehydrogenase is used in the coupled assay, the lag period is essentially abolished.

Finally, even though a zero time point or a reaction mixture lacking dihydrofolate is usually used as a control, omission of either 6-phosphogluconate dehydrogenase, dihydrofolate reductase, NADPH, or dihydrofolate from the complete reaction mixture also results in no net oxidative decarboxylation of D-6-phospho[1-^{14}C]gluconate.[5] Addition of 10 μM

---

[4] F. Garcia-Carmona, F. Garcia-Canovas, and J. A. Lozano, *Anal. Biochem.* **113**, 286 (1981).

[5] P. K. Rathod and P. Reyes, *Anal. Biochem.* **133**, 425 (1983).

DIRECT COMPARISON OF COUPLED AND SPECTROPHOTOMETRIC ASSAYS FOR
DIHYDROFOLATE REDUCTASE[a]

	Enzyme activity (nmol/hr · ml of reaction mixture)	
Enzyme source	Measured by coupled assay	Measured by spectrophotometric assay
Bovine liver	23.8	21.7
Plasmodium falciparum extract	11.2	10.8

[a] From Rathod and Reyes.[5]

methotrexate in the assay system also prevents release of radiolabeled CO_2.[6]

Comparison of the Coupled Radiometric Assay with Other Assay Procedures for Dihydrofolate Reductase

A direct comparison of this radiometric assay with the standard spectrophotometric assay for dihydrofolate reductase shows excellent agreement. This is true both with a purified enzyme preparation from bovine liver and with a crude extract of the human malarial parasite, *Plasmodium falciparum* (see table).

Comments. The coupled radiometric assay for dihydrofolate reductase described here incorporates several features which make it superior to the spectrophotometric assay. First, it is about 50 times more sensitive than the spectrophotometric assay. Second, relatively large numbers of samples can be run simultaneously, in contrast to a maximum of only two to four in a standard spectrophotometer. Third, compared to the spectrophotometric assay, product formation remains linear for longer periods of time.[6] A practical advantage of this feature is that a single end-point assay can be used to determine a wide range of initial rates with ease and accuracy.

The present coupled assay also offers a number of advantages over previously described radiometric assays for dihydrofolate reductase. As noted above, the present assay can easily accommodate multiple samples. Some of the earlier methods, however, require a separate column or paper chromatographic strip for each sample.[7,8] Furthermore, the coupled

[6] P. K. Rathod and P. Reyes, unpublished observations.
[7] S. P. Rothenberg, *Anal. Biochem.* **13**, 530 (1965).
[8] S. F. Zakrzewski and J. J. Himberg, *Anal. Biochem.* **40**, 336 (1971).

assay is inherently more specific than earlier radiometric methods because the product being measured (radiolabeled CO_2) is released in a biospecific and stoichiometric manner. On the other hand, some of the earlier procedures either relied on differential precipitation to separate folate or dihydrofolate from tetrahydrofolate[9-12] or relied on the release of tritiated water by chemical oxidation followed by absorption of degradation products and residual substrate onto charcoal.[13] The main disadvantage of batch techniques such as these is that they tend to be less reproducible. Finally, the coupled assay described here does not suffer from the disadvantage of depending on the synthesis and use of radiolabeled dihydrofolate, a rather unstable compound.

[9] S. P. Rothenberg, *Anal. Biochem.* **16,** 176 (1966).
[10] J. W. Littlefield, *Proc. Natl. Acad. Sci. U.S.A.* **62,** 88 (1969).
[11] R. Hayman, R. McGready, and M. B. Van Der Weyden, *Anal. Biochem.* **87,** 460 (1978).
[12] S. P. Rotherberg, M. P. Iqbal, and M. Da Costa, *Anal. Biochem.* **103,** 152 (1980).
[13] D. Roberts, *Biochemistry* **5,** 3549 (1966).

[57] Purification of Folate Oligoglutamate : Amino Acid Transpeptidase

By Tom Brody and E. L. R. Stokstad

Assay Method

Folate oligoglutamate : amino acid transpeptidase (EC 2.3.2.–) and an associated folate monoglutamate : amino acid ligase activity were detected in rat liver extracts. The substrates for the transpeptidase reaction are a ^{14}C-labeled amino acid such as [^{14}C]glutamic acid and a folate polyglutamate such as $H_4PteGlu_n$ or $PteGlu_n$. (The number of γ-glutamyl residues is indicated by n.) p-Aminobenzoyloligo(γ-Glu_n) also serves as a substrate. The substrates used to measure the ligase reaction are [^{14}C]glutamic acid and a folate monoglutamate such as $H_4PteGlu$.

The ^{14}C-labeled products formed during incubations of enzyme with $PteGlu_2$ or $PteGlu_7$ (transpeptidase reaction) or with $H_4PteGlu$ (ligase) were identified by chromatography on DEAE-cellulose,[1] Sephadex G-25,[2]

[1] T. Brody and E. L. R. Stokstad, *in* "Folic Acid in Neurology, Psychiatry and Internal Medicine" (M. I. Botez and E. H. Reynolds, eds.), p. 55. Raven Press, New York, 1979.
[2] T. Brody and E. L. R. Stokstad, *J. Biol. Chem.* **257,** 14271 (1982).

and BioGel P-4.[2] Identification was also based on release of incorporated radioactivity with exposure of the labeled folate products to γ-glutamyl hydrolase at pH 4.7.[1,2] When transpeptidase reactions were conducted with either PteGlu$_2$ or PteGlu$_7$ the radioactive product was identified in both cases as the folate diglutamate, PteGlu-γ-[^{14}C]Glu. The nonradioactive products formed in incubations containing PteGlu$_2$ or PteGlu$_7$ have not yet been isolated. These products are proposed to be glutamic acid and oligo(γ-Glu$_6$), respectively. The radioactive product of the ligase reaction was identified as the folate diglutamate, H$_4$PteGlu-γ-[^{14}C]Glu. The reduced state shown here is a reflection of the reduced state of the substrate rather than of a requirement for expression of ligase activity. When ^{14}C-labeled amino acids other than [^{14}C]glutamate are used as substrates enzyme activity should be measured with BioGel P-2 rather than the Dowex resin.[2]

Measurement of radioactivity incorporated into the folate substrate is dependent on the cleavage of the folate product to the respective labeled p-aminobenzoyloligoglutamate followed by conversion to the respective ^{14}C-labeled azo dye. The azo dye is easily purified from unreacted [^{14}C]glutamic acid using Dowex resin followed by scintillation counting. The products of both transpeptidase and ligase reactions were also determined directly by purification of ^{14}C-labeled products on DEAE-cellulose followed by scintillation counting. This method is similar to that of Taylor and Hanna.[3] The assay method using DEAE-cellulose is a bit more convenient than the azo dye method. However, the azo dye method has the advantage of lower backgrounds or blanks. This advantage is needed for detection of the low levels of transpeptidase and ligase activities present in rat liver.

Cleavage of tetrahydrofolates such as H$_4$PteGlu or H$_4$PteGlu$_2$ to p-aminobenzoylGlu$_n$ requires storage overnight in acid, such as HCl or trichloroacetic acid. Prolonged exposure to acid is needed when thiol reagents are present as components of the enzyme assay mixture. More rapid cleavage may be effected by removing the thiol by adding HgCl$_2$. However, the HgCl$_2$ may interfere with subsequent steps by binding directly to the azo bond or perhaps by saturating the binding sites on the Dowex resin. Cleavage of nonreduced folates such as PteGlu or PteGlu$_2$ requires brief exposure to zinc dust in acid with gelatin. Generally, the use of trichloroacetic acid should be avoided here as it converts the zinc dust to zinc ions which may interfere with Dowex chromatography.

[3] R. T. Taylor and M. L. Hanna, *Arch. Biochem. Biophys.* **181,** 331 (1977).

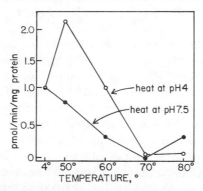

FIG. 1. Specific activity versus heat treatment. The specific activity of transpeptidase in the unheated 5-fold purified solution was 1.0 pmol/min · mg protein. Heat treatment at pH 4 (○) and pH 7.5 (●).

Purification Procedure

The purification scheme and the standard assay methods have been published.[2] The copurification of transpeptidase and ligase activities was achieved by precipitating inactive proteins with ammonium sulfate at pH 7.5, heat treatment at pH 4.0, followed by three types of column chromatography.[2] Transpeptidase activity is quite stable to storage for 1–2 weeks at 4° when of lesser purity. However, the 39,000-fold purified enzyme, as well as the γ-glutamyl hydrolase activity detected in the same preparation, lost nearly all activity with storage for 24 hr at 4°. Details of technics used in the first two steps of enzyme purification are as follows.

Heat Treatment. Rat liver was homogenized in three volumes of 10 mM Tris base, 10 mM 2-mercaptoethanol at 4° and centrifuged. The supernatant was adjusted to pH 4.0 at 4° using a small amount of ice-cold 1 M acetic acid and rapid stirring. Precipitated material was removed by centrifugation yielding a 5-fold increase in specific activity. A portion of the supernatant was then adjusted to pH 7.5 with Tris base powder. Aliquots of material at pH 4.0 or 7.5 were then warmed in water baths at the indicated temperatures (Fig. 1) for 10 min. The heat-treated extracts were then cooled, centrifuged to remove precipitated protein, adjusted to pH 7.5 where appropriate, and centrifuged again.

Results from assays of dialyzed protein were as follows. The initial adjustment of protein to pH 4 at 4° yielded a 5-fold increase in purity. When the protein was warmed at 50° and pH 4, an additional 2-fold increase resulted (Fig. 1). The recovery of enzyme was over 90% with heating at 50° for 10–20 min at pH 4. Warming at 60° at pH 4 or 7.5 destroyed activity.

Ammonium Sulfate Fractionation. Rat liver was homogenized in three volumes of 50 mM sodium phosphate (pH 7.5), 10 mM 2-mercaptoethanol at 4°. After centrifugation, 30-ml aliquots of the supernatant were brought to the indicated (Fig. 2, A and B) percent saturation of ammonium sulfate[4] over the course of 30–45 min. Solutions were centrifuged and the precipitates and supernatants dialyzed for determination of specific activities using 100 μM [^{14}C]glutamic acid and 10 μM PteGlu$_2$ or 1000 μM *dl*-H$_4$PteGlu as the substrates. The greatest increases in specific activities occurred with use of 65% ammonium sulfate and were 15-fold for transpeptidase (Fig. 2A) and 13-fold for the ligase reaction (Fig. 2B). In both cases the recovery of activity was over 90%. The recoveries of activity in the precipitates were quite low.

Activity versus PteGlu$_n$ Concentration

An unusual dependence on folate concentration was found when incubation mixtures contained 0 to ~600 μM PteGlu$_n$, 60 μg of 370-fold purified protein, with other conditions being standard. The results show that maximal activity occurred in the presence of 100–200 μM folate polyglutamate with lesser activities at higher concentrations (Fig. 3). At the highest activity values the amount of product formed was about 2 nmol, indicating conversion of 2% of the total folate substrate. After terminating reaction mixtures, the complete conversion to the corresponding azo dye derivatives was checked and confirmed by measuring the absorbance of 556 nm. The activity of the folate ligase reaction in the presence of 0–1000 μM *dl*-H$_4$PteGlu was hyperbolic.[5]

Comments

The physiological function of folate transpeptidase activity, if any, is not known. The major forms of folates in rat liver are the folate penta- and hexaglutamates. Catalytic action of transpeptidase on these cellular folates might be expected to produce a variety of compounds such as the reduced forms of pteroylglutamylglutamate, pteroylglutamylmethionine, and pteroylglutamylglutamine. The latter two compounds would not be retained by the liver and would enter the bloodstream and perhaps be excreted. It was proposed by the authors that folate transpeptidase is

---

[4] F. di Jeso, *J. Biol. Chem.* **243**, 2022 (1968).
[5] T. Brody, Doctoral Dissertation, p. 99. University of California at Berkeley (1980).

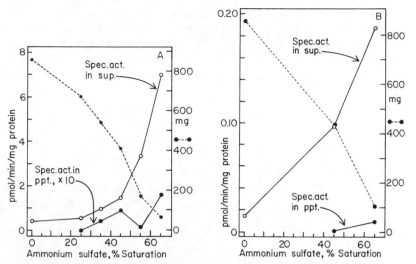

FIG. 2. Specific activity versus ammonium sulfate. Prior to addition of ammonium sulfate the specific activities of transpeptidase (A) and ligase (B) were 0.4 and 0.08 pmol/min · mg protein, respectively. After fractionation of protein with ammonium sulfate and dialysis of soluble and precipitated material, the specific activities in the supernatants (O—O), precipitates (●—●) as well as the total protein remaining in the supernatants (●--●) were measured.

identical with γ-glutamyl hydrolase, an enzyme localized in the lysosomal fraction. This proposal was based on the fact that the most highly purified preparation of transpeptidase had γ-glutamyl hydrolase activity[2] and because in nature peptidase, ligase, and transpeptidase activities are all

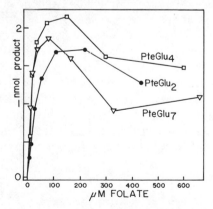

FIG. 3. Activity versus pteroyl di-, tetra-, and heptaglutamate concentration. The specific activity of transpeptidase determined with 10 μM PteGlu$_2$ was 458 pmol/min · mg protein.

commonly catalyzed by the same enzyme. Preliminary results revealed that the γ-glutamyl hydrolase activity of the purest preparations of transpeptidase displayed both neutral and acidic pH optima.

It might be added that 10 mM dithiothreitol should replace 20 mM 2-mercaptoethanol as a standard component in all future transpeptidase assays.[2]

[58] Methylenetetrahydrofolate Reductase from Pig Liver

By Rowena G. Matthews

Although mammals do not synthesize the carbon skeleton of methionine or its immediate precursor, homocysteine, they are able to regenerate methionine from homocysteine. The methyl group for this methylation comes from 5-methyltetrahydrofolate, which is in turn formed by reduction of 5,10-methylenetetrahydrofolate. Substantial quantities of methionine are converted to S-adenosylmethionine, which serves as a methyl donor for a wide variety of biological methylation reactions, and the demethylated product, S-adenosylhomocysteine, is hydrolyzed to form adenosine and homocysteine.

Methylenetetrahydrofolate reductase (NADPH–CH$_2$–H$_4$folate oxidoreductase, EC 1.5.1.20) catalyzes the reaction which commits tetrahydrofolate-bound one-carbon units to the pathway leading to homocysteine regeneration, and its activity is regulated by S-adenosylmethionine which acts as an allosteric inhibitor of the physiological reaction. These relationships are summarized in Scheme 1.[1]

Methylenetetrahydrofolate reductase from pig liver was originally characterized by Kutzbach and Stokstad[2] and was subsequently purified to homogeneity by Daubner and Matthews.[3] The mammalian enzyme differs quite substantially in its properties from the bacterial enzyme (EC 1.7.99.5). Methylenetetrahydrofolate reductase has been partially purified from *E. coli*.[4] Enzyme from this source utilizes reduced FAD as the source of reducing equivalents for the reduction of methylene- to methyl-

[1] J. Ross, J. Green, C. M. Baugh, R. E. MacKenzie, and R. G. Matthews, *Biochemistry* **23**, 1796 (1984).
[2] C. Kutzbach and E. L. R. Stokstad, *Biochim. Biophys. Acta* **250**, 459 (1971).
[3] S. C. Daubner and R. G. Matthews, *J. Biol. Chem.* **257**, 140 (1982).
[4] H. M. Katzen and J. M. Buchanan, *J. Biol. Chem.* **240**, 823 (1965).

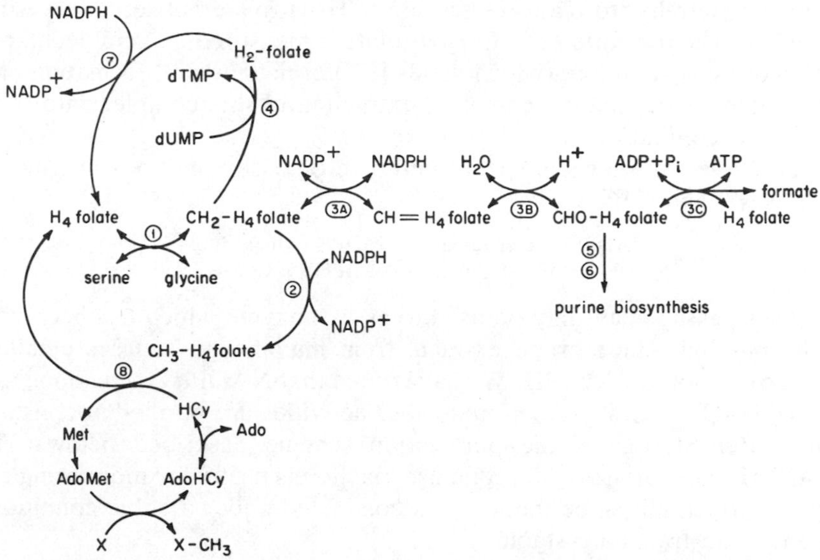

SCHEME 1. Outline of the major folate-dependent pathways in mammalian cells. The enzymes involved are (1) serine hydroxymethyltransferase, (2) methylenetetrahydrofolate reductase, (3) the trifunctional enzyme with methlenetetrahydrofolate dehydrogenase (3A), methenyltetrahydrofolate cyclohydrolase (3B), and formyltetrahydrofolate synthetase (3C) activities, (4) thymidylate synthase, (5) GAR transformylase, (6) AICAR transformylase, (7) dihydrofolate reductase, and (8) methyltetrahydrofolate–homocysteine methyltransferase (methionine synthase). [Reprinted, with permission, from ref. 1. Copyright (1984) American Chemical Society.]

H_4folate and does not oxidize NADH or NADPH. Methylenetetrahydro-folate reductase has also been purified to homogeneity from *Clostridium formicoaceticum*.[5] The enzyme from this source contains FAD and iron–sulfur centers, and utilizes reduced ferredoxin or $FADH_2$ as a source of reducing equivalents. It, too, shows no activity with pyridine nucleotides.

Assay Methods

Principle. Mammalian enzyme can be assayed by three methods. The first method, shown in Eq. (1),

$$CH_3\text{-}H_4\text{folate} + \text{menadione} \rightarrow CH_2\text{-}H_4\text{folate} + \text{menadiol} \tag{1}$$

is suitable for assays of crude extracts and can also be used for the assay of the enzyme from bacterial sources. This assay is highly specific for

[5] J. E. Clark and L. G. Ljungdahl, *J. Biol. Chem.* **259**, 10845 (1984).

methylenetetrahydrofolate reductase. $^{14}CH_3$-H$_4$folate is used as the substrate and the formation of $^{14}CH_2$-H$_4$folate is measured by acid decomposition of $^{14}CH_2$-H$_4$folate (which yields [^{14}C]formaldehyde), formation of a formaldehyde–dimedone complex, extraction of this complex into toluene, and scintillation counting.

Principles of alternate spectrophotometric assay methods are shown in Eqs. (2) and (3).

$$\text{NADPH} + \text{menadione} \rightarrow \text{NADPH} + \text{menadiol} \tag{2}$$
$$\text{NADPH} + \text{CH}_2\text{-H}_4\text{folate} \rightarrow \text{NADPH}^+ + \text{CH}_3\text{-H}_4\text{folate} \tag{3}$$

These assays can only be used to assay enzyme which has been partially purified, since crude extracts from mammalian sources catalyze rapid oxidation of NADPH. We have found that NADPH–menadione and NADPH–CH$_2$-H$_4$folate oxidoreductase activities are purified at constant ratio after Step 2 of the purification scheme described below. The NADPH–menadione oxidoreductase reaction is useful for monitoring enzyme purification since the assay is conducted under aerobic conditions and the substrates are stable.

CH$_3$–H$_4$folate-Menadione Oxidoreductase Assay

Reagents

A. 4.2 mM (6R,S)-[$methyl$-^{14}C]H$_4$folate (2000 dpm/nmol)
B. Saturated solution of menadione in 20% methanol/80% H$_2$O
C. 100 mM potassium phosphate buffer (pH 6.7), 0.6 mM EDTA, 0.67% (w/v) bovine serum albumin.
D. Dimedone, 3 mg/ml, in 1 M sodium acetate buffer (pH 4.5)
E. Toluene, analytical reagent grade

Procedure. The reaction mixture in a total volume of 1 ml consists of 0.5 ml phosphate buffer (C), 0.06 ml CH$_3$-H$_4$folate, 0.1 ml menadione. The solution is incubated for 2 min at 37° and the reaction is then initiated by the addition of enzyme. After incubation for 15 min at 37°, the reaction is terminated by addition of 0.3 ml of dimedone solution and by placing the tube in a heater block at 100° for 2 min.

The tube is cooled on ice, 3 ml of toluene is added, and the contents mixed by vortexing. The toluene layer is separated by centrifugation at 1500 g in a clinical desk-top centrifuge for 5 min. A 1-ml aliquot of the toluene (upper) phase is added to 8 ml aqueous scintillation fluid and counted. All assays are corrected for the counts found in a sample to which enzyme was added after quenching with dimedone and heating for 1 min.

NADPH–Menadione Oxidoreductase Assay

Reagents

A. 10 mM NADPH in 0.02 M unneutralized Tris (base)
B. Saturated solution of menadione in 20% methanol/80% H_2O
C. Potassium phosphate buffer, 100 mM (pH 7.2), 0.6 mM in EDTA

Procedure. The reaction mixture in a total volume of 2 ml consists of 1.0 ml phosphate buffer, 0.025 ml NADPH, and enzyme. This mixture is incubated at 25° for 5 min in a 3-ml cuvette, and then the reaction is initiated by addition of 0.2 ml menadione. The reaction is monitored by measuring the change in absorbance at 343 nm, where menadione and menadiol are isosbestic. The extinction coefficient associated with the oxidation of NADPH is 6220 M^{-1} cm^{-1} at 343 nm.[3]

NADPH–CH₂–H₄folate Oxidoreductase Assay

Reagents

A. 10 mM NADPH in 0.02 M unneutralized Tris (base)
B. Potassium phosphate buffer, 100 mM (pH 7.2), 0.3 mM in EDTA, and 100 mM in formaldehyde
C. 10 mM (6R,S)-H₄folate, in 0.5 M 2-mercaptoethanol/0.25 M tri- ethanolamine chloride buffer (pH 7)

Procedure. (6R,S)-H₄Folate is prepared by catalytic hydrogenation of a neutral aqueous solution of folic acid as described by Blakley,[6] and is purified on DEAE-cellulose-52 by elution with 0.25 M triethanolamine- chloride buffer (pH 7) and 0.5 M 2-mercaptoethanol. Stock solutions of H₄folate were stored under nitrogen at −70°. CH₂-H₄Folate is generated nonenzymatically in the assay buffer during preincubation. The reaction mixture consisting of 0.025 ml NADPH and 1 ml of potassium phosphate/ formaldehyde in a total volume of 2 ml is equilibrated with nitrogen in a 3- ml cuvette by passage of a fine stream of nitrogen gas through the solution for 5 min at 25°. The cuvette is sealed with Parafilm, and 0.02 ml of H₄folate is added to the solution by injection through the Parafilm using a Hamilton syringe. After 5 min at 25°, the reaction is initiated by injection of enzyme and the reaction is monitored at 340 nm in a spectrophotome- ter. An extinction coefficient of 6230 M^{-1} cm^{-1} at 340 nm is used to calculate the extent of product formation. The highest enzyme activities are observed if the enzyme solutions are preincubated aerobically in 125 μM NADPH for 10 min prior to addition to the assays.

[6] R. L. Blakley, *Biochem. J.* **65**, 331 (1957).

Purification of Methylenetetrahydrofolate Reductase from Pig Liver[3]

Methylenetetrahydrofolate reductase was purified in four steps consisting of (1) homogenization and centrifugation followed by (2) chromatography on DEAE-52, (3) Matrex Gel Blue A (Amicon), and (4) DEAE-Sephadex A-50. The following procedure is described for preparation of enzyme from 3 kg of pig liver and is scaled according to the amount of liver actually used.

Preparation of Crude Extract. Frozen pig liver was thawed under cold running water for 45 min, and then cut into cubes. Homogenization was performed in a Waring blender, using 4 liters of buffer containing the following protease inhibitors: phenylmethylsulfonyl fluoride, 100 mg; N-α-p-tosyl-L-lysine chloromethylketone, 12 mg; soybean trypsin inhibitor, 50 mg; and aprotinin, 5 ml. The homogenizing buffer was 50 mM potassium phosphate (pH 5.9), 0.3 mM EDTA, and phenylmethylsulfonyl fluoride was added from a stock solution in isopropyl alcohol (20 mg/ml) immediately prior to homogenization. The homogenate was centrifuged at 20,000 g for 60 min. The supernatant was filtered through cheesecloth, and its pH was adjusted to 7.0 with 1 M NH$_4$OH.

Fractionation on DEAE-52. The supernatant from Step 1 was adsorbed onto DEAE-52 which had previously been equilibrated with 50 mM phosphate buffer (pH 7.2), 0.3 mM EDTA. Settled DEAE (720 ml) was added to the supernatant and the slurry was stirred for 30 min at 4° and then filtered on a Büchner funnel and the DEAE was washed twice with 1-liter portions of 50 mM phosphate buffer (pH 7.2), 0.3 mM EDTA. The DEAE cake was extracted twice with 1-liter portions of 0.3 M phosphate buffer (pH 7.2), 0.3 mM EDTA. Each extraction was performed by stirring the slurry for 30 min at 4° and then collecting the filtrate by vacuum filtration. The two filtrates were combined and utilized for the next step.

Fractionation on Matrex Gel Blue A. Amicon Matrex Gel Blue A, 70 ml of settled beads which had previously been equilibrated with 50 mM potassium phosphate buffer (pH 7.2), 0.3 mM EDTA, was added to the pooled filtrates. The suspension was stirred overnight at 4° and then the beads were removed from this suspension by vacuum filtration on a Büchner funnel. Care was taken to keep the beads from going dry. The beads were slurried in 10 mM Tris buffer (pH 7.0), and then filtered and resuspended in Tris buffer several times to remove residual phosphate buffer. The beads were then stirred for 30 min at 4° with 70 ml of 5 mM NADH in 10 mM Tris (pH 7.0). This treatment removes contaminating proteins without releasing methylenetetrahydrofolate reductase. The beads were collected by filtration and washed with Tris buffer, and then were suspended in a small volume of 50 mM phosphate buffer (pH 7.2), 0.3 mM

EDTA, 10% glycerol. A column (110 × 28 mm) was poured, and the beads were rinsed with 1 bed volume of 0.5 M NaCl in 50 mM phosphate buffer (pH 6.7), 0.3 mM EDTA, 10% glycerol. (The solution is prepared using pH 7.2 buffer, and the pH of 6.7 is measured in the presence of 0.5 M NaCl.) Finally, the enzyme was released from the column with 2 M NaCl in 50 mM phosphate, 0.3 mM EDTA, 10% glycerol (pH 6.1 in the presence of the NaCl). Active fractions were pooled and diluted 8-fold with 50 mM phosphate buffer (pH 7.2), 0.3 mM EDTA, 10% glycerol.

Chromatography on DEAE-Sephadex. The diluted fractions from Step 3 were applied to three columns (50 × 10 mm) of DEAE-Sephadex A-50 which had previously been equilibrated with 50 mM phosphate buffer (pH 6.9), 0.3 mM EDTA, 0.25 M NaCl, 10% glycerol. The columns were loaded at room temperature with a pressure head of about 1 m. As loading progressed, sharp, bright yellow bands formed at the top of the columns. On completion of loading these yellow bands were transferred with a disposable pipet to the top of a column (150 × 10 mm) of DEAE-Sephadex A-50 equilibrated with the same buffer. After washing with 1 bed volume of buffer, the enzyme was eluted with a 200-ml linear gradient of 0.25–1.0 M NaCl in 50 mM phosphate buffer (initially pH 6.9 in the presence of 0.25 M NaCl) containing 0.3 mM EDTA and 10% glycerol. The enzyme usually eluted at about 0.37 M NaCl. Active fractions were concentrated and desalted in an Amicon ultrafiltration cell using an XM-50 membrane and 50 mM phosphate buffer (pH 7.2), 0.3 mM EDTA, 10% glycerol. Visible electronic absorbance spectra of active fractions eluting from the DEAE-Sephadex column were recorded and only fractions with 380 nm/450 nm absorbance ratios of 1.0 or less were pooled.

The purification scheme is summarized in Table I. At the completion of the purification, the purity of the enzyme is assessed by polyacrylamide gel electrophoresis in the presence of sodium dodecyl sulfate. The enzyme is usually homogeneous,[7] showing a single band with an apparent subunit molecular weight of 77,300.

Properties of Methylenetetrahydrofolate Reductase from Pig Liver

Stability. The activity of purified enzyme is stable for several months upon storage at $-70°$, provided that 10% glycerol is present.

Physical Properties. Polyacrylamide gel electrophoresis in the presence of sodium dodecyl sulfate results in a single band with an apparent molecular weight of 77,300.[3] Amino acid analysis of an aliquot of enzyme of known flavin content gives a molecular weight per flavin of 74,500,

[7] M. A. Vanoni, D. P. Ballou, and R. G. Matthews, *J. Biol. Chem.* **258**, 11510 (1983).

TABLE I

PURIFICATION OF METHYLENETETRAHYDROFOLATE REDUCTASE FROM PIG LIVER[a]

| Step | Volume (ml) | Activity | | Protein[c] | Specific activity | Purification (-fold) | Yield (%) |
		NADPH–menadione oxidoreductase (μmol/min)[b]	CH$_3$–H$_4$folate– menadione oxido- reductase (μmol/min)[b]				
1. Supernatant from 11,000 g centrifugation	3960	ND[d]	160	263,000	0.00061	1	100
2. DEAE–52 chromatography	1980	682	159	14,200	0.0112	18	99
3. Chromatography on Amicon Matrex Gel Blue A	133	345	79	217	0.332	540	49
4. Chromatography on DEAE-Sephadex A-50	8.5	118	27	1.4	19.4	32,000	17

[a] The starting material was 3.0 kg of pig liver.
[b] Units are at 25°.
[c] Protein determinations were performed using a Coomassie Blue dye-binding method (Bio-Rad protein assay), and were compared with standard curves based on reaction of protein assay mix with bovine serum albumin. Determination of the protein concentration of methylenetetrahydrofolate reductase by amino acid analysis gave a value 76.5% of that obtained using the Bio-Rad protein assay. Therefore, a factor of 0.765 was used to correct values obtained by the Bio-Rad protein assay for the sample of enzyme after Step 4.
[d] ND, Not determined.

TABLE II
AMINO ACID COMPOSITION OF
METHYLENETETRAHYDROFOLATE REDUCTASE[a]

Amino acid	Residues/mol FAD[b]
Aspartic acid	58.1
Threonine	22.5
Serine	34.6
Glutamic acid	93.9
Proline	48.4
Glycine	67.7
Alanine	36.8
Valine	39.1
Isoleucine	38.7
Methionine	19.9
Leucine	75.1
Tyrosine	15.0
Phenylalanine	30.5
Lysine	45.2
Histidine	18.4
Arginine	30.9
Tryptophan	ND[b]
Half-cystine	ND[b]

[a] Reprinted, with permission, from ref. 3.
[b] These values are the means of duplicate analysis after 24 hr of hydrolysis. ND, Not determined.

excluding contributions to the molecular weight from tryptophan, half-cystine and cysteine (Table II). The flavin has been identified as FAD, and the extinction coefficient of the noncovalently bound flavin is 12,100 M^{-1} cm^{-1} at 450 nm.[3] The molecular mass of the native enzyme has been determined by scanning transmission electron microscopy[8] and was found to be 136 ± 29 kDa. In the presence of the allosteric inhibitor, the molecular mass of the enzyme was 144 ± 29 kDa. Thus both active and inhibited native enzymes appear to consist of dimers of identical 77,300-Da subunits, each containing noncovalently bound FAD.

Kinetic Properties. Methylenetetrahydrofolate reductase exhibits ping-pong kinetics in the catalysis of the reactions shown in Eqs. (1), (2), and (3). At pH 7.2, the kinetic parameters for NADPH–CH$_2$-H$_4$folate oxidoreductase activity are K_m (NADPH), 16 μM; K_m (CH$_2$-H$_4$folate), 88

[8] R. G. Matthews, M. A. Vanoni, J. F. Hainfeld, and J. Wall, *J. Biol. Chem.* **259** 11647 (1984).

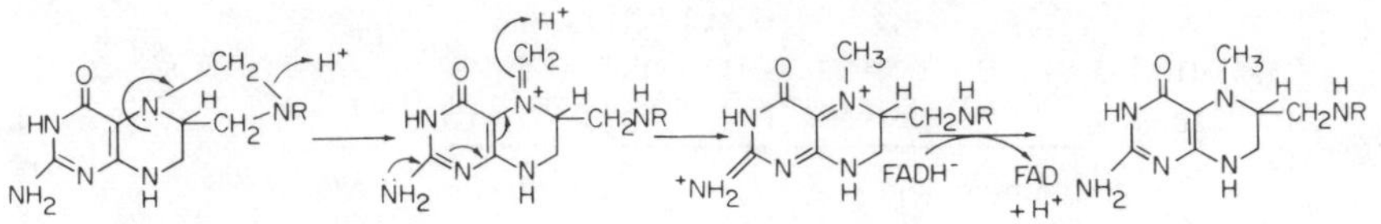

SCHEME 2. Proposed mechanism for reduction of CH_2-H_4folate by methylenetetrahydrofolate reductase. (Reprinted, with permission, from ref. 9.)

μM; V_{max}/E_T, 1700 min^{-1}.[7] At pH 6.7, values for V_{max}/E_T are unchanged, but the K_m for CH_2-H_4folate is substantially lower, 19 μM.[3]

Rapid reaction kinetics at pH 7.2 are completely consistent with a ping-pong mechanism, although only about 75–85% of the enzyme appears to be active.[7] The extrapolated first-order rate constant for reduction of oxidized enzyme by NADPH is 9600 min^{-1}, while that for oxidation of photoreduced enzyme by CH_2-H_4folate is 3000 min^{-1}. Also in agreement with a ping-pong mechanism, the enzyme catalyzes isotopic exchange between [methyl-^{14}C]H_4folate and CH_2-H_4folate in the absence of pyridine nucleotides.[7]

Methylenetetrahydrofolate reductase also catalyzes the NADPH-linked reduction of quinoid dihydrofolate and dihydropterin derivatives.[9] This activity is thought to mimic the normal pathway for reduction of CH_2-H_4folate, which involves ring opening to form a 5-iminium cation, tautomerization to form quinonoid 5-CH_3-H_2folate, and then reduction to form CH_3-H_4folate, as shown in Scheme 2.

Inhibition Studies. Methylenetetrahydrofolate reductase is inhibited by dihydrofolate.[10] Dihydropteroylpolyglutamates are potent inhibitors, with the most potent being the dihydropteroylhexaglutamate (K_i = 13 nM).[11] In all cases, inhibition by dihydropteroylpolyglutamates is linearly competitive with respect to CH_2-H_4folate.

Methylenetetrahydrofolate reductase is also inhibited by *S*-adenosylmethionine (AdoMet).[2,12] The onset of inhibition on introduction of AdoMet into an assay is slow, although the lag in inhibition can be abolished by 10-min preincubation of the enzyme with inhibitor prior to addition of substrates. Inhibition can be relieved by *S*-adenosylhomocysteine, which does not itself activate the enzyme.

[9] R. G. Matthews and S. Kaufman, *J. Biol. Chem.* **255,** 6014 (1980).
[10] R. G. Matthews and B. J. Haywood, *Biochemistry* **18,** 4845 (1979).
[11] R. G. Matthews and C. M. Baugh, *Biochemistry* **19,** 2040 (1980).
[12] M. A. Vanoni, S. C. Daubner, D. P. Ballou, and R. G. Matthews, *in* "Chemistry and Biology of Pteridines" (J. A. Blair, ed.), p. 235. de Gruyter, Berlin, 1983.

Methylenetetrahydrofolate reductase is extremely sensitive to proteolysis, and limited proteolysis of the native enzyme with trypsin results in complete desensitization of the enzyme toward inhibition by AdoMet, with no accompanying loss of catalytic activity. This desensitization corresponds to the cleavage of the 77-kDa subunit into two fragments of 39 and 36 kDa, which remain associated noncovalently. Scanning transmission electron microscopy has shown that these fragments correspond to preexisting spatially distinct domains in the intact subunit.[8]

[59] Enzymatic Assay of 5-Methyl-L-tetrahydrofolate

By Philip A. Kattchee and Robert W. Guynn

5-Methyl-L-tetrahydrofolate (L-CH$_3$THF) is a major form of folate in tissue and blood. The low levels of this compound in biological material can be assayed by protein-binding radiotracer methods[1,2] or bioassays[3] (ability to support bacterial growth). Such methods are very sensitive, but apart from any concern about specificity, they can be only as accurate as the solutions used to standardize them. A common practice has been to calculate the concentration of a given standard solution of L-CH$_3$THF from its absorption at 290 nm, the absorption maximum of the pure compound.[4] Other than the method reported below, the only approach to actually standardize a solution for biologically, enzymatically active L-CH$_3$THF has been a previously reported spectrophotometric assay utilizing 5-methyltetrahydrofolate–homocysteine methyltransferase (methionine synthase, EC 2.1.1.13) purified from *Escherichia coli* B.[5] Because of the requirement of the latter method for an enzyme from a source not conveniently available to all laboratories, we have developed alternative spectrophotometric and fluorometric enzymatic assays for L-CH$_3$THF which can be used routinely. The enzymes needed are relatively stable and can be partially purified from convenient sources in reasonably large quantities.

[1] S. P. Rothenberg, M. daCosta, and Z. Rosenberg, *N. Engl. J. Med.* **286,** 1335 (1972).
[2] S. Waxman and C. Schreiber, *Blood* **42,** 281 (1973).
[3] E. Usdin, P. M. Phillips, and G. Toennies, *J. Biol. Chem.* **221,** 865 (1956).
[4] V. S. Gupta and F. M. Huennekens, *Arch. Biochem. Biophys.* **120,** 712 (1967).
[5] L. Jaenicke, this series, Vol. 18 Part B, p. 605.

METHODS IN ENZYMOLOGY, VOL. 122

Assay Methods

Principle. The methods are based upon the enzyme sequence:

$$5\text{-Methyl-L-tetrahydrofolate} \xrightarrow[\text{menadione}]{\text{FAD}} 5,10\text{-methylene-L-tetrahydrofolate} \qquad (1)$$

$$5,10\text{-Methylene-L-tetrahydrofolate} + NADP^+$$
$$\rightarrow 5,10\text{-methenyl-L-tetrahydrofolate} + NADPH \qquad (2)$$

$$NADPH \xrightarrow[\text{menadione}]{\text{FAD}} NADP^+ \qquad (3)$$

$$\text{NET:} \quad 5\text{-Methyl-L-tetrahydrofolate} \xrightarrow[\text{menadione}]{\text{FAD}} 5,10\text{-methenyl-L-tetrahydrofolate} \qquad (4)$$

The 5,10-methenyl-L-tetrahydrofolate formed in a preincubation step is measured spectrophotometrically or fluorometrically after acidification of the reaction mixture.

The assay depends upon two sequential oxidations of 5-methyl-L-tetrahydrofolate to 5,10-methenyl-L-tetrahydrofolate by 5,10-methylene-L-tetrahydrofolate reductase (EC 1.5.1.20) and 5,10-methylene-L-tetrahydrofolate dehydrogenase (EC 1.5.1.5). The stoichiometry (and even feasibility) of the assay depends on the facts that menadione is a much stronger oxidant[6] than either 5,10-methylene-L-tetrahydrofolate or $NADP^+$, and that reduced menadione is itself readily reoxidized by oxygen. The assay would not be possible using $NADP^+$ alone because of the unfavorable equilibrium constant of the reaction.[7] Formaldehyde is included in the reaction mixture to minimize the nonenzymatic, reversible dissociation of the intermediate, 5-methylene-L-tetrahydrofolate.

The methods are linear at least within the ranges given. In purified samples of L-CH₃THF the enzymatic assay agrees within 1–2% of the value predicted from the UV absorption (corrected for small contaminants of dihydrofolate and tetrahydrofolate).

Enzymes and Reagents

Enzymes. Beef liver 5,10-methylene-L-tetrahydrofolate reductase (specific activity 0.8–1.2 units/mg protein at 25°) and 5,10-methylene-L-tetrahydrofolate dehydrogenase (specific activity 0.6–0.8 units/mg protein) are partially purified from beef and chicken liver, respectively, by the methods of Kattchee and Guynn.[8] The dehydrogenase isolated from yeast is also available commercially but has not been studied in this assay system.

---

[6] N. R. Trenner and F. A. Bacher, *J. Biol. Chem.* **137**, 745 (1941).

[7] L. Jaenicke and H. Rudiger, *FEBS Lett.* **4**, 316 (1969).

[8] P. A. Kattchee and R. W. Guynn, *Anal. Biochem.* **118**, 85 (1981).

Reagents. All solutions are prepared in distilled water and are stable for at least 1 month unless otherwise noted.

Stored below −15°
  Flavin adeninedinucleotide (FAD), 0.5 mM
  NADP$^+$, 12 mM
  Bovine serum albumin (BSA), 15 mg/ml
Stored at 0–4°
  Potassium phosphate, 0.2 M (pH 7.2)
  EDTA, 0.1 M (pH 7.0)
  Potassium perchlorate, saturated solution
  Menadione (prepared fresh daily), saturated (~2 mM) solution in
    20% (v/v) methanol
  Sodium ascorbate (prepared fresh daily), 1 M (pH 7)
Stored at room temperature
  HCHO, 1.2 M
  HCl, 5 M

Instruments. The spectrophotometric assay can be performed with either a single- or dual-beam instrument holding cuvettes of 1-cm light path. The fluorometric procedure is designed for an instrument capable of handling 1.0-ml volumes. The incident and emission wavelengths for 5,10-methenyl-L-tetrahydrofolate are 350 and 470 nm, respectively.[9] For the fluorometric assay quinine sulfate in H_2SO_4 is used as a reference standard.[10]

Spectrophotometric Procedure

Into 12 × 75-mm disposable tubes is placed 2.0 ml of a reaction mixture containing in final concentration 100 mM potassium phosphate (pH 7.2), 0.5 mM EDTA, 0.2 mM NADP$^+$, 2.5 μM FAD, 40 mM sodium ascorbate, ~0.02 mM menadione, 0.05 unit/ml 5,10-methylene-L-tetrahydrofolate reductase, 0.2 unit/ml 5,10-methylene-L-tetrahydrofolate dehydrogenase, 225 μg/ml BSA, 0.5 mM formaldehyde, and 3–60 nmol L-CH$_3$THF. The blank contains all reagents including sample except for 5,10-methylene-L-tetrahydrofolate reductase. The reaction mixtures are incubated at 37° for 20 min and then terminated with 0.35 ml of 5 N HCl/ml of reaction mixture. The precipitate is removed by centrifugation at room temperature, and the absorption is measured at 350 nm. The content of 5-methyl-L-tetrahydrofolate in each cuvette is calculated from the net

[9] R. H. Himes and J. C. Rabinowitz, *J. Biol. Chem.* **237,** 2903 (1962).
[10] O. H. Lowry and J. V. Passonneau, "A Flexible System of Enzymatic Analysis." Academic Press, New York, 1972.

ΔOD (corrected for the blank and the absorption of the cuvette itself) and the extinction coefficient of 5,10-methenyl-L-tetrahydrofolate of 25.1 $\times$ 10^3 M^{-1} cm^{-1}.[11]

$$\mu\text{mol L-CH}_3\text{THF/cuvette} = (\Delta\text{OD})(\text{final volume in cuvette})/25.1$$

Fluorometric Procedure

A reagent mixture is prepared to contain, in final concentrations, 50 mM potassium phosphate (pH 7.2), 40 μM NADP$^+$, 0.1 mM EDTA, 40 mM sodium ascorbate, 0.50 μM FAD, 0.50 mM formaldehyde, 225 μg/ml BSA, and 4 μM menadione. The reagent mixture (0.95 ml) plus 0.5–3.0 nmol of L-CH$_3$THF, 0.025 unit of 5,10-methylene-L-tetrahydrofolate reductase, and 0.1 unit of 5,10-methylene-L-tetrahydrofolate dehydrogenase in a final volume of 1.0–1.1 ml is placed in an fluorometer 10 $\times$ 75-mm tube. The mixture is incubated for 10 min at 37°, after which the reaction is terminated with 0.35 ml 5 N HCl. The fluorescence is measured after centrifugation, if necessary, to remove precipitated protein. A standard curve of DL- or L-CH$_3$THF assayed by the spectrophotometric method is run simultaneously, as are blanks in which the 5,10-methylene-L-tetrahydrofolate reductase has been omitted.

The fluorometric assay can be made more sensitive, allowing the determination of 50–250 pmol L-CH$_3$THF in very pure samples. In such cases the concentrations of FAD and menadione are halved, and an enzyme mixture is prepared containing (in final concentration), 50 mM potassium phosphate (pH 7.2), 225 μg/ml BSA, 0.05 units/ml 5,10-methylene-L-tetrahydrofolate reductase, and 0.2 units/ml 5,10-methylene-L-tetrahydrofolate dehydrogenase. Into a 10 $\times$ 75-mm fluorometer tube is placed 200 μl of reagent cocktail, 20 μl enzyme mixture, and 50–200 pmol L-CH$_3$THF in 100 μl or less. After incubation at 37° for 10 min, 0.8 ml of 1 N HCl is added and the fluorescence measured and compared with a standard curve as above.

Variations of the Procedures

Sulfhydryl groups should not be present in concentrations greater than 10 mM because of their ability to bind HCHO and thus promote dissociation of 5,10-methylene-L-tetrahydrofolate, the intermediate product of the assay. The times suggested for the preincubations are more than sufficient for the systems we have examined. The time needed for a complete reaction would need to be determined for each new application, however.

[11] B. V. Ramasastri and R. L. Blakley, *J. Biol. Chem.* **239**, 106 (1964).

Discussion

CH₃THF is an unstable compound that is simple neither to prepare nor to maintain pure. It is almost inevitably contaminated with both precursors and a variety of breakdown products. The presence of UV-absorbing contaminations tends to invalidate the common practice to date of standardizing solutions of CH_3THF by measuring the absorption at 290 nm. We have found that relying on the UV absorption alone can result in up to a 40% overestimation of DL-CH_3THF in random, commercially obtained samples and as much as a 100% error in the estimate of the concentration of DL-CH_3THF synthesized by borohydride reduction of 5-formyl-DL-tetrahydrofolate[12] (both corrected for the L isomer). Paper chromatography of these latter two preparations, in fact, demonstrated three to four distinct spots. Even enzymatically synthesized and chromatographically purified L-CH_3THF[4] in our hands contained 94% L-CH_3THF and 5% THF by enzymatic assays.

These findings underscore the difficulties of trying to depend solely upon UV absorption to assay accurately solutions containing L-CH_3THF and demonstrate the use of the specific assay described. Although microbiological and protein-binding assays for L-CH_3THF are significantly more sensitive than the methods described above, the accuracy and dependability of these other methods can only be as good as the standard curves used. The enzymatic assays described here permit the L-CH_3THF content of such standards to be determined accurately.

[12] I. Chanarin and J. Perry, *Biochem. J.* **105,** 633 (1967).

[60] Purification of 5,10-Methenyltetrahydrofolate Cyclohydrolase from *Clostridium formicoaceticum*

By LARS G. LJUNGDAHL and JOAN E. CLARK

10-Formyltetrahydrofolate + H^+ ⇌ 5,10-methenyltetrahydrofolate + H_2O

Clostridium formicoaceticum, described by Andreesen *et al.*[1], belongs to the acetogenic bacteria, which synthesize acetate from one-carbon precursors via a tetrahydrofolate–corrinoid pathway.[2,3] These bacteria

[1] J. R. Andreesen, G. Gottschalk, and H. G. Schlegel, *Arch. Mikrobiol.* **72,** 154 (1970).
[2] L. G. Ljungdahl and H. G. Wood, *in* "B₁₂" (D. Dolphin, ed.), Vol. II, p. 165. Wiley (Interscience), New York, 1982.

are rich sources of tetrahydrofolate enzymes needed to convert and reduce formate to 5-methyltetrahydrofolate, which is the precursor of the methyl group of acetate.

Methenyltetrahydrofolate cyclohydrolase (EC 3.5.4.9) activity has been demonstrated in animal and plant tissues and in many bacteria. In eukaryotic cells such as yeast[4] and liver,[5,6] the cyclohydrolase activity resides in trifunctional proteins, which in addition to the cyclohydrolase catalyze formate–tetrahydrofolate ligase (10-formyltetrahydrofolate synthetase, EC 6.3.4.3) and the 5,10-methylenetetrahydrofolate dehydrogenase (EC 1.5.1.5) reactions. In some bacteria including *Escherichia coli*[7] and *Clostridium thermoaceticum*[8] the cyclohydrolase is associated with 5,10-methylenetetrahydrofolate dehydrogenase in enzymes that are bifunctional. In these bacteria the 10-formyltetrahydrofolate synthetase is a separate enzyme. In other bacteria all three activities are catalyzed by separate enzymes. These bacteria include *C. formicoaceticum,* from which 10-formyltetrahydrofolate synthetase,[9] 5,10-methylenetetrahydrofolate dehydrogenase,[8] and 5,10-methenyltetrahydrofolate cyclohydrolase[10] have been obtained as separate homogeneous proteins. The purification and properties of the latter enzyme are dealt with in this chapter.

Assay Method

Principle. The 5,10-methenyltetrahydrofolate cyclohydrolase activity is assayed by following the decrease in absorbance at 350 nm of 5,10-methenyltetrahydrofolate as it is hydrolyzed to 10-formyltetrahydrofolate.[8,11] The enzyme is most active around pH 7. At this pH 5,10-methenyltetrahydrofolate is hydrolyzed nonenzymatically at a measurable rate, which is deducted from the rate of hydrolysis in the presence of the enzyme. The enzymatic and the nonenzymatic reactions are strongly affected by the nature of the buffer system that is used. Phosphate and imidazole buffers increase the rate of nonenzymatic hydrolysis,[11] whereas

[3] L. G. Ljungdahl, *in* "Organic Chemicals from Biomass" (D. Wise, ed.), p. 219. Benjamin/Cummings, Menlo Park, California, 1983.
[4] J. L. Paukert, G. R. Williams, and J. C. Rabinowitz, *Biochem. Biophys. Res. Commun.* **77,** 147 (1977).
[5] R. E. MacKenzie and L. V. L. Tan, this series, Vol. 66, p. 609.
[6] J. L. Paukert and J. C. Rabinowitz, this series, Vol. 66, p. 616.
[7] I. K. Dew and R. J. Harvey, *J. Biol. Chem.* **253,** 4245 (1978).
[8] L. G. Ljungdahl, W. E. O'Brien, M. R. Moore, and M.-T. Liu, this series, Vol. 66, p. 599.
[9] W. E. O'Brien, J. M. Brewer, and L. G. Ljungdahl, *Experientia, Suppl.* **26,** 249 (1976).
[10] J. E. Clark and L. G. Ljungdahl, *J. Biol. Chem.* **257,** 3833 (1982).
[11] D. M. Greenberg, this series, Vol. 6, p. 386.

they decrease the enzymatic rate. The enzymatic rate is high with tri-ethanolamine, morpholinopropane, and maleate buffers, in which the nonenzymatic rate is relatively slow.

Reagents

A. Triethanolamine–HCl (pH 7.5), 0.5 M

B. (l)-L-5,10-Methenyltetrahydrofolate, 1.7 mM, is synthesized enzymatically from formate, tetrahydrofolate, and ATP with formyltetrahydrofolate synthetase from *C. formicoaceticum*[9] or *C. thermoaceticum*.[12] In this synthesis the incubation mixture contains the components used for assay of formyltetrahydrofolate synthetase in the forward reaction.[12] At the completion of the synthetase reaction the reaction mixture is applied to a DE-23 cellulose (Whatman) column previously equilibrated with oxygen-free 50 mM potassium maleate (pH 8), containing 0.1 M 2-mercaptoethanol. The column is washed with the equilibration buffer; then the 10-formyltetrahydrofolate is eluted with 0.2 M potassium maleate (pH 7.5), containing 0.1 M 2-mercaptoethanol. Fractions of 2 ml are collected under anaerobic conditions in tubes containing 0.1 ml of 4 N HCl. The acid causes the convertion of 10-formyltetrahydrofolate to 5,10-methenyltetrahydrofolate. The absorbance is measured at 350 and 305 nm. Fractions with the ratio A_{350}/A_{305} greater than 2.5 are combined, and this solution of 5,10-methenyltetrahydrofolate is directly used as substrate. It may be stored under N_2 at $-20°$ for at least 2 months. The amount of 5,10-methenyltetrahydrofolate is determined by its absorbance at 350 nm ($\varepsilon = 24.9 \times 10^3 \ M^{-1} \ cm^{-1}$).

Procedure. Two cuvettes, light path of 1 cm, containing 0.45 ml of the assay buffer (A) are equilibrated at 35°. The nonenzymatic reaction is started by the addition of 50 μl of the 5,10-methenyltetrahydrofolate solution (B) to one of the cuvettes. The addition of the substrate lowers the pH to 7.2, which is the pH at which the reaction is run. To the second cuvette the enzyme (5 μl) is first added and the reaction is then started by the addition of 50 μl of the substrate. Both reactions are followed by measuring the decrease in absorption at 350 nm. The enzymatic rate is obtained by deducting the nonenzymatic rate of the first cuvette from the rate of the second cuvette. Units of enzyme activity can be calculated as previously suggested.[8] They are expressed as moles of 5,10-methenyltetrahydrofolate hydrolyzed per minute and specific activity is in units per milligram of protein.

[12] W. T. Shoaf, S. H. Neece, and L. G. Ljungdahl, *Biochim. Biophys. Acta* **334**, 448 (1974).

In both the nonenzymatic and the enzymatic reactions a significant lag period is observed at the start of the reactions before a linear rate is observed. This lag period is about 5 times longer in the nonenzymatic than in the enzymatic reaction and it is dependent on the pH. Thus the lower the pH the longer the lag period. At pH 7.8 or higher, no lag is observed and the nonenzymatic reaction rate is extremely rapid, making it difficult to observe the enzymatic reaction. A mechanism for the nonenzymatic hydrolysis has been proposed by Robinson and Jencks.[13]

Cell Material

Clostridium formicoaceticum (ATCC 23439) is grown at 37° under anaerobic conditions under an atmosphere of N_2 and at a pH of 7.5 maintained by the addition of 5 N KOH during the growth. The medium is prepared in three solutions, which are sterilized separately. Solution A contains 10 g fructose in 50 ml water. Solution B contains 900 ml water, 10 g Na_2HCO_3, 5 g tryptone (Difco), 5 g yeast extract (Difco), 10 g K_2HPO_4, 1 mg pyridoxine · HCl, 1 g NH_4Cl, and 5 ml of a mineral solution.[14] Solution C contains 0.125 g sodium thioglycolate dissolved in 30 ml anaerobic water. Solution B is first autoclaved and then bubbled with sterile CO_2 until the pH is about 7.5. Solutions A and C are first made anaerobic by boiling and bubbling with CO_2 and then autoclaved. At the time of inoculation solutions A and C are added to solution B. The culture is kept under an atmosphere of CO_2 until growth is visible; at this point the culture is bubbled with N_2, to continue throughout the remainder of the growth phase. Cell yields of about 6 g per liter of wet cells are obtained. The bacterial cells can be stored frozen at −16° for at least 1 year without significant loss of cyclohydrolase activity.

Purification Procedure

The purification is that given by Clark and Ljungdahl.[10] All work is performed at 5° with potassium phosphate buffers of pH 7.2. A summary of a purification is given in Table I.

Step 1. Extraction. Frozen cells (200 g, wet weight) are suspended and allowed to thaw in 600 ml of 50 mM buffer. The suspension is centrifuged at 57,000 g for 90 min. The pellet is discarded. The supernatant solution contains about 90% of the cyclohydrolase. The freezing and thawing operation to release the cyclohydrolase is preferred over the use of nonfrozen, freshly grown cells homogenized using a French pressure cell. The freeze–thaw procedure yields extracts with specific activities about twice

[13] D. R. Robinson and W. P. Jencks, *J. Am. Chem. Soc.* **89**, 7098 (1967).
[14] E. A. Wolin, M. J. Wolin, and R. S. Wolfe, *J. Biol. Chem.* **238**, 2882 (1963).

TABLE I

PURIFICATION OF 5,10-METHENYLTETRAFOLATE
CYCLOHYDROLASE FROM *C. formicoaceticum*[a]

Step	Total protein (mg)	Total units (μmol/min)	Specific activity (units/mg)
1. Extract	21,100	50,900	2.83
2. DEAE-cellulose	7,500	48,000	6.36
3. Blue Sepharose	520	37,900	73
4. Ultrogel	117	28,000	240
5. DEAE-Sephadex	51	20,200	396
6. Blue agarose	23	10,900	469

[a] Frozen cells, 200 gm.

of that of extracts obtained with the French pressure cell. The enzyme is not released from freshly grown cells (unless frozen) by simply suspending them in buffer.

Step 2. DEAE-Cellulose Chromatography. The extract of Step 1 is applied to DE-23 cellulose column (3.8 × 35 cm) equilibrated with 50 mM buffer, which also is used to elute the enzyme. Fractions of 20 ml are collected. This step removes acidic proteins including ferredoxin and rubredoxin.

Step 3. Blue Sepharose Chromatography. The active fractions of Step 2 are combined and the solution is applied to a Blue Sepharose (prepared as described by Böhme[15]) column (5.5 × 40 cm), which previously had been washed with first 8 M urea and then with 50 mM buffer. After application of the enzyme solution, the column is washed with 50 mM buffer until the protein content in the eluate is negligible. The cyclohydrolase is then eluted with 50 mM buffer containing 0.25 M KCl. Fractions of 10 ml are collected and those of high specific activity are combined. This solution is concentrated to 20 ml with an Amicon Diaflo apparatus using a PM10 membrane.

Step 4. Ultrogel AcA Gel Filtration. The concentrated enzyme solution from Step 3 is applied to an Ultrogel AcA 44 (LKB, Bromma, Sweden) column (3.2 × 81 cm) connected in tandem with an Ultrogel AcA 54 column (2.6 × 96 cm). The solvent is 50 mM buffer containing 0.2 M KCl. Fractions of 5 ml are collected. The active fractions are combined and concentrated in the Diaflo apparatus described in Step 3. The solution is then diluted with 2 mM buffer and again concentrated. This process is repeated several times to desalt the protein solution.

[15] H. J. Böhme, *J. Chromatography* **69**, 209 (1972).

Step 5. DEAE-Sephadex Chromatography. The desalted enzyme solution of Step 4 is applied to a DEAE-Sephadex (Pharmacia) column (1.6 × 14 cm) previously equilibrated with 10 mM buffer. The column is washed with 2 column volumes of 10 mM buffer, after which the enzyme is eluted with a 200-ml linear gradient from 0 to 0.1 M KCl in 10 mM buffer. Fractions of 4 ml are collected. The enzyme elutes at about 0.05 M KCl. The active fractions are combined.

Step 6. Blue A Agarose Chromatography. The enzyme solution from Step 5 is applied to a 1.6 × 10-cm column of Blue A Agarose (Amicon Corp., Lexington, MA) previously equilibrated with 10 mM buffer. The column is washed with 100 ml of 10 mM buffer and the cyclohydrolase is then eluted with 10 mM buffer containing 0.29 M KCl. Fractions of 1 ml are collected and those with specific activities of 460 ± 5% are combined. They contain apparent homogeneous enzyme as was demonstrated by a single symmetrical Schlieren peak in sedimentation velocity experiments, linear concentration versus r^2 plots from sedimentation equilibrium experiments, and single protein bands from sodium dodecyl sulfate and standard gel electrophoresis.

In the above purification procedure the cyclohydrolase is separated from 10-formyltetrahydrofolate synthetase[9] in Step 2 and from 5,10-methylenetetrahydrofolate dehydrogenase[8] in Step 3.

The purified enzyme is stable for at least 1 week when stored at 5° in 50 mM phosphate buffer (pH 7.0). When in this buffer the enzyme can be frozen in an acetone–dry ice bath and lyophilized without loss of activity. In other buffers such as maleate and Tris–HCl at pH 7 and 5°, about 50% of the activity is lost within a week.

Properties

Several properties of 5,10-methenyltetrahydrofolate cyclohydrolase from *C. formicoaceticum* are listed in Table II. In addition the amino acid composition has been determined.[10] The enzyme apparently is a dimer of two identical subunits. Thermostability studies of the enzyme reveal that it is rapidly inactivated at 55° or higher temperatures. At 55° 70% of the activity is lost within 2 min.

Comment

Methenyltetrahydrofolate cyclohydrolase activity has been found in many different types of cells. It now appears that in eukaryotic cells this activity is associated with a protein which is trifunctional,[5,6] containing

TABLE II
PROPERTIES OF 5,10-METHENYLTETRAHYDROFOLATE
CYCLOHYDROLASE FROM *C. formicoaceticum*

Property	Value
Specific activity of purified enzyme at 35°, pH 7.2	470
Molecular weight[a]	41,000
Molecular weight of subunits[b]	25,500
s°_{20w}	3.9
Stokes radius	29.6 Å
Absorbance at 280 nm of 0.1% solution at pH 7.0	0.47
Apparent K_m for methenyl-H_4folate	0.19 mM
pH	
Range	5.8–8
Optimum[c]	6.6–7.6
Apparent activation energy	7,500 cal/mol

[a] Determined by sedimentation equilibrium centrifugation.
[b] Determined by sodium dodecyl sulfate–gel electrophoresis. The data suggest two identical subunits.
[c] In triethanolamine or morpholinopropane sulfonate buffers.

two additional enzyme activities concerned with tetrahydrofolate metabolism. In bacteria the cyclohydrolase activity is catalyzed either by a bifunctional enzyme with methylenetetrahydrofolate dehydrogenase activity or by a monofunctional enzyme. Among the acetogens *Clostridium thermoautotrophicum* and *C. thermoaceticum* have a bifunctional cyclohydrolase–(NADP-dependent) dehydrogenase, whereas *C. formicoaceticum* and *Acetobacterium woodii* have separate cyclohydrolase and dehydrogenase enzymes.[16] The latter monofunctional enzyme in the two acetogens is NAD dependent. The existence of trifunctional, difunctional, and monofunctional enzymes converting tetrahydrofolates leads to interesting speculation about how the tetrahydrofolate pathway has evolved. Genetic studies may perhaps give an answer to this question.

---

[16] J. E. Clark, Ph.D. Dissertation, University of Georgia, Athens (1983).

[61] Purification and Properties of 5,10-Methylenetetrahydrofolate Reductase from *Clostridium formicoaceticum*

By JOAN E. CLARK and LARS G. LJUNGDAHL

5,10-Methylenetetrahydrofolate + $2H^+$ + 2 ferredoxin$_{red}$
$$\rightleftharpoons \text{5-methyltetrahydrofolate} + 2 \text{ ferredoxin}_{ox} \quad (1)$$

Acetogenic bacteria, including *Clostridium formicoaceticum,* produce acetate from carbon dioxide via a unique tetrahydrofolate–corrinoid pathway.[1,2] Part of this pathway involves the reduction of carbon dioxide to 5-methyltetrahydrofolate, with methylenetetrahydrofolate reductase catalyzing the final reduction [Eq. (1)] in this sequence of reactions. Methyltetrahydrofolate then serves as a precursor of the methyl group of acetate to which it is converted in a series of reactions involving either CO, CO_2, or pyruvate as a source of the carboxyl group.[3]

It is likely that the reductase from *C. formicoaceticum* under physiological conditions catalyzes the reduction of methylenetetrahydrofolate with reduced ferredoxin as shown in Eq. (1).[4,5] In addition reductase can utilize $FADH_2$ in the forward reaction [as written in Eq. (1)] whereas in the reverse reaction FAD, rubredoxin, menadione, benzyl viologen, and methylene blue serve as electron acceptors.[4] The reaction with benzyl viologen allows for a convenient spectrophotometric assay of the reductase activity. Pyridine nucleotides are not utilized by the reductase from *C. formicoaceticum*. Similarly, the enzyme from *Escherichia coli* does not use pyridine nucleotides. It is considered a FAD-dependent enzyme.[6] In contrast, NADP is the electron carrier for the methylenetetrahydrofolate reductase from pig and rat liver.[7,8]

[1] L. G. Ljungdahl and H. G. Wood, *in* "B$_{12}$" (D. Dolphin, ed.), Vol. II, p. 165. Wiley (Interscience), New York, 1982.

[2] L. G. Ljungdahl, *in* "Organic Chemicals from Biomass" (D. Wise, ed.), p. 219. Benjamin/Cummings, Menlo Park, California, 1983.

[3] S.-I. Hu, E. Pezacka, and H. G. Wood, *J. Biol. Chem.* **259**, 8892 (1984).

[4] J. E. Clark and L. G. Ljungdahl, *J. Biol. Chem.* **259**, 10845 (1984).

[5] J. E. Clark, L. G. Ljungdahl, and D. V. DerVartanian, *Fed. Proc., Fed. Am. Soc. Exp. Biol.* **43**, 2061 (1984).

[6] H. M. Katzen and J. M. Buchanan, *J. Biol. Chem.* **240**, 825 (1965).

[7] C. Kutzbach and E. L. R. Stokstad, *Biochim. Biophys. Acta* **250**, 459 (1971).

[8] S. C. Daubner and R. G. Matthews, *J. Biol. Chem.* **257**, 140 (1982).

General Anaerobic Procedures

The 5,10-methylenetetrahydrofolate reductase from *C. formicoaceti-cum* is oxygen labile and must be handled under anaerobic conditions. In addition, the assay described below for the reductase activity with benzyl viologen must be conducted anaerobically to prevent reoxidation of reduced benzyl viologen by air.[9] The anaerobic procedures are those described by Yamamoto *et al.*[10] Buffers are boiled to remove oxygen; then bubbled continuously with O_2-free argon or nitrogen for 20 min while chilled in an ice bath. Nitrogen and argon are purified of oxygen by passage through a copper-filled furnace at 450° (Sargent-Welch Scientific Co., Skokie, IL). Additional compounds are added as solids to the anaerobic buffer. Most work is performed inside an anaerobic chamber (Coy Laboratory Products, Ann Arbor, MI) which is filled with a gas mixture containing 5% H_2 and 95% N_2. Columns too big for the chamber are connected with stainless steel or Iso-Versinic butyl rubber tubings to a fraction collector inside the anaerobic chamber. Ordinary plastic or rubber tubing allows oxygen to penetrate and destroy the enzyme and should be avoided.

Assay Method

Principle. The *C. formicoaceticum* methylenetetrahydrofolate reductase can be assayed by two different methods. One method, originally described by Katzen and Buchanan,[6] is a sampling method which utilizes menadione as the electron carrier in the oxidation of [*methyl*-^{14}C]methyltetrahydrofolate to [*methylene*-^{14}C]methylenetetrahydrofolate. The [*methylene*-^{14}C]methylenetetrahydrofolate is in nonenzymatic equilibrium with tetrahydrofolate and H^{14}CHO. The latter compound is trapped and extracted with 5,5-dimethyl-1,3-cyclohexanedione and toluene. The second assay, which is used in our laboratory, is a continuous method which spectrophotometrically monitors the one-electron reduction of the artificial electron carrier, benzyl viologen[9] (BV), at 555 nm ($\varepsilon = 12 \times 10^3$ M^{-1} cm^{-1})[11] according to Eq. (2):

5,10-Methyltetrahydrofolate $\times$ 2 BV$_{ox}$
$$\rightleftharpoons 5,10\text{-methylenetetrahydrofolate} + 2\ BV_{red} + 2H^+ \quad (2)$$

[9] L. Michaelis and E. S. Hill, *J. Gen. Physiol.* **16,** 859 (1933).
[10] I. Yamamoto, T. Saiki, S.-I. Liu, and L. G. Ljungdahl, *J. Biol. Chem.* **258,** 1826 (1983).
[11] J. M. Brewer, A. J. Pesce, and R. B. Ashworth, "Experimental Techniques in Biochemistry." Prentice-Hall, Englewood Cliffs, New Jersey, 1974.

Reagents

   A. Potassium phosphate (pH 7.2), 100 mM; sodium ascorbate, 20 mM; benzyl viologen, 20 mM

   B. (*dl*)-L-5-Methyltetrahydrofolate, 13 mM

Solution A is prepared by adding sodium ascorbate and benzyl viologen as solids to the anaerobically prepared phosphate buffer. This solution can be stored at room temperature in the dark and under an argon atmosphere for about 2 weeks.

Solution B is prepared by dissolving solid 5-methyltetrahydrofolate in anaerobically prepared 10 mM sodium ascorbate. The concentration of the methyltetrahydrofolate is determined using the extinction coefficient at 290 nm[12] ($\varepsilon = 31.7 \times 10^3\,M^{-1}\,cm^{-1}$). This solution can be stored at $-20°$ under argon for 2–4 weeks. 5-Methyltetrahydrofolate is prepared by the method of Blair and Saunders[13] as modified by Parker *et al.*[14] except that all solutions are prepared anaerobically as described above. The methyltetrahydrofolate is purified using an DE-23 cellulose (Whatman) column run anaerobically with ammonium acetate as elution solvent. Fractions having a 290/245 nm absorbance ratio greater than 3.5 are pooled and lyophilized. The methyltetrahydrofolate is stored as a solid at $-20°$ in an argon atmosphere.

Procedure. A semimicro cuvette having a 1-cm light path and a 7-mm circular opening is fitted with a 7-mm-diameter serum stopper. Two needles (2 inch, $22\frac{1}{2}$ gauge) are placed through the serum stopper into the cuvette. A stream of O_2-free argon is passed through the inlet needle for 5 min, displacing air in the cuvette through the outlet needle. A 0.6-ml aliquot of solution A is added via syringe to the argon-filled cuvette. This solution is bubbled with argon for 5 min. The cuvette is placed in a spectrophotometer and equilibrated to 37° in 5 min. The reaction is initiated either by first adding 6 μl of solution B and then enzyme (usually 1 μl) via syringes or by first adding the enzyme and then solution B. The progress of the reaction is recorded continuously at 555 nm with an absorbance scale of either 0.1 or 0.2. Enzyme units can be calculated from the equation, enzyme units $= A \times 0.607/24$, where A is the change in absorbance per minute, 0.607 is the assay volume, and 24 is twice the millimolar extinction coefficient for the one-electron reduction of benzyl viologen. Enzyme units are expressed as 2 μmol benzyl viologen reduced per minute, which is equivalent to 1 μmol methyltetrahydrofolate oxi-

---

[12] R. L. Blakley, "The Biochemistry of Folic Acid and Related Pteridines." North-Holland Publ., Amsterdam, 1969.

[13] J. A. Blair and K. J. Saunders, *Anal. Biochem.* **34,** 376 (1970).

[14] D. J. Parker, T.-F. Wu, and H. G. Wood, *J. Bacteriol.* **198,** 770 (1971).

dized per minute. Specific activity is in units per milligram of protein. Protein is determined by a colorimetric method with 1 mg/ml rose bengal.[15]

Cell Material

Clostridium formicoaceticum, strain ATCC 23439, described by Andreesen *et al.*[16] is grown according to the procedure outlined by Ljungdahl and Clark[17] using their medium prepared in three solutions A, B, and C. However, the addition of 1 ml per liter medium of a fourth solution D containing 0.882 g $Co(NO_3)_2 \cdot 6H_2O$, 0.0172 g Na_2SeO_3, 0.333 g $Na_2WO_4 \cdot 2H_2O$, 0.22 g $Na_2MoO_4 \cdot 2H_2O$, and 0.0242 g $NiCl_2$ dissolved in 101 ml of water adjusted to pH 2 with 2 N H_2SO_4 proved to increase the activity of 5-methyltetrahydrofolate reductase in the cells about 20-fold. Solution C is also altered by the addition of 1.9 mg of riboflavin. Solution D is added to solution B concomitantly with solutions A and C. Cell yield with this (high-metal) medium is about 7.5 g wet weight per liter.

Purification Procedure

The purification is summarized in Table I and follows the procedure of Clark and Ljungdahl.[4] The work is carried out at 10° with all solutions prepared anaerobically as described above. With the exception of the lysate preparation, Step 1, the procedure is performed in a anaerobic chamber with a nitrogen/hydrogen (95/5%) atmosphere. The basic buffer is anaerobically prepared 50 mM Tris–HCl (pH 7.4), 20% glycerol, 2 mM sodium dithionite, and 5 μM FAD. Methyl viologen (0.2 mM) is included in the equilibration buffers of the chromatography steps as an indicator of anaerobicity.

Step 1. Extraction. Frozen cells (120 g wet weight) are suspended in 240 ml basic buffer, minus glycerol and containing 0.2 mM methyl viologen and a few crystals of deoxyribonuclease I. The cell suspension is passed through a French pressure cell at 10,000 psi. The lysate is collected in argon-filled vials then centrifuged in capped stainless-steel centrifuge tubes for 65 min at 57,000 g. The resulting supernatant is the crude extract.

Step 2. Ammonium Sulfate Fractionation. Four successive ammonium sulfate fractionations are performed on the crude extract with am-

[15] J. I. Elliott and J. M. Brewer, *Arch. Biochem. Biophys.* **190,** 351 (1978).
[16] J. R. Andreesen, G. Gottschalk, and H. G. Schlegel, *Arch. Mikrobiol.* **72,** 154 (1970).
[17] L. G. Ljungdahl and J. E. Clark, this volume [60].

TABLE I

PURIFICATION OF 5,10-METHYLENETETRAHYDROFOLATE REDUCTASE
FROM *C. formicoaceticum*[a]

Step	Total protein (mg)	Total units (μmol/min)	Specific activity (units/mg)
1. Extract	12,700	12,300	1
2. Ammonium sulfate	1,600	8,400	5
3. Phenyl-Sepharose	115	4,900	43
4. BioGel HTP	45	4,500	102
5. DEAE–Sephadex	13	1,800	139

[a] From 120 g wet cells.

monium sulfate added to 55, 65, 75, then 85% saturation. Solid ammonium sulfate is added over a period of 30 min with constant stirring. The solution is centrifuged in capped stainless-steel tubes for 30 min at 31,000 *g*. The 65–75 and 75–85% saturation precipitates are combined and dissolved in the basic buffer. The final ammonium sulfate fractionation (75–85%) can be omitted if the solution having ammonium sulfate to 75% saturation is centrifuged for 60 min instead of 30 min.

Step 3. Phenyl-Sepharose Chromatography. Ammonium sulfate, to 1.3 *M*, is added to the dissolved precipitate from Step 2. This enzyme solution is applied to a phenyl-Sepharose (Sigma) column (5 × 9 cm) which had been equilibrated with the basic buffer minus glycerol and containing 1.3 *M* ammonium sulfate. The column is washed with 1 column volume of the equilibration buffer. The reductase is then eluted with a 1600 ml (total volume) linear gradient of the basic buffer, minus glycerol from 1.3 *M* ammonium sulfate to 40% glycerol. The reductase elutes in 90 ml at approximately 0.65 *M* ammonium sulfate and 20% glycerol. The combined fractions are desalted with the basic buffer and an Amicon ultrafiltration device containing a PM10 membrane. It is necessary to desalt the enzyme to a final concentration of approximately 10 m*M* ammonium sulfate to obtain binding of the reductase to the HTP column of Step 4.

Step 4. BioGel HTP Chromatography. The desalted fraction from Step 3 is applied to a hydroxylapatite column (Bio-Rad) (1.5 × 10 cm) which had been equilibrated with 50 m*M* potassium phosphate (pH 7.2), 20% glycerol, 2 m*M* dithionite, and 5 μ*M* FAD. The column is washed with 3 column volumes of the equilibration buffer. The reductase is eluted with 150 m*M* potassium phosphate containing 20% glycerol, 2 m*M* dithionite, and 5 μ*M* FAD. The combined fractions are immediately desalted as in

Step 3. This entire step is performed within a few hours to prevent the substantial loss of reductase activity which occurs with time in phosphate buffers.

Step 5. DEAE-Sephadex. The enzyme obtained from Step 4 is applied to a DEAE-Sephadex (Pharmacia) column (1.5 × 7 cm) equilibrated with the basic buffer minus FAD. The column is washed with 3 column volumes of 150 mM Tris–HCl (pH 7.4), 20% glycerol, 2 mM dithionite. A 400 ml (total volume) linear gradient with 20% glycerol and 2 mM dithionite is run from 150 to 250 mM Tris–HCl (pH 7.4). The reductase elutes in 80 ml at approximately 180 mM buffer. The combined fractions are immediately concentrated.

For removal of exogenously added FAD and dithionite, the concentrated enzyme solution (0.5 ml) is applied to an anaerobic Sephadex G-25 (Pharmacia) superfine column (1 × 25 cm) equilibrated with 50 mM Tris–HCl (pH 7.4) and 20% glycerol.

Enzyme Properties

The methylenetetrahydrofolate reductase from *C. formicoaceticum* is oxygen labile, having a halftime of inactivation of less than 1 hr in aerobic buffer (0.1 M Tris–HCl, pH 7.4). The anaerobic procedures described above and the inclusion of sodium dithionite in enzyme solutions prevent the inactivation of the enzyme by oxygen. The only other tetrahydrofolate-dependent enzyme which we have encountered to be sensitive to oxygen is the 5,10-methylenetetrahydrofolate dehydrogenase from *Acetobacterium woodii*.[18] This enzyme must be purified with anaerobic procedures in the presence of dithiothreitol. The *C. formicoaceticum* methylenetetrahydrofolate reductase is additionally stabilized during the purification procedure by the presence of 5 μM FAD. When the purification is performed in the absence of the exogenously added FAD, both the FAD content and the final specific activity are 2-fold lower than reported here for the purification conducted in the presence of 5 μM FAD. Exogenously added FAD has also been reported to stabilize the reductase from pig liver.[8]

Table II lists some properties of the 5-methyltetrahydrofolate reductase.[19,20] The enzyme, which contains four each of two different subunits, has an $\alpha_4\beta_4$ subunit composition. In contrast, the pig liver enzyme is either a dimer or trimer of identical subunits.[8]

[18] S. W. Ragsdale and L. G. Ljungdahl, *J. Biol. Chem.* **259**, 3499 (1984).
[19] J. B. Jones, Jr., *Commun. Soil Sci. Plant Anal.* **8**, 349 (1977).
[20] J. C. Rabinowitz, this series, Vol. 53, p. 275.

TABLE II

PROPERTIES OF
5,10-METHYLENETETRAHYDROFOLATE
REDUCTASE FROM *C. formicoaceticum*

Property	Value
Specific activity of purified enzyme at 37°, pH 7.4	140
Molecular weight[a]	
Active enzyme	237,000
α subunit	26,000
β subunit	35,000
Metals and cofactors[b]	
Iron[c]	15.2 ± 0.3
Acid-labile sulfide[d]	19.5 ± 1.3
Zinc[c]	2.3 ± 0.2
FAD[e]	1.7
K_m	
dl-5-Methyltetrahydrofolate	0.12 mM
Benzyl viologen	11.1 mM

[a] The molecular weights of the active enzyme and of the subunits were determined using gel filtration and polyacrylamide gel electrophoresis in the presence of sodium dodecyl sulfate, respectively.

[b] Metals and cofactors are given in moles per mole enzyme assuming an M_r of 237,000.

[c] Metals were determined using plasma emission spectroscopy.[19]

[d] Acid-labile sulfide determined according to Rabinowitz.[20]

[e] FAD was identified by thin-layer chromatography and quantified using an extension coefficient ε_{445} of 13×10^3 M^{-1} cm^{-1}.

The *C. formicoaceticum* reductase contains per 237,000 molecular weight, 15 molecules iron, 19 inorganic sulfur, 2 zinc, and 2 FAD. The iron and inorganic sulfur are present as iron–sulfur clusters which, along with the enzyme-bound FAD, are evident in the visible absorbance spectrum of the enzyme[4] with a peak at 385 nm and a shoulder at 430 nm. The enzyme is reduced upon addition of the substrate methyltetrahydrofolate. The electron paramagnetic resonance spectrum[5] of the substrate-reduced enzyme has g values at 1.93 and 2.02 for the reduced iron–sulfur clusters and at 2.00 for the semiquinone form of the enzyme-bound FAD. The presence of the semiquinone intermediate suggests that the reductase transfers electrons one at a time during catalysis. Thus, the catalytic process of the *C. formicoaceticum* enzyme seems to differ from that of the enzyme from pig liver. This enzyme is also a flavoenzyme, but does not appear to have a semiquinone intermediate nor is there evidence for the involvement of metals.[8]

Assays have been performed in the physiological direction [Eq. (1)] of the *C. formicoaceticum* reductase to determine the ability of the enzyme to transfer electrons from various reduced carriers to 5,10-methylenetetrahydrofolate.[4] The enzyme is able to transfer reducing equivalents from $FADH_2$ and reduced ferredoxin to methylenetetrahydrofolate in spectrophotometrically monitored reactions. However, the enzyme does not transfer electrons from reduced rubredoxin or from NADH and NADPH to methylene-H_4folate. The fact that reduced ferredoxin and $FADH_2$ serve as electron donors and rubredoxin does not is compatible with their respective electron potentials (E_0'). Although ferredoxin and rubredoxin have been purified from *C. formicoaceticum*,[21] their electron potentials have not been determined. However, ferredoxins normally have potentials at or lower than -360 mV, and those for rubredoxins are from -60 to $+50$ mV. The potential for $FAD/FADH_2$ is -219 mV. The potential for 5,10-methylenetetrahydrofolate/5-methyltetrahydrofolate is -120 mV[4] as was calculated from data published by Katzen and Buchanan.[6]

---

[21] S. W. Ragsdale and L. G. Ljungdahl, *J. Bacteriol.* **157,** 1 (1984).

[62] Assay and Detection of the Molybdenum Cofactor

By R. V. HAGEMAN and K. V. RAJAGOPALAN

All of the known biological functions of molybdenum are carried out by enzymes which contain the metal in tight association. All the molybdoenzymes characterized to date (listed in Table I) are in fact complex proteins containing not only Mo but other prosthetic groups as well. Among these, xanthine dehydrogenase, aldehyde oxidase, purine hydroxylase, and pyridoxal oxidase each catalyze the oxidative hydroxylation of a wide variety of substrates such as purines, pteridines, pyridines, and aliphatic and aromatic aldehydes, using the elements of water, and are commonly referred to as molybdenum hydroxylases. The same type of reaction is catalyzed by sulfite oxidase, formate dehydrogenase, carbon monoxide oxidase, and nicotinic acid hydroxylase, but these en-

METHODS IN ENZYMOLOGY, VOL. 122

TABLE I

MOLYBDENUM-CONTAINING ENZYMES FROM DIVERSE SOURCES

Enzyme	Source	Reaction catalyzed	Prosthetic groups
Xanthine dehydrogenase	Animals	$RH + H_2O \rightarrow ROH + 2e + 2H^+$	Mo, MPT,[a] FAD, Fe/S
	Plants		
	Microorganisms		
Aldehyde oxidase	Animals	$RH + H_2O \rightarrow ROH + 2e + 2H^+$	Mo, MPT, FAD, Fe/S
	Drosophila melanogaster	$RH + H_2O \rightarrow ROH + 2e + 2H^+$	
Pyridoxal oxidase	Animals	$RH + H_2O \rightarrow ROH + 2e + 2H^+$	Mo, MPT, FAD, Fe/S
	Drosophila melanogaster	$RH + H_2O \rightarrow ROH + 2e + 2H^+$	
Nicotinic acid hydroxylase	Bacteria	$RH + H_2O \rightarrow ROH + 2e + 2H^+$	Mo, MPT, FAD, Fe/S, Se
Purine hydroxylase	Aspergillus nidulans	$RH + H_2O \rightarrow ROH + 2e + 2H^+$	Mo, MPT, FAD, Fe/S
Sulfite oxidase	Animals	$SO_3^{2-} + H_2O \rightarrow SO_4^{2-} + 2e + 2H^+$	Mo, MPT, heme
	Plants		
	Bacteria		
CO dehydrogenase	Bacteria	$CO + H_2O \rightarrow CO_2 + 2e + 2H^+$	Mo, MPT, FAD, Fe/S
Nitrate reductase, assimilatory	Plants	$NO_3^- + 2e + 2H^+ \rightarrow NO_2^- + 2H_2O$	Mo, MPT, FAD, heme
	Microorganisms		
Nitrate reductase, respiratory	Bacteria	$NO_3^- + 2e + 2H^+ \rightarrow NO_2^- + H_2O$	Mo, MPT, Fe/S
Formate dehydrogenase	Bacteria	$HCOOH + H_2O^* \rightarrow O{=}C{=}O^* + 2e + 2H^+$	Mo (or W), MPT, Se, Fe/S
Biotin sulfoxide reductase	Bacteria	Biotin sulfoxide $+ 2e + 2H^+ \rightarrow$ biotin $+ H_2O$	Mo, MPT
Trimethylamine N-oxide reductase	Bacteria	$[(CH_3)_3N \rightarrow O] + 2e + 2H^+ \rightarrow (CH_3)_3N + H_2O$	Mo, MPT
Nitrogenase	Microorganisms	$N_2 + 6e + 8H^+ \rightarrow 2NH_4^+$	Fe/S, Fe-Mo-cofactor

[a] MPT is molybdopterin.

zymes show rigorous specificity for their substrates. The reactions catalyzed by nitrate reductases, biotin sulfoxide reductase, and N-oxide reductases can be considered to be the reverse of the hydroxylase reaction. Finally, a fundamentally different type of reaction is carried out by nitrogenase, an enzyme which is markedly different from the other molybdoenzymes in other respects as well.

The strength of association of Mo to the enzyme proteins has long been evident from the stoichiometric retention of the metal in purified molybdoproteins. Yet, until recently there has been no information on the nature of the forces involved in the stable interaction between the metal and the various apoproteins. The initial breakthrough in this area was reported in 1963 when Cove and Pateman first suggested the existence of a molybdenum-containing organic cofactor common to several molybdoenzymes. The hypothesis arose from their characterization of a group of mutants of *Aspergillus nidulans* which lacked nitrate reductase and xanthine dehydrogenase activities.[1] Nason and co-workers substantiated the hypothesis in studies of the *Neurospora crassa* mutant *nit-1,* which similarly lacked nitrate reductase and xanthine dehydrogenase activities. These investigators showed that nitrate reductase activity could be reconstituted in extracts of *nit-1* by a variety of bacterial, plant, and animal molybdenum cofactor sources.[2] Many systems involving cofactor mutations or demolybdoenzymes which can be reconstituted have since been described.[3] The ability to reconstitute molybdoenzymes using such a wide variety of cofactor sources strongly supported the existence of a universal molybdenum cofactor common to all molybdoenzymes except nitrogenase.

Due to the extreme lability of the active cofactor, the nature of the organic moiety eluded characterization until 1980, when Rajagopalan and co-workers succeeded in isolating a stable, catalytically inactive form derived from the cofactor.[4] Further studies indicated that the cofactor is a novel 6-alkylpterin containing sulfur and a monophosphate ester.[5] Recent work has established the structures of two stable derivatives, Form A and Form B (Fig. 1, A and B), which result from oxidative modification of the

[1] J. A. Pateman, D. J. Cove, B. M. Rever, and D. B. Roberts, *Nature (London)* **201,** 58 (1964).
[2] A. Nason, K.-Y. Lee, S.-S. Pan, P. A. Ketchum, A. Lamberti, and J. DeVries, *Proc. Natl. Acad. Sci. U.S.A.* **68,** 3242 (1971).
[3] J. L. Johnson, *in* "Molybdenum and Molybdenum-Containing Enzymes" (M. P. Coughlan, ed.), p. 347. Pergamon, Oxford, 1980.
[4] J. L. Johnson, B. E. Hainline, and K. V. Rajagopalan, *J. Biol. Chem.* **255,** 1783 (1980).
[5] J. L. Johnson and K. V. Rajagopalan, *Proc. Natl. Acad. Sci. U.S.A.* **79,** 6856 (1982).

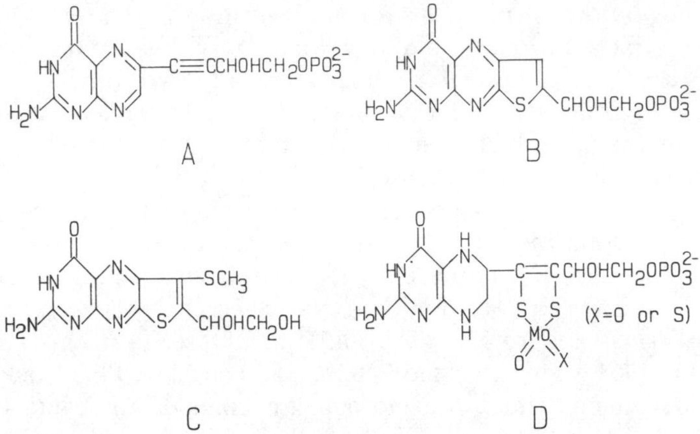

FIG. 1. Structures of Form (A), Form B (B), and urothione (C) and proposed structure of Mo cofactor (D).

cofactor.[6] Form B is structurally related to urothione,[5] a sulfur-containing pterin found in human urine (Fig. 1C).[7] The above evidence has led to the proposed model for the active molybdenum–pterin complex shown in Fig. 1D.[5]

Unlike other tightly bound prosthetic groups such as thiamine pyrophosphate, flavin coenzymes, pyridoxal phosphate, biotin, heme, and cobamide coenzymes, the reduced, metal-free pterin of the molybdenum cofactor, termed molybdopterin, is extremely unstable and has not been detected in the free state. When separated from molybdoenzymes or carrier molecules it undergoes rapid chemical modification and inactivation within a short period. Because of this lability neither the intact Mo cofactor nor molybdopterin has yet been isolated in a pure state. The lability presumably stems from the tendency of the pterin ring to become oxidized, and from the proposed enedithiol group in the side chain.

This instability imposes some constraints on the methods for quantitative measurement of the cofactor. The procedures currently available include biological assay of activity by reconstitution of the apoprotein of nitrate reductase in extracts of *Neurospora crassa nit-1* and chemical assay by oxidative degradation of molybdopterin to Form A or Form B mentioned above. The optimum conditions for these assays are described in this chapter.

[6] J. L. Johnson, B. E. Hainline, K. V. Rajagopalan, and B. H. Arison, *J. Biol. Chem.* **259**, 5414 (1984).

[7] M. Goto, A. Sakurai, K. Ohta, and H. Yamakami, *J. Biochem.* (*Tokyo*) **65**, 611 (1969).

Reconstitution Assay

Growth of Neurospora. *Neurospora crassa* mutant strain *nit-1*, allele 34547, obtained from the Fungal Genetics Stock Center, Humboldt State University Foundation, Arcata, CA, is grown on Fries' basal medium[8] containing (in g/liter) sucrose (20), NH_4Cl (4.3), sodium potassium tartrate (7.7), KH_2PO_4 (1.0), $MgSO_4 \cdot 7H_2O$ (0.5), NaCl (0.1), and $CaCl_2$ (0.1). Biotin (5 μg/liter) and a trace element solution (1.0 ml/l) consisting of (per liter) 8.8 g $ZnSO_4 \cdot 7H_2O$, 960 mg $FeCl_3 \cdot 6H_2O$, 270 mg $CuCl_2$, 88 mg $Na_2B_4O_7 \cdot 10H_2O$, 87 mg $Na_2MoO_4 \cdot 2H_2O$, and 72 mg $MnCl_2 \cdot 4H_2O$ are also added. For slants, the above medium is supplemented with 0.5% casein hydrolysate (Sigma) and 2.0% agar (Difco). For nitrate induction media, the NH_4Cl is replaced with 6.4 g $NaNO_3$ per liter of medium.

Slants are either started from silica gel stocks or transferred from existing cultures, kept at 30° for 1 day, and then at room temperature in the light for 7–21 days. Inocula are prepared by adding 5 ml water to a 10- to 20-day-old slant, followed by vigorous shaking to suspend the conidia. The suspension is decanted into a sterile tube, and standardized by measuring the A_{600} after appropriate dilution. For growth of mycelia, 4 to 8 A_{600} units of the conidial suspension (about 0.2 ml) were used to inoculate 800 ml of medium in a 2.8-liter Fernbach flask. Flasks are incubated at room temperature for 41 hr with vigorous shaking for aeration. Cultures are harvested by filtration, washed with distilled water, and resuspended in nitrate induction medium. After 5 more hours, the mycelial mat is harvested, washed in cold water, broken into small pieces, frozen in liquid nitrogen, and stored at −70°. These methods are derived from Amy and Rajagopalan.[9]

Although most work on nitrate reductase in *N. crassa* has used mycelia grown on Fries' medium, Vogel's salts has been recommended for its convenience in use.[8] It is our experience that the reconstitution activity of the extract prepared from mycelia grown on Vogel's salts is much less stable than that of the extract of mycelia grown on Fries' medium. The latter extract retains much of its activity after storage for 4 hr on ice while the former extract loses essentially all of its activity. This presumably results from the destruction of the apo nitrate reductase in the extract.

Preparation of Cell-Free Extracts. Crude extracts are prepared by homogenizing mycelia in 2 volumes of 0.1 *M* potassium phosphate buffer, pH 7.0 (KP_i buffer), 1 m*M* EDTA, 1 m*M* phenyl methyl sulfonyl fluoride (PMSF), 1 m*M* dithiothreitol (DTT). DTT and PMSF are both degraded slowly in this buffer and thus should be added fresh for best results. All

[8] R. H. Davis and F. J. deSerres, this series, Vol. 17 Part A, p. 19.
[9] N. K. Amy and K. V. Rajagopalan, *J. Bacteriol.* **140**, 114 (1979).

operations are carried out at 0–4°. Homogenization is performed in a cold Duall tissue grinder (Kontes), and the resulting brei is centrifuged at 27,000 g for 15 minutes. The supernatant is decanted and is designated the *nit-1* extract.

Reconstitution of Nitrate Reductase in nit-1 Extracts. Reconstitution experiments are carried out by mixing 100 μl of reconstitution buffer (25 mM KP$_i$ buffer, 25 mM Na$_2$MoO$_4$, and 0.25 mM EDTA) with 1–50 μl of cofactor source, 100 μl of *nit-1* extract, and water to final volume of 250 μl. Reconstitution is allowed to proceed either for 10 min at room temperature (standard assay) or for 5 hr at 15° (quantitative assay). The presence of molybdate in the reconstitution mixture ensures that uncomplexed molybdopterin present in the cofactor sample will also be measured.

Assay of Nitrate Reductase Activity. An aliquot of the reconstitution mixture is used for the assay of nitrate reductase activity.[9,10] Assays are started by the addition of NADPH, and usually stopped by the addition of 1% sulfanilamide in 3 N HCl, followed by the addition of 0.02% N-1-naphthylethylenediamine di-HCl. We have found that color development in this reaction is inhibited by reducing agents in general, and NADPH in particular. For routine work, the interference by NADPH can be corrected for by the use of an appropriate standard curve. However, when accurate values for the activity of the nitrate reductase are required, the barium acetate–ethanol precipitation procedure of Lafferty and Garrett[11] is used to remove the NADPH. DTT and ascorbate do not interfere with the diazo coupling at the levels usually present in the nitrate reductase assay as performed.

We have examined the effect of growth phase on the reconstitution activity of the *nit-1* mycelia by varying the size of the inoculum to control the growth stage of the mycelium. This was technically more convenient than varying the time of harvest and induction. A conidial suspension in water from a 10- to 20-day-old slant was calibrated by measuring A_{600}, and liquid cultures were inoculated with various volumes of the suspension. After 41 hr of growth at room temperature, the mycelia were harvested, washed, and resuspended in nitrate induction medium for 5 hr with vigorous shaking. Extracts of the mycelia were then compared for activity in reconstitution and stability experiments. The highest reconstitution was obtained with mycelia grown with an inoculum of 4–8 A_{600} units of conidial suspension (0.7–1.5 × 10^7 conidia). This corresponds to a yield of 5–6 g mycelia (dry weight) per liter. Allowing the growth to attain stationary phase by using larger inocula caused a decrease in the stability of the *nit-1*

[10] R. H. Garrett and A. Nason, *J. Biol. Chem.* **255,** 2870 (1969).
[11] M. A. Lafferty and R. H. Garrett, *J. Biol. Chem.* **249,** 7555 (1974).

extracts. Our findings indicate that some of the inconsistencies that have been reported in utilizing the *nit-1* system for measurement of molybdenum cofactor can be traced to the growth conditions used for the *nit-1* mycelia. It is apparent that if the inoculum is not standardized, some batches of *nit-1* mycelia could be in the optimal region of the growth curve, while other batches would not be, thus leading to erratic results.

The efficiency of the reconstitution of the *nit-1* nitrate reductase depends on the temperature at which the reaction is carried out. The time courses of activation at several different temperatures are shown in Fig. 2. As can be seen, the rate of the reconstitution reaction and the extent of reconstitution both vary with temperature. The maximum activity attained is essentially constant up to 20°, but decreases rapidly at higher temperatures. For quantitative assay of the cofactor the reconstitution is carried out at 15° for 5 hr, conditions in which the reaction proceeds at a reasonable rate and to the maximum extent.

Release of Mo Cofactor from Molybdoenzymes. Crude extracts of fungi, bacteria, and animal tissues contain storage forms of the cofactor in which the cofactor is loosely attached to stabilizing macromolecules, and which can reconstitute the nitrate reductase in *nit-1* extracts without any treatment to release the cofactor.[3] In contrast, the cofactor bound to purified molybdoenzymes is unavailable for reconstitution unless the do-

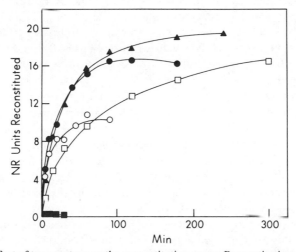

FIG. 2. Effect of temperature on the reconstitution assay. Reconstitution was carried out at the indicated temperatures with 100 μl of *nit-1* extract, 50 μl of water, 100 μl of reconstitution buffer, and 1 μl of cofactor source (1.38 μM sulfite oxidase in 6 M GuHCl). Aliquots were removed at intervals and assayed for nitrate reductase activity. (□) 10°; (▲) 15°; (●) 20°; (○) 25°; and (■) 37°.

nor proteins are denatured before mixing with *nit-1* extracts. Acid treatment,[12] organic solvents,[13] heat,[9] and SDS[3] have been utilized as means of denaturing the enzymes. Little comparative data are available on the relative efficiencies of different denaturation methods for releasing molybdenum cofactor from molybdoenzymes. We have examined the relative efficiency of different denaturation procedures on the release of reconstitutively active molybdenum cofactor from sulfite oxidase.

Reconstitutions were carried out either for 10 min at room temperature or for 5 hr at 15°. The room temperature reconstitution measures the initial rate of reconstitution, while the 15° reconstitution measures the quantitative efficiency of transfer of the cofactor from the donor sulfite oxidase to the *nit-1* nitrate reductase. The sulfite oxidase stock solution was 75 μM in 10 mM Tris–HCl (pH 8.0) and was diluted 50-fold into the appropriate buffer for denaturation. Aliquots (1 μl) were withdrawn after denaturation and used for reconstitution as described. Assays of the 15° reconstitutions were stopped with Ba–EtOH, whereas assays of the room temperature reconstitutions were stopped with the acid–sulfanilamide. The various denaturation treatments were carried out by dilution of 10 μl sulfite oxidase into 490 μl of (1) 6 M GuHCl, 10 mM Tris (pH 8.0) for 10 min at 0°; (2) 1% SDS, 10 mM Tris (pH 8.0) at room temperature for 10 min; (3) 98% formamide, 2% water, 10 mM Tris (pH 8.0) for 10 min at 0°; (4) 10 mM Tris (pH 8.0), heating at 100° for 30 sec, then cooling on ice. Tris treatment (untreated control) consisted simply of dilution into 10 mM Tris (pH 8.0) at 0°. Acid treatment was performed by diluting 10 μl of the sulfite oxidase stock with 140 μl of water at 0°, and adding 50 μl of ice-cold 0.1 M H$_3$PO$_4$ (final pH 2.5). After 1 min, 40 μl of 0.2 M Tris base was added (final pH 7.5). All solutions for denaturation contained 10 mM sodium ascorbate.

Table II shows the differences in the initial rate of reconstitution, as measured at room temperature, and the maximum extent of reconstitution, as measured at 15°, among the different denaturation methods. GuHCl and SDS appear to have equal effectiveness for releasing cofactor. Formamide is considerably less effective, while heat and acid treatment yield much less activity. The undenatured sulfite oxidase appears to release some cofactor to the nitrate reductase of *nit-1*, but this could well represent denatured protein resulting from freezing and thawing. For reasons of convenience and efficiency, denaturation with 6 M GuHCl is the preferred procedure for release of Mo cofactor from molybdoenzymes in the biological assay.

[12] P. A. Ketchum, H. Y. Cambier, W. A. Frazier, III, G. H. Madansky, and A. Nason, *Proc. Natl. Acad. Sci. U.S.A.* **66,** 1016 (1970).
[13] P. T. Pienkos, V. K. Shah, and W. J. Brill, *Proc. Natl. Acad. Sci. U.S.A.* **74,** 5468 (1977).

TABLE II
EFFECT OF DENATURATION CONDITIONS ON
RELEASE OF MO COFACTOR ACTIVITY FROM
SULFITE OXIDASE

Denaturation conditions	Units of nitrate reductase formed/pmol sulfite oxidase	
	Standard assay	Quantitative assay
GuHCl	8.6	18.8
SDS	9.5	15.6
Formamide	2.8	7.5
Heat	2.2	4.9
Acid	0.4	3.2
Tris	0.1	1.0

Quantitative Transfer of Molybdenum Cofactor to nit-1. We have examined a number of different molybdoenzymes for their ability to reconstitute the apo nitrate reductase of *nit-1* extracts after GuHCl denaturation. Reconstitutions were carried out in the presence of 10 mM Na_2MoO_4 at 15°, utilizing a stable *nit-1* extract to maximize the efficiency of transfer. The results are presented in Table III. The amount of nitrate reductase formed after reconstitution is expressed as nanomoles NO_2^- formed per minute per nanomole cofactor in the donor protein, assuming that each subunit contains a cofactor molecule. To determine the efficiency of transfer, the turnover number of the *N. crassa* nitrate reductase was taken to be 20,000 nmol NO_2^- formed per minute per nanomole Mo, reported by Jacob and Orme-Johnson.[14] Using this number, it can be seen that the transfer of the molybdenum cofactor from a donor molecule to the *nit-1* nitrate reductase is a highly efficient process. Some enzymes appear to give somewhat lower transfer efficiencies, but this may well reflect some active site heterogeneity of the purified proteins. The similar efficiencies observed with the different proteins provide an indication that the molybdenum cofactor from these different enzymes is in fact highly related structurally, and perhaps identical. Also, the essentially 100% efficiency observed with the different enzymes allows this reconstitution method to quantitatively estimate the amount of cofactor present in a sample.

[14] G. S. Jacob and W. H. Orme-Johnson, *in* "Molybdenum and Molybdenum-Containing Enzymes" (M. P. Coughlan, ed.), p. 327. Pergamon, Oxford, 1980.

TABLE III

AMOUNT OF COFACTOR ACTIVITY RELEASED FROM DIFFERENT
MOLYBDENUM-CONTAINING ENZYMES[a]

Enzyme	Source	Units of nitrate reductase/nmol Mo	Efficiency (% of theory)
Sulfite oxidase	Chicken liver	22,000 ± 1000 (18)	110
Xanthine dehydrogenase	Chicken liver	22,700 ± 800 (4)	113
Xanthine oxidase	Bovine milk	18,400 ± 800 (3)	92
Purine hydroxylase II[b]	Aspergillus nidulans	16,800 (2)	84
Carbon monoxide oxidase[c]	Pseudomonas carboxydovorans	15,100 ± 1500 (8)	75
Nitrate reductase[d]	Chlorella vulgaris	16,600 ± 1500 (4)	83

[a] Proteins were denatured by diluting a concentrated stock solution into 6 M GuHCl at 0° for 10–20 min. Aliquots of 1 μl were removed and mixed with 100 μl of reconstitution buffer and 150 μl *nit-1* extract for 5 hr at 15°. Aliquots were then removed and assayed for nitrate reductase activity. Activity is shown with its associated standard error, with the number of observations in parentheses.

[b] Gift of M. P. Coughlan.

[c] Gift of O. Meyer.

[d] Gift of L. Solomonson.

Preparation of Oxidized Molybdopterin Derivatives

Even though direct chemical quantitation of molybdopterin has not been possible because of its lability, qualitative detection has been accomplished by oxidative modification to stable fluorescent derivatives. Two well-characterized products, Form A and Form B (see above), can be readily obtained and identified.[5,6] The procedures for preparing, assaying, and isolating these derivatives are described below.

For generation and isolation of Form A from a purified molybdoenzyme, a protein sample (2 mg/ml in 0.01 M Tris–HCl, pH 7.0) is adjusted to pH 2.5 with HCl. A 1% I_2/2% KI solution is prepared by dissolving the solids in a minimal volume of water and then diluting to the required final volume. The iodine solution is added to the acidified protein at a ratio of 1 : 20 (v/v), and the sample is heated for 20 min in a boiling water bath, cooled, and centrifuged at 35,000 g for 10 min to remove any precipitate.

Form A is the predominant product of this modification procedure, and can be quantitated at this stage by its fluorescence spectrum in 1 N NH$_4$OH. Even though the conversion of molybdopterin to Form A is only about 50% efficient,[6] it has been shown that this yield is reproducible even with different molybdoenzymes.[15]

[15] O. Meyer and K. V. Rajagopalan, *J. Bacteriol.* **157**, 643 (1984).

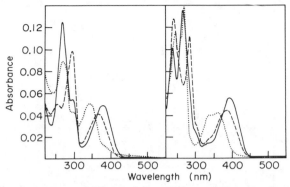

FIG. 3. Absorption spectra of Form A (left) and Form B (right) at pH 13 (——), pH 6.8 (---), and pH 1 (···). From Johnson *et al.*[6]

For isolation of Form A, the clarified solution is applied to a column (2.5 × 34 cm) of Sephadex G-25 (fine) equilibrated with 0.01 M acetic acid and is eluted with the same solvent. The fluorescent fractions are pooled, adjusted to pH 8 with concentrated NH_4OH, and applied to a column of QAE-Sephadex (acetate form) equilibrated with H_2O. The column is washed with H_2O and 0.01 M acetic acid, and the fluorescent material is eluted with 0.01 M HCl. The product of this isolation procedure, Form A, can be dephosphorylated using alkaline phosphatase.[6] Isolation of Form B of the molybdenum cofactor is carried out under identical conditions but without the addition of the iodine solution during the heat step. The absorption spectra and fluorescent properties of purified Forms A and B are shown in Figs. 3 and 4, respectively.

The method described above can be adapted for converting molybdopterin in intact bacterial or tissue cells to its oxidized, fluorescent degradation products, Forms A and B. The procedure used for obtaining Form A from *Escherichia coli* is as follows.[16] *Escherichia coli* cells grown under test conditions are harvested by centrifugation, and resuspended in 5 ml water per gram wet weight cells. The suspension is adjusted to pH 2.35 with 1 N HCl and mixed with 1 ml of 1% I_2/2% KI. The sample is then placed in a boiling water bath, and subsequently shielded from light throughout the following procedures. After 10 min boiling, an additional 1 ml I_2/KI is added, and boiling continued for 20 min more. After cooling, the boiled cell suspension is centrifuged at 12,000 g for 5 min. The supernatant is applied to a Sephadex G-25 column (1.7 × 62 cm), previously equilibrated with 0.01 N acetic acid, and eluted with the same solution. The broad fluorescent peak eluting near the included volume is pooled, adjusted to pH 8.5 with 1 M ammonium hydroxide, and made 7.5 mM in

16 M. E. Johnson and K. V. Rajagopalan, unpublished data.

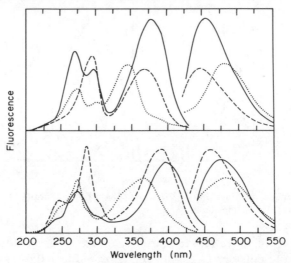

FIG. 4. Fluorescence spectra of Form A (top) and Form B (bottom) at pH 13 (———), pH 6.8 (---), and pH 1 (⋯). The relative instrument settings were 30 (———), 100 (---), and 10 (⋯) in the top panel and 30 (———), 100 (---), and 3 (⋯) in the bottom panel. From Johnson *et al.*[6]

Mg^{2+} with 1.0 M magnesium chloride. A 0.50-ml aliquot of alkaline phosphatase, 1 mg/ml in 0.1 M ammonium bicarbonate, is added and the mixture incubated for 12 hr at 22°. The sample is then applied to a 0.5 × 2.5-cm column of QAE-Sephadex (acetate form) equlibrated with water, and eluted with 0.01 N acetic acid. Fluorescence of column fractions is

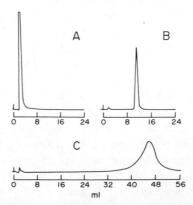

FIG. 5. HPLC elution properties of Form B (A), dephospho Form B (B), and the periodate-cleavage product of the latter (C) from a C_{18} reverse-phase column (4.6 × 250 mm) with 20% methanol as the solvent at a flow rate of 1 ml/min. From Johnson and Rajagopalan.[5]

TABLE IV
HPLC ELUTION BEHAVIOR OF OXIDIZED
MOLYBDOPTERIN DERIVATIVES

Derivative	Elution volume (ml)	
	Form A	Form B
Phospho	4[a]	4[a]
Dephospho[b]	8	11
Periodate-treated dephospho[c]	15	46

[a] Void volume of the column.
[b] Obtained by treatment of Form A or Form B with calf intestinal alkaline phosphatase.
[c] Prepared by treatment of dephospho Forms A and B with 10 mM sodium metaperiodate in 0.01 N acetic acid for 1 min.

monitored with excitation at 380 nm and emission at 460 nm. The fluorescent peak fractions are pooled and used for further analysis by HPLC. Form B and its dephospho derivative are prepared by a similar method but without addition of iodine during the extraction step.

The procedure used for whole cells yields a mixture of fluorescent compounds, and the fluorescence spectrum of the sample usually does not resemble those of pure Form A or Form B. The presence of the molybdopterin derivatives in such preparations can be established by HPLC chromatography. The HPLC elution behavior of these compounds is summarized in Fig. 5 and Table IV. Forms A and B, which are phosphorylated, are eluted in the void volume of the C_{18} reverse-phase column using 20% methanol as the solvent, whereas the dephospho forms are significantly retarded.

It is seen from Fig. 1 that the dephospho derivatives of Forms A and B contain vicinal hydroxyls, making them susceptible to cleavage by periodate. This periodate sensitivity is also depicted in Fig. 5, and is an important means of verification of the identity of these molybdopterin derivatives.

Summary

This chapter describes procedures for the biological assay of the molybdenum cofactor and for the preparation of stable, easily identifiable derivatives of molybdopterin, the organic moiety of the cofactor. Using these procedures or context-dependent modifications thereof, it has been shown that *N. crassa nit-1* cells are deficient in molybdopterin and that reconstitution of extracts of induced cells of the organism leads to specific incorporation of the pterin into nitrate reductase.[17] The labile existence of a molybdopterin–Mo complex and its sensitivity to sulfhydryl reagents have also been demonstrated.[18] Withal the biochemistry of the cofactor is still at its evolving stage, these procedures should be helpful in obtaining additional information on this extremely unstable compound.

[17] S. Kramer, R. V. Hageman, and K. V. Rajagopalan, *Arch. Biochem. Biophys.* **233**, 821 (1984).
[18] R. C. Wahl, R. V. Hageman, and K. V. Rajagopalan, *Arch. Biochem. Biophys.* **230**, 264 (1984).

[63] Methanopterin and Tetrahydromethanopterin Derivatives: Isolation, Synthesis, and Identification by High-Performance Liquid Chromatography

By Jan T. Keltjens, Gerda C. Caerteling, and Godfried D. Vogels

Methanopterin, N-[1'-(2''-amino-4''-hydroxy-7''-methyl-6''-pteridinyl)-ethyl]-4-[2',3',4',5'-tetrahydroxypent-1'-yl-(5' → 1'')-O-α-ribofuranosyl-5''-phosphoric acid]aniline, in which the phosphate group is esterified with α-hydroxyglutaric acid, is a complex pterin derivative structurally related to folic acid (Fig. 1).[1,2] The compound is present in methanogenic bacteria. *Methanosarcina barkeri* contains a derivative with an additional glutamic acid bounded to the α-hydroxyglutaric residue.[2] In the process of methanogenesis the reduced form of methanopterin (5,6,7,8-tetrahydromethanopterin, H$_4$MPT) functions as a carrier of one-carbon units at three consecutive reduction levels.[3] As in tetrahydrofolate biochemistry[4,5] the

[1] P. van Beelen, A. P. M. Stassen, J. W. G. Bosch, G. D. Vogels, W. Guijt, and C. A. G. Haasnoot, *Eur. J. Biochem.* **138**, 563 (1984).
[2] P. van Beelen, J. F. A. Labro, J. T. Keltjens, W. J. Geerts, G. D. Vogels, W. H. Laarhoven, W. Guijt, and C. A. G. Haasnoot, *Eur. J. Biochem.* **139**, 359 (1984).
[3] J. C. Escalante-Semerena, K. L. Rinehart, Jr., and R. S. Wolfe, *J. Biol. Chem.* **259**, 9447 (1984).
[4] J. C. Rabinowitz, *in* "The Enzymes" (P. D. Boyer, H. Lardy, and K. Myrbäck, eds.), 2nd Ed., Vol. 2, Part A, p. 185. Academic Press, New York, 1960.
[5] R. L. Blakley, "The Biochemistry of Folic Acid and Related Pteridines." North-Holland Publ., Amsterdam, 1969.

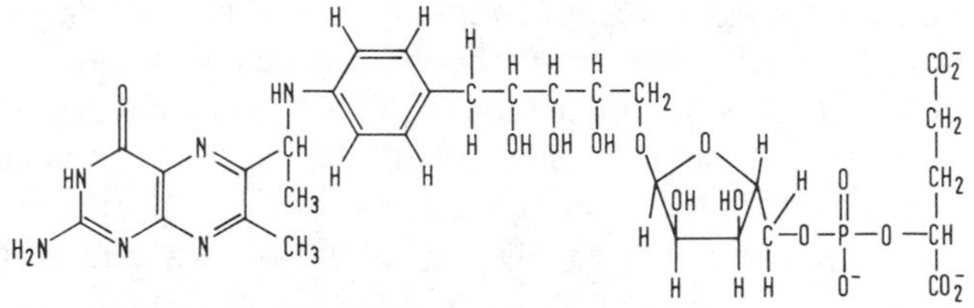

FIG. 1. Structure of methanopterin established by van Beelen *et al.*[1]

following intermediates are encountered: 5- and 10-formyl-5,6,7,8-tetrahydromethanopterin (5- and 10-formyl-H$_4$MPT), 5,10-methenyl-5,6,7,8-tetrahydromethanopterin (5,10-methenyl-H$_4$MPT), 5,10-methylene-5,6,7,8-tetrahydromethanopterin (5,10-methylene-H$_4$PMT), and 5-methyl-5,6,7,8-tetrahydromethanopterin (5-methyl-H$_4$MPT).

Here we describe the isolation and purification procedures for methanopterin, H$_4$MPT, and 5,10-methenyl-H$_4$MPT, the synthesis of formyl-H$_4$MPT derivatives, 5,10-methylene-H$_4$MPT, and 5-methyl-H$_4$MPT, and the identification of the compounds by high-performance liquid chromatography (HPLC).

Materials

Mass Culture of Cells

Procedures for maintenance, mass culturing, harvesting, and storage of methanogenic bacteria have been described in a preceding volume of this series.[6]

The amount of 5,10-methenyl-H$_4$MPT may be increased—at the expense of the other (tetrahydro)methanopterin derivatives—by incubating a thick slurry of cells with 10^5 Pa hydrogen for 15 min at the optimal growth temperature in a bottle under vigorous shaking.[7] Subsequently, an anoxic solution of 3 ml 1 M Na$_2$CO$_3$ per kilogram wet cells is injected, the bottle is evacuated and gassed with 10^5 Pa hydrogen, and the incubation is continued. After about 0.2 mmol CH$_4$ per kilogram wet cells has been formed, the incubation is stopped by cooling the cells to 0°.

Extraction of Methanopterin Derivatives

One kilogram of wet cells is mixed with 200 ml distilled water, the pH is adjusted to 4.5 with acetic acid, and subsequently 600 ml ethanol is added.

[6] J. A. Romesser and W. E. Balch, this series, Vol. 67, p. 545.
[7] P. van Beelen, J. W. van Neck, R. M. de Cock, G. D. Vogels, W. Guijt, and C. A. G. Haasnoot, *Biochemistry* **23**, 4498 (1984).

The suspension is heated in an evacuated bottle under vigorous shaking at 80° for 30 min. After cooling to 0° the cells are centrifuged (9,000 g, 30 min), the supernatant is decanted, and the pellet is extracted twice more with 60% aqueous ethanol. The combined supernatants are flash evaporated to about 5% of the original volume. Before and after each step, the pH is checked and adjusted to pH 4.5 either with acetic acid or with KOH.

Desalting Procedure

Salt-free preparations of methanopterin derivatives are conveniently obtained by use of Sep-Pak C_{18} cartridges (Waters Associates). Prior to use the cartridge is activated with 100% methanol and subsequently washed with 25 mM potassium acetate buffer (pH 4.5) or with 25 mM potassium phosphate buffer (pH 3) when H_4MPT is treated. Methanopterin derivatives adhere to the C_{18} material and are eluted from the cartridge with 50% aqueous methanol acidified to pH 4.5 with acetic acid, or to pH 3 in the treatment of H_4MPT. Methanol is removed by flash evaporation.

Enzyme Preparations

The conversion and assay of tetrahydromethanopterin derivatives have been studied only with crude enzyme preparations of *Methanobacterium thermoautotrophicum*.[3,7,8] Two types of coenzyme-free crude extracts are used starting from crude cell-free extract prepared as described by Romesser and Balch[6]:

G-25 Enzyme Mixture.[9,10] To crude cell-free extract, $MnCl_2$ is added in an anaerobic chamber to a final concentration of 50 mM. The solution is allowed to stand with periodic swirling for 15 min at 4° and is subsequently centrifuged in sealed stainless-steel centrifuge tubes (27,000 g, 15 min). An amount of the supernatant equal to about 5% (v/v) of the bed volume is passed at 4° through a column packed with Sephadex G-25 and placed in an anaerobic chamber. Sephadex G-25 is preswollen in anoxic 50 mM Tris–HCl buffer (pH 7.2) containing 30 mM $MgCl_2$, 1 mM dithiothreitol, and 10% (v/v) glycerol. Equilibration and elution are performed with the same buffer. The enzyme fraction elutes in the void volume separated from the low molecular weight coenzymes. An amount

[8] P. van Beelen, H. L. Thiemessen, R. M. de Cock, and G. D. Vogels, *FEMS Microbiol. Lett.* **18**, 135 (1983).
[9] J. A. Romesser and R. S. Wolfe, *Zentralbl. Bakteriol., Mikrobiol. Hyg., Abt. I, Orig. C* **3**, 271 (1982).
[10] J. A. Leigh and R. S. Wolfe, *J. Biol. Chem.* **258**, 7536 (1984).

of the enzyme fraction equal to the amount applied to the column is collected, divided in 1.5 to 2-ml portions, and stored in 5-ml serum bottles closed with butyl rubber stoppers and crimped aluminum seal caps under a positive pressure of hydrogen at −70°.

Ammonium Sulfate-Treated Extract.[3] A total of 43.6 g of finely ground ammonium sulfate is added over a period of 1 hr to 100 ml crude cell-free extract containing 1 mM dithiothreitol and present inside an anaerobic chamber at 4°. After the salt has been dissolved the solution is allowed to stand for 15 min and subsequently centrifuged (27,000 g, 15 min). The supernatant is dialyzed twice for 18 hr at 4° in 40 volumes each of anoxic 20 mM potassium phosphate buffer (pH 7) that contains 1 mM dithiothreitol. Aliquots (1.5–2 ml) of the enzyme mixture are stored as described above.

Identification and Purification of Methanopterin and 5,10-Methenyl-H₄MPT

Identification

Routinely the presence of methanopterin and 5,10-methenyl-H₄MPT is checked by thin-layer chromatography (TLC) (Table I).[11,12] Methanopterin and 5,10-methenyl-H₄MPT are identified after developing and thoroughly drying the plates; the former compound appears upon irradiation with long-wave (365 nm) ultraviolet light as a blue fluorescent spot, whereas the latter is bright-yellow fluorescent.

The analysis by means of HPLC is described in the last part of this section.

Purification

The purification[1,11–13] of methanopterin and 5,10-methenyl-H₄MPT is performed under aerobic conditions at acidic pH and 4°. Since 5,10-methenyl-H₄MPT is sensitive to light, all steps must be carried out under subdued light. Since the compounds behave quite differently on anion-exchange chromatography a common procedure for the first purification step can be used.

Anion-Exchange Column Chromatography. The ethanol extract of cells is desalted, diluted in 25 mM potassium acetate buffer (pH 4.5), and

[11] J. T. Keltjens, M. J. Huberts, W. H. Laarhoven, and G. D. Vogels, *Eur. J. Biochem.* **130,** 537 (1983).

[12] J. T. Keltjens, L. Daniels, H. G. Janssen, P. J. Borm, and G. D. Vogels, *Eur. J. Biochem.* **130,** 545 (1983).

[13] P. van Beelen, R. M. de Cock, W. Guyt, C. A. G. Haasnoot, and G. D. Vogels, *FEMS Microbiol. Lett.* **21,** 159 (1983).

TABLE I
PROPERTIES OF METHANOPTERIN, 5,10-METHENYL-H$_4$MPT, DEGRADATION PRODUCTS OF
METHANOPTERIN DERIVATIVES, AND SOME COENZYMES ENCOUNTERED DURING
PURIFICATION PROCEDURES

Compound	R_f values[a]	Retention time HPLC (min)[b]	Color of solutions	Fluorescence	
				In neutral solutions	On TLC plate
Methanopterin	0.22	12.12	None, pale yellow	None	Blue
5,10-Methenyl-H$_4$MPT	0.10	12.86	None, pale yellow	None	Yellow
7-Methylpterin	0.54	10.04	None	Blue	Blue
6-Ethyl-7-methylpterin	0.70	16.50	None	Blue	Blue
6-Ethyl-7-methyl-7,8-dihydropterin	ND[c]	13.0	None	None	None
6-Acetyl-7-methyl-7,8-dihydropterin	0.85	13.27	Yellow	Yellow	Green-yellow
Factor F$_{430}$	0.38–0.39	7.88	Yellow	None	None
Factor F$_{560}$	0.37	9.00	Purple	None	None
Coenzyme F$_{420}$	0.32	9.53	Green-yellow	Blue	Blue
FAD	0.64	13.90	Yellow	Yellow	Yellow
Formyl-methanofuran	0.42	ND	None	None	None[d]

[a] Kieselgel-60 plates (0.25 mm; Merck) developed with 1-butanol/acetic acid/water (33/12/15, v/v).
[b] Solvent system 2.
[c] ND, Not detectable.
[d] A yellow spot is obtained after spraying with 0.2% (w/v) ninhydrin in methanol.

then applied to a column packed with QAE-Sephadex A-25 (acetate form) and equilibrated with the same acetate buffer; 2 ml preswollen column material is used for each gram wet weight of cells. After application of the sample the column is washed with 1 bed volume acetate buffer and next a linear gradient (7 bed volumes: 25–500 mM potassium acetate, pH 4.5) is started. Elution is continued with 500 mM potassium acetate (pH 4.5) until methanopterin is removed from the column. Fractions are checked for the presence of 5,10-methenyl-H$_4$MPT and methanopterin by means of TLC and HPLC; the former compound is eluted at about 300 mM potassium acetate, whereas methanopterin is found at the end of the gradient.

Both compounds are pooled separately, flash evaporated, and desalted as described above.

Second Anion-Exchange Column Chromatography of 5,10-Methenyl-H₄MPT. The pool which contains 5,10-methenyl-H₄MPT is diluted in 25 mM potassium acetate buffer (pH 4.5) and applied to a column packed with about 200 ml DEAE-Sephadex A-25 (acetate form) equilibrated with the same acetate buffer. In order to remove the poorly soluble 7-methylpterin,[14] a blue fluorescent degradation product of H₄MPT, the column is thoroughly rinsed with 10 bed volumes 25 mM potassium acetate (pH 4.5). Next, the column is eluted with 20 bed volumes 200 mM potassium acetate (pH 4.5). The fractions that contain only 5,10-methenyl-H₄MPT as judged by TLC, HPLC, and ultraviolet-visible light spectroscopy are pooled, flash evaporated to 5% of the original volume, and desalted by using Sep-Pak C$_{18}$ cartridges.

Second Anion-Exchange Column Chromatography of Methanopterin. The methanopterin-containing pool is still contaminated with a number of compounds among which are factor F_{430}[15] and methanofuran (CDR factor).[10] Purification may be accomplished by chromatography using a column packed with DEAE-Sephadex A-25 (HCO$_3$⁻ form) and equilibrated with 50 mM NH₄HCO₃ (pH 8.5): a bed volume of 0.4 ml is used for each gram wet cells. The desalted methanopterin-containing pool is diluted with 50 mM NH₄HCO₃ (pH 8.5) and applied to the column. Next the column is washed with 1 bed volume equilibration buffer, followed by a linear gradient (5 bed volumes, 50–500 mM NH₄HCO₃, pH 8.5), and elution is continued with 500 mM NH₄HCO₃ until methanopterin is removed from the column. The fractions which contain only methanopterin as judged by TLC, HPLC, and ultraviolet-visible light spectroscopy are pooled, flash evaporated, and desalted as described above.

Properties of Methanopterin and 5,10-Methenyl-H₄MPT

The ultraviolet spectroscopic properties of methanopterin and 5,10-methenyl-H₄MPT are listed in Table II.

Methanopterin (M_r 772 of the acid form) is a pale yellow hygroscopic compound. Solutions show a blue fluorescence in the region of pH 2–5. The compound is stable in slightly acid, neutral, and basic solutions. At pH values below 3 the glycosidic binding between the tetrahydroxypentityl and the ribofuranosyl moieties is hydrolyzed.[2] This holds for all methanopterin derivatives.

[14] J. T. Keltjens, P. van Beelen, A. M. Stassen, and G. D. Vogels, *FEMS Microbiol. Lett.* **20,** 259 (1983).
[15] J. T. Keltjens, C. G. Caerteling, A. M. van Kooten, H. F. van Dijk, and G. D. Vogels, *Arch. Biochem. Biophys.* **223,** 235 (1983).

TABLE II
ULTRAVIOLET-VISIBLE LIGHT SPECTRAL PROPERTIES OF METHANOPTERIN DERIVATIVES

Compound	pH	λ_{max} (ε), in nm (mM^{-1} cm^{-1})			Ref.
Methanopterin	1.0	320 (10.4)			2
	3.1	339 (7.2)	277 (14.6)		
	6.0	342 (7.4)	274 (16.9)	235 (26.7)	
	11.3	358 (8.3)	251 (33.3)		
H$_4$MPT	7.0	302 (15.2)	247 (22.5)	220 (32.9)	3
5-Formyl-H$_4$MPT	8.0	283 (17.5)	245 (23.6)	220 (40.9)	18
10-Formyl-H$_4$MPT	6.0	295	245	220	18
5,10-Methenyl-H$_4$MPT	7.0	335 (21.6)	287 (13.3)	225 (30.7)	3
5,10-Methylene-H$_4$MPT	7.0	287 (16.9)	250 (32.1)	220 (34.3)	3
5-Methyl-H$_4$MPT	7.0	295 (8.6)	238 sh (15.4)	218 sh (22.9)	3

5,10-Methenyl-H$_4$MPT (M_r 787 of the acid form) is a pale yellow hygroscopic compound that fluoresces only weakly in solutions. In an acidic environment the compound is protected from oxidation by the presence of the 5,10-methenyl bridge. At higher pH values, the compound is in equilibrium with formyl-H$_4$MPT (see below) and the reduced pterin moiety is then susceptible to oxidation accompanied by degradation. The green-yellow fluorescent 6-acetyl-7-methyl-7,8-dihydropterin is one of the degradation products.[16] 5,10-Methenyl-H$_4$MPT is also sensitive to ammonia ions.[12]

Identification and Purification of 5,6,7,8-Tetrahydromethanopterin

Identification and Quantitation

The analysis of H$_4$MPT by HPLC is described in the last part of this section. In addition, the compound may be identified enzymatically as described by Escalante-Semerena et al.[3]

Principle. Formaldehyde binds nonenzymatically to H$_4$MPT, yielding 5,10-methylene-H$_4$MPT. An oxidoreductase present in G-25 enzyme mixture and in ammonium sulfate-treated extract converts 5,10-methylene-H$_4$MPT quantitatively into hydrogen and 5,10-methenyl-H$_4$MPT, which has a strong absorption band in the 340-nm region (Table II).

Reagents

Anoxic 20 mM potassium phosphate buffer (pH 7) containing 1 mM dithiothreitol and 0.3 mM formaldehyde

[16] J. T. Keltjens, H. J. Rozie, and G. D. Vogels, *Arch. Biochem. Biophys.* **229,** 532 (1984).

G-25 enzyme mixture of ammonium sulfate-treated extract. The enzyme preparations are 50-fold diluted with anoxic 20 mM potassium phosphate (pH 7) which contains 1 mM dithiothreitol

Procedure. Inside an anaerobic chamber the potassium phosphate–formaldehyde solution and 10- to 200-μl of the H$_4$MPT-containing sample are added together in a cuvette to a final volume of 3.0 ml. The cuvette is closed with a rubber stopper and transferred out of the chamber and the solution is sparged with oxygen-free nitrogen for 5 min. After preincubation for 10 min at either 37 or 60°, depending on the mesophilic or thermophilic character of the methanogenic bacteria, the reaction is initiated by adding 25 μl of diluted G-25 enzyme mixture or ammonium sulfate-treated extract by means of a gas-tight syringe. The absorbance at 340 nm is recorded until no further increase occurs. The concentration of H$_4$MPT is calculated from the increase of the absorbance, corrected for the absorbance of the enzyme mixture, and from the molar extinction coefficient $\varepsilon_{340} = 20,600 \ M^{-1} \ cm^{-1}$ of 5,10-methenyl-H$_4$MPT.

Purification of H$_4$MPT

The purification of H$_4$MPT is based with minor modifications on the method developed by Escalante-Semerena *et al.*[17] For all the preceding operations and the steps described in this section, contact of H$_4$MPT-containing solutions with air should be strictly avoided; column chromatography is performed inside an oxygen-free chamber. In addition the pH should be controlled carefully: the compound is purified at pH 3–4.5, where protonation at N-5 of the reduced pterin moiety stabilizes the compound against oxidation.

Anion-Exchange Column Chromatography. The first step involves column chromatography performed with DEAE-Sephadex A-25 (acetate form); 1 ml swollen column material is used for each gram of wet cells. The column is equilibrated with 30 mM potassium acetate buffer (pH 4) that contains 1 mM dithiothreitol. After application of the desalted ethanol extract and subsequent dilution of the extract with equilibration buffer, the column is rinsed with 1 bed volume of equilibration buffer. Elution is continued with 1 bed volume equilibration buffer containing 160 mM NaCl, followed by a linear gradient (3 bed volumes; 160–300 mM NaCl) in equilibration buffer. H$_4$MPT elutes at about 200 mM NaCl. The fractions are assayed enzymatically and those containing H$_4$MPT are pooled, concentrated by rotary evaporation, desalted, again concentrated to nearly dryness, and diluted in anoxic HPLC eluant.

[17] J. C. Escalante-Semerena, J. A. Leigh, K. L. Rinehart, Jr., and R. S. Wolfe, *Proc. Natl. Acad. Sci. U.S.A.* **81,** 1976 (1984).

High-Performance Liquid Chromatography. Preparative HPLC is performed isocratically on a reverse-phase system using C_{18} material. Detection takes place either at 254 or 260 nm. The column is eluted with 15% (v/v) methanol in 25 mM sodium formate that is acidified to pH 3 with concentrated HCl and 1 mM dithiothreitol. Prior to injection of the samples, oxygen-free nitrogen is sparged through the HPLC solvent for 40 min and sparging is continued throughout the whole operation. In addition, oxygen-leaking parts of the system are covered with a blanket of nitrogen. H_4MPT-containing samples are stored protected from oxygen and light in serum bottles, and are removed from the bottles and injected by means of a gas-tight syringe. From preliminary runs the retention time of H_4MPT is determined, together with the amount of sample which allows optimal separation from contaminating materials. The position of H_4MPT in the HPLC pattern is found after exposing a small amount of sample to air or by adding formaldehyde to a sample; in both cases the peak that has disappeared may be attributed to H_4MPT. In our system as specified below and eluted with 1 ml/min, the compound shows a retention time of 10.6 min.

H_4MPT is collected through a needle connected to the outlet and placed either in an anaerobic serum bottle or directly into a serum bottle that is flushed with nitrogen. Methanol is removed by rotary evaporation and desalting of the preparates is performed as described above.

Properties of H_4MPT

H_4MPT (M_r 776 of the acid form) is a pale yellow hygroscopic compound. Solutions of the pure compound do not show visible fluorescence. The ultraviolet spectral characteristics are described in Table II.

As noted above the compound is quite sensitive to oxygen and presumably also to light, notably at neutral and basic conditions. Oxidation results into the formation of 7,8-dihydromethanopterin and a number of blue fluorescent degradation products, e.g., 7-methylpterin, 6-ethyl-7-methylpterin, and 6-ethyl-7-methyl-7,8-dihydropterin, and the green-yellow fluorescent 6-acetyl-7-methyl-7,8-dihydropterin.

Preparation and Properties of the Other Physiologically Active Tetrahydromethanopterin Derivatives

Formyl-H_4MPT Derivatives

Two distinct formyl-H_4MPT derivatives play a role in chemical and enzymatic conversions of 5,10-methenyl-H_4MPT: an oxygen-stable compound, tentatively identified as 5-formyl-H_4MPT and an oxygen-sensitive compound, presumably 10-formyl-H_4MPT.[18] Details of these conversions are not yet fully established. When the pH of a solution of 5,10-methenyl-H_4MPT is raised to pH 12–13, the compound is quantitatively converted to the oxygen-stable 5-formyl-H_4MPT. Upon neutralization an equilibrium between 5-formyl-H_4MPT and 5,10-methenyl-H_4MPT is slowly established. 5-Formyl-H_4MPT may also be prepared enzymatically under nitrogen atmosphere from 5,10-methenyl-H_4MPT by a cyclohydrolase present in G-25 enzyme mixture from *M. barkeri*. The cyclohydrolase shows optimal activity at pH 8 in 20 mM potassium phosphate buffer. The reaction can be followed by the decrease of the absorbance at 340 nm. The ultraviolet spectroscopic properties of the oxygen-stable 5-formyl-H_4MPT are listed in Table II.

Treatment of 5,10-methenyl-H_4MPT in 20 mM potassium phosphate buffer (pH 8) with G-25 enzyme mixture from *M. thermoautotrophicum* under nitrogen yields an equilibrium mixture of the starting compound and both the oxygen-stable and the oxygen-labile formyl-H_4MPT. By using formylmethanofuran[19] as the one-carbon donor to H_4MPT in an enzymatic reaction performed with G-25 enzyme mixture of *M. thermoautotrophicum* under nitrogen gas in 20 mM potassium phosphate buffer (pH 7), only the oxygen-labile 10-formyl-H_4MPT is observed as an intermediate that is subsequently converted to 5,10-methenyl-H_4MPT.

5,10-Methylene-H_4MPT

Formaldehyde binds nonenzymatically in an equilibrium reaction and in a 1:1 stoichiometry, yielding 5,10-methylene-H_4MPT.[3] When formaldehyde is added in a more than 2.5-fold excess, the equilibrium is shifted completely toward the latter compound. This is the most convenient way to prepare 5,10-methylene-H_4MPT. All handling should be performed anaerobically, and excess formaldehyde is removed by lyophilization.[3]

5,10-Methenyl-H_4MPT is reduced to 5,10-methylene-H_4MPT with an equimolar amount of KBH_4 under anoxic conditions at 60° in 20 mM potassium phosphate buffer (pH 7). The course of the reaction is followed by measuring the decrease of the absorbance at 340 nm. After completion anoxic HCl is added to pH 3–4.5 and 5,10-methylene is recovered from the reaction mixture as described under Desalting Procedure.

Neither G-25 enzyme mixture nor ammonium sulfate-treated extract is suited to prepare 5,10-methylene-H_4MPT from 5,10-methenyl-H_4MPT under reducing conditions, since both types of enzyme preparations contain

---

[18] G. C. Caerteling, J. T. Keltjens, and G. D. Vogels, unpublished results (1985).
[19] J. A. Leigh, K. L. Rinehart, Jr., and R. S. Wolfe, *Biochemistry* **24**, 995 (1985).

the enzyme that catalyzes the reduction to 5-methyl-H_4MPT at considerable rate. In the reverse way, 5,10-methenyl-H_4MPT may be prepared from 5,10-methylene-H_4MPT under nonreducing conditions.[3] 5,10-Methylene-H_4MPT, 2 μmol in 2 ml anoxic 20 mM potassium phosphate buffer, is incubated under nitrogen atmosphere in a serum bottle with 1 mg protein from ammonium sulfate-treated extract at optimal temperature (either 37 or 60°). After completion of the reaction (10–20 min), the reaction mixture is cooled and carefully acidified to pH 4.5 with anoxic acetic acid. The insoluble material is removed by centrifugation and the supernatant is passed through a Sep-Pak C_{18} cartridge; 5,10-methenyl-H_4MPT is recovered as described under Desalting Procedure.

The ultraviolet spectroscopic properties of 5,10-methylene-H_4MPT (M_r 788 of the acid form) are listed in Table II. The compound is sensitive to oxygen.

5-Methyl-H_4MPT

Chemically, 5-methyl-H_4MPT may be prepared both from 5,10-methylene-H_4MPT[3] or from 5,10-methenyl-H_4MPT by reduction under anaerobic conditions with a large (more than 10-fold) excess of KBH_4 at pH 7 in potassium phosphate (25 mM). The reaction is performed at 60° for 10 min. After completion the reaction mixture is acidified to pH 3–4.5 and salts are removed by passage over a Sep-Pak C_{18} cartridge as described above.

Enzymatically, 5-methyl-H_4MPT is also prepared either from 5,10-methylene-H_4MPT or from 5,10-methenyl-H_4MPT under hydrogen atmosphere by the use of G-25 enzyme mixture. The reaction and purification steps are carried out as described in the preceding subsection.

5-Methyl-H_4MPT (M_r 790 of the acid form) is somewhat more stable to oxygen due to the protection of the N-5 of the pterin moiety by the methyl group. However, prolonged direct exposure of solutions to air, more than 1 hr, causes decomposition. The ultraviolet spectral characteristics are described in Table II. Solutions do not show visible fluorescence.

Identification of Methanopterin Derivatives by Reverse-Phase High-Performance Liquid Chromatography

Methanopterin derivatives may be identified by reverse-phase HPLC using a column packed with octadecylsilane bonded to microparticulate silica and gradient elution as specified below.

Apparatus and Materials

HPLC Equipment. In the authors' laboratory a precolumn (50 × 2.1 mm) of 37- to 50-μm C_{18} Corasil (Waters Associates) was either connected

to 10-μm LiChrosorb RP-18 (Merck) in a column with a length of 25 cm ×
4.0 mm i.d. (solvent system 1 and 2) or to 5-μm LiChrosorb RP-8 (Merck)
in a column with a length of 10 cm × 4.6 mm i.d. (solvent system 3). The
columns were eluted at room temperature. The solvent delivery system
consisted of Waters M6000 and M45 pumps with a 660 programmer and a
U6K injector. The detector was a Waters 450 variable-wavelength detector coupled with a Hewlett-Packard 3390 A integrator.

Solvent Systems. Routinely three solvent systems are used:

1. Solution (a) consisting of 25 mM sodium formate in glass-distilled
water acidified to pH 3.0 with concentrated HCl and (b) the same formate
buffer, which contains 50% (v/v) methanol. A linear gradient of 0–50%
solution (b) is applied in 10 min. Flow rate is kept constant at 2.0 ml/min
at 10 to 14 MPa.

2. Solution (a) consisting of 25 mM acetic acid in glass-distilled water
adjusted with KOH to pH 6.0 and (b) the same buffer containing 50%
methanol.[20] Elution is performed as under system 1.

3. Solution (a) consisting of 20 mM potassium phosphate buffer (pH
5.0) and (b) the same buffer containing 20% acetonitrile. A linear gradient
of 15–60% solution (b) is applied in 25 min. Flow rate is kept constant at
1.0 ml/min.

The solutions are made anoxic by sparging with oxygen-free nitrogen
for 40 min; sparging is continued throughout the whole operation.

Samples. Samples that were obtained in studying chemical and enzymatic conversions or from column chromatographic separations were injected without pretreatment. Samples containing H$_4$MPT derivatives
were stored anaerobically and carefully protected from light. Aliquots
(25–100 μl) that contained at least 0.1 nmol of the component were recorded at 0.04 AUFS deflection.

For the determination of methanopterin derivatives present in whole
cells, cells were extracted as described above. The pooled ethanol extracts were concentrated by rotary evaporation to 30% of the original
volume. Aliquots of 25 μl were used for analysis. This procedure can be
carried out with about 10 mg wet cells, equivalent to 1.5 mg dry weight.[20]

Operating Procedure. The integrator was started 0.5 min after injection of the sample; the gradient was started 2.0 min after injection of the
sample. After 20–30 min the integrator was stopped.

Comments

Results of HPLC separations of methanopterin derivatives achieved
by three solvent systems are given in Fig. 2,A–C and Table III. In system

[20] P. van Beelen, W. J. Geerts, A. Pol, and G. D. Vogels, *Anal. Biochem.* **131,** 285 (1983).

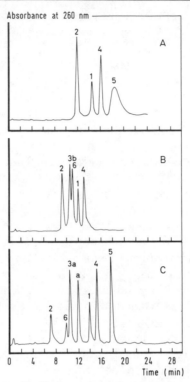

Fig. 2. High-performance liquid chromatography separation of methanopterin derivatives. A mixture of methanopterin derivatives (0.7–2.5 nmol per compound) was separated
using system 1 (A), system 2 (B), and system 3 (C) as specified in the text. The effluent was
monitored by ultraviolet absorption at 260 nm and 0.04 AUFS deflection. Identification of
peaks: 1, methanopterin; 2, H₄MPT; 3a, 5-formyl-H₄MPT; 3b, 10-formyl-H₄MPT; 4, 5,10-
methenyl-H₄MPT; 5, 5,10-methylene-H₄MPT; 6, 5-methyl-H₄MPT; a, 6-acetyl-7-methyl-
7,S-dihydropterin.

1 methanopterin, 5-methyl-H₄MPT, and 10-formyl-H₄MPT are not resolved, but the compounds are separated in solvent system 2 (Fig. 2,A
and B, Table III). However, in the latter system, 5,10-methenyl-H₄MPT
and 5,10-methylene-H₄MPT comigrate. Most methanopterin derivatives
discussed here are resolved by using a 5-μm column eluted with a linear
gradient of 3–12% acetonitrile in 20 mM potassium phosphate buffer (pH
5.0) (solvent system 3). This system is somewhat more time-consuming
compared to the two former ones.

   With the compounds tested a linear relationship between the peak
areas and the applied amounts was obtained. By increasing the sensitivity
of the detector and recording at 240 nm the method allows the detection of
1–10 pmol of a compound. When the system is operated under carefully

TABLE III

HIGH-PERFORMANCE LIQUID CHROMATOGRAPHIC PROPERTIES
OF METHANOPTERIN DERIVATIVES

Compound	Retention time (min)		
	System 1	System 2	System 3
Methanopterin	14.52	12.12	13.94
H₄MPT	11.86	9.36	7.27
5,10-Methenyl-H₄MPT	16.12	12.86	15.22
5-Formyl-H₄MPT	12.83	9.86	10.49
10-Formyl-H₄MPT	14.60	10.74	9.99
5,10-Methylene-H₄MPT	18.50	12.92	17.64
5-Methyl-H₄MPT	14.65	10.33	9.89

anaerobic conditions and when the samples are handled properly, no noticeable degradation was observed during HPLC of the oxygen-labile H₄MPT derivatives.

In our hands HPLC has proved to be a powerful tool for the study of chemical and enzymatic conversions. As described by van Beelen et al.,[20] HPLC using solvent system 2 is well suited for the quantification of oxygen-stable methanopterin derivatives and other coenzymes in whole cells of methanogenic bacteria. Though no extensive studies were performed, we feel that the method also may be extended to the oxygen-labile methanopterin derivatives. However, results obtained with one solvent system must be interpreted in combination with the results from separate runs using another solvent system, since other coenzymes may comigrate with methanopterin derivatives. It must be pointed out here that the results described above apply to those derivatives of methanopterin, whose structure is given in Fig. 1. However, *M. barkeri* contains a methanopterin derivative with an additional glutamate moiety. The presence of the latter moiety is reflected in aberrant chromatographic properties not discussed here.

Author Index

Numbers in parentheses are footnote reference numbers and indicate that an author's work is referred to although the name is not cited in the text.

Subject Index

L

L1210 mouse leukemia cells, folate transport protein, 267–269
Lactobacillus, folylpolyglutamates, 329
Lactobacillus 30a
　growth
　　crude medium for, 130
　　for enzyme purification, 129–130
　　in stock culture, 129
　histidine decarboxylase, 128–135
　mutant 3
　　culture, 136
　　histidine decarboxylase, 133–134
　prohistidine decarboxylase, 135–138
Lactobacillus casei
　culture, 351
　folate transport protein, 261–266
　folate uptake system, 262
　folylpolyglutamate synthase, 350
　transport-defective subline, folate transport protein, 266–267
Leigh's disease, 24, 29
Leucovorin, 309
Luciferase, bacterial, 147
Lumazine, reverse-phase HPLC, 274–293

M

Malonyl transacylase, 58
Melanin, biosynthesis, 273
Methanobacterium thermoautotrophicum
　crude enzyme preparations, 414
　5-deazaflavins, 205–206
Methanococcus vannielii NADPH : F_{420} oxidoreductase, 208
Methanogen 5-deazaflavin, synthetic analogs, 208
Methanogenic bacteria, mass culture of cells, 413
Methanol oxidase, flavin coenzyme, 240
Methanopterin
　identification, 415
　properties, 416–418
　purification, 415–417
　reduced form, 412–413
　reverse-phase HPLC, 424–425
　structure, 412–413
Methanopterin derivatives, 412
　degradation products, 416

desalting, 414
extraction, 413–414
reverse-phase HPLC, 422–425
UV-visible light spectral properties, 418
Methanosarcina barkeri, 206, 412
Methenyltetrahydrofolate cyclohydrolase, 373
　assay, 386–388
　bifunctional protein, 386, 391
Methenyltetrahydrofolate cyclohydrolase distribution, 386, 390–391
　in eukaryotic cells, association with trifunctional protein, 386, 390–391
　monofunctional enzyme, 386, 391
　properties, 390–391
　purification, procedure, 388–390
5,10-Methenyltetrahydrofolate cyclohydrolase, *C. formicoaceticum,* purification, 385–391
5,10-Methenyltetrahydromethanopterin, 413
　identification, 415
　properties, 416–418
　purification, 415–417
5,10-Methenyl-5,6,7,8-tetrahydromethanopterin, reverse-phase HPLC, 524–525
Methionine synthase, 373, 381
Methotrexate, 293–294, 339
Methotrexate analogs, 299
　dansyl derivatives, 297–299
　HPLC separation, 293–300
　　equipment, 295
　lysine and ornithine
　　separation, 296–300
　synthesis, 295–296
Methotrexate polyglutamate derivatives, analysis *in vivo,* 339–346
　chromatography, 342–345
　materials, 340
　methods, 340–343
　molecular chromatography, 342–345
　reverse phase separation, 344–346
Methylenetetrahydrofolate, ternary complex with tritiated 5-fluorodeoxyuridylate and thymidylate synthase, 313–319
　estimation, 315
　preparation, 314–315
Methylenetetrahydrofolate dehydrogenase, 373